Topics in
Current Physics

12

Topics in Current Physics

Founded by Helmut K. V. Lotsch

Volume 1 **Beam-Foil Spectroscopy**
Editor: S. Bashkin

Volume 2 **Modern Three-Hadron Physics**
Editor: A. W. Thomas

Volume 3 **Dynamics of Solids and Liquids by Neutron Scattering**
Editors: S. W. Lovesey and T. Springer

Volume 4 **Electron Spectroscopy for Surface Analysis**
Editor: H. Ibach

Volume 5 **Structure and Collisions of Ions and Atoms**
Editor: I. A. Sellin

Volume 6 **Neutron Diffraction**
Editor: H. Dachs

Volume 7 **Monte Carlo Methods** in Statistical Physics
Editor: K. Binder

Volume 8 **Ocean Acoustics**
Editor: J. A. DeSanto

Volume 9 **Inverse Source Problems** in Optics
Editor: H. P. Baltes

Volume 10 **Synchrotron Radiation** Techniques and Applications
Editor: C. Kunz

Volume 11 **Raman Spectroscopy** of Gases and Liquids
Editor: A. Weber

Volume 12 **Positrons in Solids**
Editor: P. Hautojärvi

Volume 13 **Computer Processing of Electron Microscope Images**
Editor: P. W. Hawkes

Volume 14 **Excitons**
Editor: K. Cho

Volume 15 Physics of **Superionic Conductors**
Editor: M. B. Salamon

Positrons in Solids

Edited by P. Hautojärvi

With Contributions by
A. Dupasquier P. Hautojärvi M. J. Manninen
P. E. Mijnarends R. M. Nieminen A. Vehanen
R. N. West

With 66 Figures

Springer-Verlag Berlin Heidelberg New York 1979

Professor Dr. Pekka Hautojärvi

Department of Technical Physics, Helsinki University of Technology,
SF-02150 Espoo 15, Finland

ISBN 3-540-09271-4 Springer-Verlag Berlin Heidelberg New York
ISBN 0-387-09271-4 Springer-Verlag New York Heidelberg Berlin

Library of Congress Cataloging in Publication Data. Main entry under title: Positrons in solids. (Topics in current physics; v. 12), Includes bibliographical references and index.
1. Positrons. 2. Positron annihilation. 3. Solids. I. Hautojärvi, Pekka. II. Dupasquier, Alfredo E.
III. Series. QC793.5.P622P67 539.7'214 79–1191

Offset printing and bookbinding: Konrad Triltsch, Graphischer Betrieb, Würzburg.
2153/3130-543210

Preface

In condensed matter initially fast positrons annihilate after having reached equilibrium with the surroundings. The interaction of positrons with matter is governed by the laws of ordinary quantum mechanics. Field theory and antiparticle properties enter only in the annihilation process leading to the emergence of energetic photons. The monitoring of annihilation radiation by nuclear spectroscopic methods provides valuable information on the electron-positron system which can directly be related to the electronic structure of the medium. Since the positron is a positive electron its behavior in matter is especially interesting to solid-state and atomic physicists. The small mass quarantees that the positron is really a quantum mechanical particle and completely different from any other particles and atoms. Positron physics started about 25 years ago but discoveries of new features in its interaction with matter have maintained continuous interest and increasing activity in the field. Nowadays it is becoming part of the "stock-in-trade" of experimental physics.

A striking feature is the great diversity of fields in which positron annihilation is applied. In addition to solid-state physics, which is the topic of this book, there are intensive activities in atomic physics, and positronium chemistry is the traditional name for the chemical applications. There exists plenty of earlier reference material in the form of review monographs and conference proceedings. However, the recent development and achievements especially in momentum density and defect studies have rapidly attracted the interest of a wider scientific community outside of the traditional positron physics thus providing justification for the publication of this topical volume.

The first chapter is an introduction devoted to readers who have no former familiarity with positron annihilation. It gives a short account of annihilation processes and of conventional experimental techniques. Also some recent topics of positron studies are very briefly discussed.

Chapter 2 is concerned with electron momentum density studies by means of angular correlation of 2γ-annihilation radiation. It discusses the state-of-the-art in the application of independent-particle theory and describes the recent development of two-dimensional detector geometries which have greatly increased the resolution and efficiency of the positron method. The latest results on metals and alloys are thoroughly reviewed.

The extreme sensitivity of positrons to crystal imperfections in metals is reflected through positron trapping. This makes the positron annihilation method capable of yielding unique information on the concentration, configuration and internal structure of lattice defects in solids. Positrons have within a few years proved to be a new microscopic tool the use of which has produced significant achievements in the physics of lattice defects. The new experimental results and progress in theoretical understanding are reviewed in Chaps.3 and 4, respectively.

Chapter 5 deals with ionic solids and serves as an example of problems in nonmetals. It offers a solution to the origin of the complex annihilation characteristics in alkali halides and reviews the interaction of positrons with various kind of defect centers.

I want to thank the authors who have not only contributed to this book but in fact created it. The very fluent cooperation has kept the editorial work at a minimum. In addition, I am especially indebted to Risto Nieminen for his comments and suggestions which have largely influenced the content of the book.

Helsinki, December 1978 *Pekka Hautojärvi*

Contents

List of Contributors

DUPASQUIER, ALFREDO

Gruppo Nazionale di Struttura della Materia des CNR,
Istituto di Fisica del Politecnico, Milano, Italy

HAUTOJÄRVI, PEKKA

Department of Technical Physics, Helsinki University of Technology.
SF-02150 Espoo 15, Finland

MANNINEN, MATTI J.

Research Institute for Theoretical Physics, University of Helsinki,
Siltavuorenpenger 20c, SF-00170 Helsinki 17, Finland

MIJNARENDS, PETER E.

Netherlands Energy Research Foundation E.C.N.,
1755 ZG Petten (N.H.), The Netherlands

NIEMINEN, RISTO M.

Department of Technical Physics, Helsinki University of Technology,
SF-02150 Espoo 15, Finland

VEHANEN, ASKO

Department of Technical Physics, Helsinki University of Technology,
SF-02150 Espoo 15, Finland

WEST, ROY N.

School of Mathematics and Physics, University of East Anglia,
Norwich NR4 7TJ, United Kingdom

1. Introduction to Positron Annihilation

P. Hautojärvi and A. Vehanen

With 15 Figures

Positron physics is concerned with the interaction of low-energy positrons with matter. The existence of the positron was predicted by Dirac and verified by Anderson more than 40 years ago. The birth and the rapid initial development of positron physics occurred in the early 1950s as it was realized that the characteristics of the quantum electrodynamic annihilation process depend almost entirely on the state of the positron-electron system in the matter. During the last decade the field has started to grow and widen strongly, as indicated in Fig.1.1 which shows the number of annually published papers [1.1]. The reason for the rapid growth lies in observations that positrons can provide unique information on a wide variety of problems in condensed matter physics. Also inexpensive experimental equipment which nowadays is commercially available has its own contribution to the popularity of positron studies.

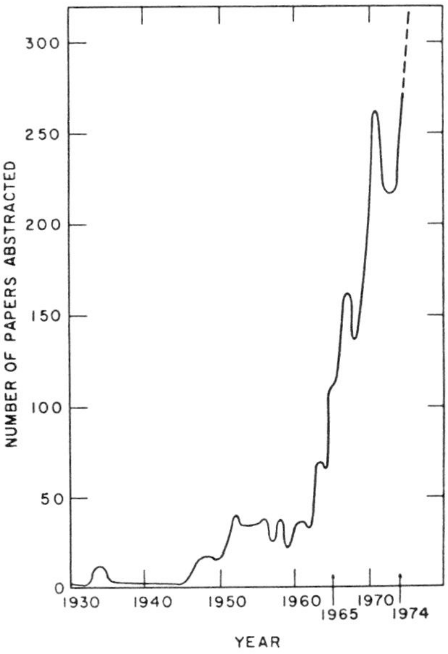

Fig.1.1. The number of annually published papers dealing with the study of low-energy positrons and positronium [1.1]

There have been several international conferences specially devoted to positron annihilation studies. The first was held in Detroit, USA in 1965 [1.2], a European meeting was organized in Saclay, France, in 1969, the second international conference was held in Kingston, Canada, in 1971 [1.3], the third in Otaniemi, Finland, in 1973 [1.4] and the fourth in Helsingør, Denmark, in 1976 [1.5]. The fifth will take place in Lake Yamanata, Japan, in 1979 and even the sixth conference has been scheduled to be held in the USA in 1982.

The positron technique has many advantages in the study of matter. It provides a nondestructive method because the information is carried out of the material by penetrating annihilation radiation. No special sample preparation is necessary and in some applications also in situ studies, e.g., on dynamic phenomena at elevated temperatures are possible. Several reviews and bibliographical surveys on the use of positrons to the study of condensed matter have been published [1.1-9].

This chapter consists of tutorial material for readers having no former familiarity with positron annihilation. The first section describes the principles of the positron method. Section 1.2 deals with the annihilation process of free positrons. The conventional measurement systems for the positron lifetime, the angular correlation of the 2γ-annihilation radiation and the annihilation line shape are discussed together with some examples in Sect.1.3. Positronium formation and its consequences in annihilation characteristics are treated in Sect.1.4. Some recent topics of positron studies are briefly discussed in Sect.1.5 and a short summary is given in Sect.1.6.

1.1 Positron Method

When energetic positrons from a radioactive source are injected into a condensed medium they first slow down to thermal energies in a very short time of the order of 1 ps. The mean implantation range varying from 10 to 1000 μm guarantees that the positrons reach the bulk of the sample material. Finally, after living in thermal equilibrium, the positron annihilates with an electron from the surrounding medium dominantly into two 511 keV gamma quanta. The average lifetime of positrons is characteristic of each material and varies from 100 to 500 ps. The picture above is distorted in molecular media, where positronium formation may occur during the slowing down process. This phenomenon, however, is treated separately in Sect.1.4.

Figure 1.2 shows schematically the positron annihilation experiment, where the most commonly used radioisotope ^{22}Na is implied. Within a few picoseconds after the positron emission the nucleus emits an energetic 1.28 MeV photon which serves as a birth signal. The lifetime of the positron can thus be measured as the time delay between the birth and annihilation gammas. The momentum of the annihilating

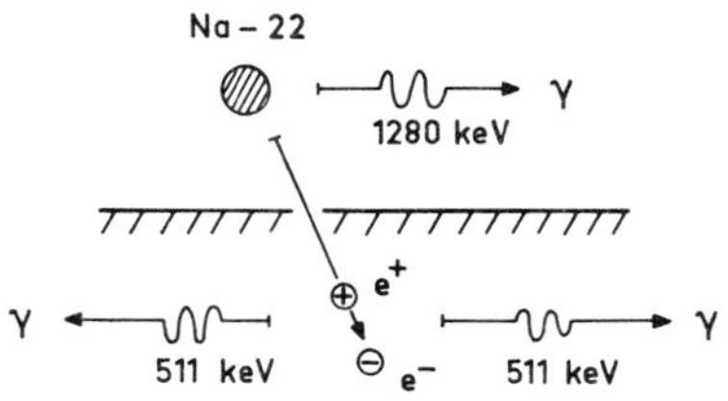

Fig.1.2. The positron experiment. Positrons from a radioactive isotope like ^{22}Na annihilate in the sample material. Positron lifetime is determined from the time delay between the birth gamma (1.28 MeV) and the two annihilation quanta. The momentum of the electron-positron pair is measured as an angle deviation between the two 511 keV quanta or as a Doppler shift in the energy of the annihilation radiation

electron-positron pair is transmitted to the annihilation quanta and it can be detected as a small angle deviation from collinearity between the two 511 keV photons. The motion of the pair also produces a Doppler shift to the annihilation radiation and this is seen in an accurate energy measurement of one of the photons.

1.2 Annihilation of Free Positrons

The positron-electron annihilation is a relativistic process where the particle masses are converted into electromagnetic energy, the annihilation photons. From the invariance properties of quantum electrodynamics several selection rules can be derived. One-gamma annihilation is possible only in the presence of a third body absorbing the recoil momentum and its relative probability is negligible. The main process is the two-gamma annihilation, since the spin-averaged cross section for the three-gamma annihilation is 0.27% of that for the two-gamma annihilation. The three-gamma annihilation is important only in a spin-correlated state like ortho-positronium, where the selection rules forbid the two-quantum process. This situation is treated in Sect.1.4.

From the nonrelativistic limit of the 2γ-annihilation cross section derived by DIRAC [1.10] one obtains the annihilation probability per unit time or the annihilation rate

$$\lambda = \pi r_0^2 c n_e \quad , \tag{1.1}$$

which is independent of the positron velocity. Here r_0 is the classical electron radius, c the velocity of light and n_e is the electron density at the site of the positron. By measuring the annihilation rate λ, the inverse of which is the mean lifetime τ, one directly obtains the electron density n_e encountered by the positron. Thus a positron can serve as a test particle for the electron density of the medium. However, because of the opposite charges, a strong Coulomb attraction exists between the positron and electrons. Consequently, the electron density n_e is enhanced from the equilibrium value in matter due to the Coulomb screening of the positron. Calculation of these positron-electron correlations is a complicated many-body problem which is well understood only in the case of electron gas.

The kinetic energy of the annihilating pair is typically a few electron volts. In their center-of-mass frame the photon energy is exactly m_0c^2 = 511 keV and the photons are moving strictly into opposite directions. Because of the nonzero momentum of the pair the photons deviate from collinearity in the laboratory frame. As illustrated in Fig.1.3 the momentum conservation yields a result

$$\theta \simeq p_T/m_0c \quad , \tag{1.2}$$

where $180^\circ - \theta$ is the angle between the two photons in the laboratory frame and P_T is the momentum component of the electron-positron pair transverse to the photon emission direction. Usually θ is very small($\theta < 1^\circ$) and (1.2) is valid. Because the momentum of the thermalized positrons is almost zero, the measured angular correlation curves describe the momentum distribution of annihilated electrons in matter.

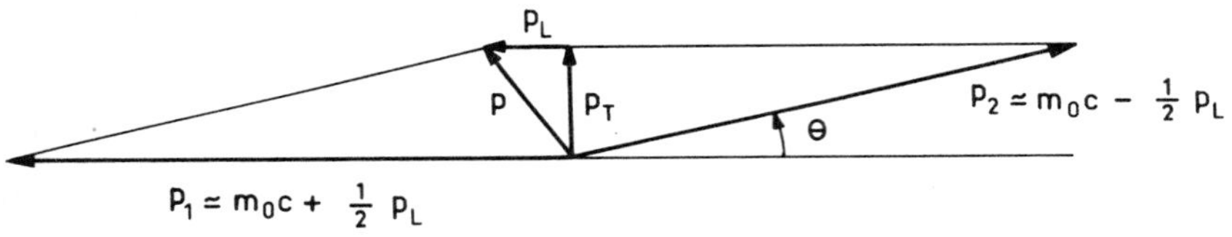

Fig.1.3. The vector diagram of the momentum conservation in the 2γ-annihilation process. The momentum of the annihilating pair is denoted by p, subscripts L and T refer to longitudinal and transverse components, respectively

The motion of the pair also causes a Doppler shift in the energy of the annihilation photons measured in the laboratory system. The frequency shift is $\Delta\nu/\nu = v_L/c$, where the longitudinal center-of-mass velocity v_L of the pair equals $p_L/2m_0$. Since the energy of a photon is proportional to its frequency, we get for the Doppler shift at the energy m_0c^2

$$\Delta E = (v_L/c)E = cp_L/2 \quad . \tag{1.3}$$

Thus also the line shape of the annihilation radiation reflects the momentum distribution of electrons in matter.

1.3 Experimental Techniques

1.3.1 Lifetime Measurements

The conventional positron source in the lifetime measurements is ^{22}Na. As indicated in Fig.1.2 the positron emission is accompanied by a suitable start signal, the 1.28 MeV photon from the de-excitation of the resulting ^{22}Ne nucleus. The positron source is usually prepared by evaporating a few microcuries of aqueous $^{22}NaCl$ onto a thin metal or plastic foil (typically 1 mg cm^{-2}) and covering it with the same foil. The source is then sandwiched by two identical pieces of sample material, the thickness of which must be high enough (>0.1 mm) to absorb all positrons.

The lifetime spectrometer is shown schematically in Fig.1.4. It is a fast-slow coincidence system conventionally used in nuclear spectroscopy. The detectors consist of fast plastic scintillators coupled to fast photomultiplier tubes. The energy windows of the single-channel analyzers (SCA) in the slow channels are adjusted so that the detectors A and B register the birth and annihilation gammas of individual positrons, respectively. The fast signals from the anodes of the photomultipliers are fed to constant-fraction-timing discriminators (CF DISC) to produce time signals. These are then led to a time-to-amplitude converter (TAC), the output of which is proportional to the time interval between the start and stop signals. The output pulses from the time-to-amplitude converter are then transferred to a multichannel analyzer (MCA) through a linear gate driven by the coincidences in the slow channel, i.e., the output pulses are accepted only if the corresponding start and stop signals have correct energy values determined by the windows of the single-channel analyzers.

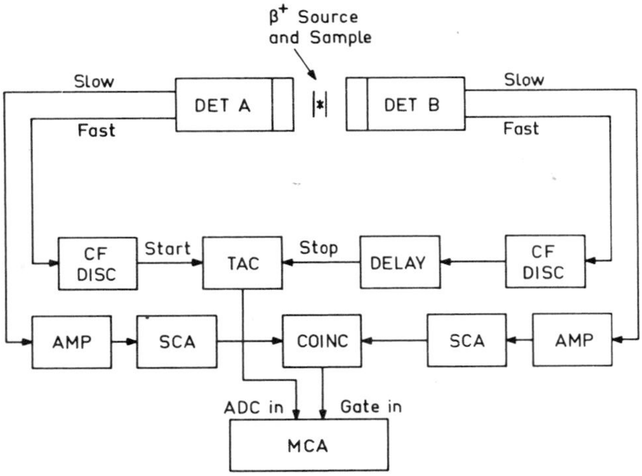

Fig.1.4. Schematic diagram of the fast-slow coincidence system used in the positron lifetime measurement. CF DISC: constant-fraction-timing discriminator; TAC: time-to-amplitude converter; AMP: amplifier; SCA: single-channel analyzer; COINC: coincidence circuit; MCA: multichannel analyzer

In practice there are random and systematic errors in the measurement and as a result we have a convolution of an ideal spectrum with the instrumental resolution function. The latter can be measured by replacing the source-sample sandwich with a ^{60}Co gamma source without touching other settings of the system. This isotope emits almost simultaneously two photons (1.17 and 1.33 MeV) in a cascade. A time resolution of 300 ps (full width at half maximum, FWHM) is typically obtained with available commercial equipment and it is of the same order of magnitude as the lifetimes to be measured. The resolution is mainly limited by the decay time of the light centers in the scintillators and by the transit time spread of electrons in the photomultipliers. The best FWHM resolution reported in positron lifetime measurements is about 170 ps (FWHM) [1.11].

An example of a lifetime spectrum is shown in Fig.1.5 [1.12] together with the instrumental resolution (prompt) curve measured with ^{60}Co. In a perfect well-annealed metal all positrons annihilate as free particles from Bloch-like states with a well-defined time-independent annihilation rate λ. The lifetime measurement then produces a single-exponential spectrum of the form $\exp(-\lambda t)$. This is well seen in Fig.1.5 for sodium. The corresponding mean lifetime τ is the inverse of λ. The constant background in the spectrum is due to random coincidences.

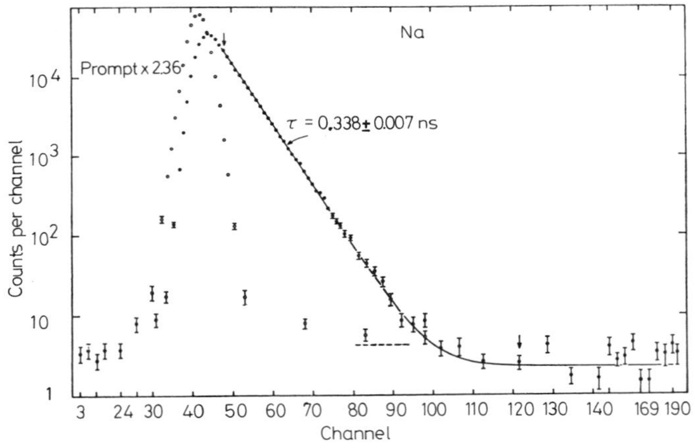

Fig.1.5. Typical positron lifetime spectrum of a metal (sodium) in a semilogarithmic plot. The prompt curve represents the instrumental resolution [1.12]

If positrons annihilate from different states in the sample the result is a multiexponential lifetime spectrum. This is normally analyzed by computers in order to extract lifetime values τ_i and relative intensities I_i associated with the different components. In cases where individual components have rather close decay values the result of the curve fitting analysis is quite ambiguous (see discussions in Sects.3.1.2 and 5.1.3). This problem does not exist, when the lifetime spectrum is characterized with the displacement of its centroid relative to the centroid of the resolution curve. The reliable determination of the centroid shift requires simultaneous measurement of the two curves, but it can be realized with high accuracy using digital stabilization and a router-mixed system [1.13]. The resulting single parameter, the average lifetime τ_{ave}, can be used to label various states of the sample, but part of the underlying physical information connected with different annihilation modes is lost.

1.3.2 Angular Correlation Measurements

The angular correlation of the 2γ-annihilation photons is measured with a typical system described in Fig.1.6. The positron source is usually ^{22}Na, ^{64}Cu, or ^{58}Co with an activity of 10 mCi-1 Ci. The positrons from the source penetrate into the sample and annihilate there. The annihilation photons are detected in coincidence by NaI scintillation counters, which are shielded from direct view of the source. The lead collimators in front of the detectors define the instrumental angular resolution being typically less than 1 mrad. Because of the small angle deviations to be measured, the distance between the detectors is several meters. To achieve adequate counting rates the detectors and slits are made in the x direction as long as possible. The single-channel analyzers (SCA) are tuned for 511 keV photons and the device simply counts the coincidence pulses as a function of the angle θ_z. The data collection for an angular correlation curve with good statistics usually requires counting over a couple of days.

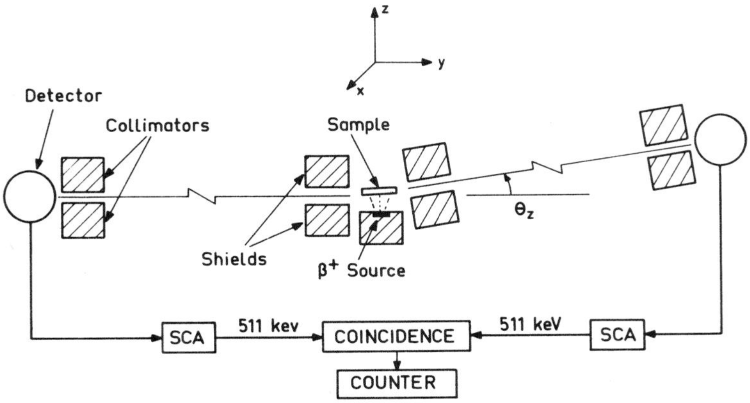

Fig.1.6. Angular correlation apparatus with long-slit geometry for 2γ-annihilation radiation

The arrangement in Fig.1.6 is called the long-slit geometry according to the shape of the collimators. The device can resolve neither the angular deviation in the x direction nor the Doppler shift in the y direction and consequently, the counting rate becomes

$$N(\theta_z) = C \int_{-\infty}^{+\infty}\int_{-\infty}^{+\infty} dp_x dp_y \rho(p_x, p_y, \theta_z m_0 c) \quad , \tag{1.4}$$

where $\rho(p_x, p_y, p_z)$ is the momentum distribution of the annihilating positron-electron pairs in the sample medium. When the positron is free and in thermal equilibrium with the medium, its momentum is negligible and then the angular correlation curve represents the p_z-distribution of the annihilated electrons (for more details see Sect.2.1).

Examples of angular correlation curves resulting from free positrons in annealed copper and aluminum are given in Fig.1.7. Because of symmetry the curves are folded around $\theta_z = 0$ and only the positive-angle portions are shown. The curves are seen to consist of two parts. The inverted parabola is due to annihilations with valence electrons. The broader, roughly Gaussian-shaped component is due to annihilations with core electrons having higher momentum values. The momentum distribution of valence electrons has a cutoff at the Fermi surface and the corresponding momentum value ($\theta_F m_0 c$) can be directly obtained from the intersection of the two components. Figure 1.7 also shows a significant difference in the core annihilation fractions and a smaller difference in the widths of the parabolas. These effects reflect different core and valence electron densities in the two metals.

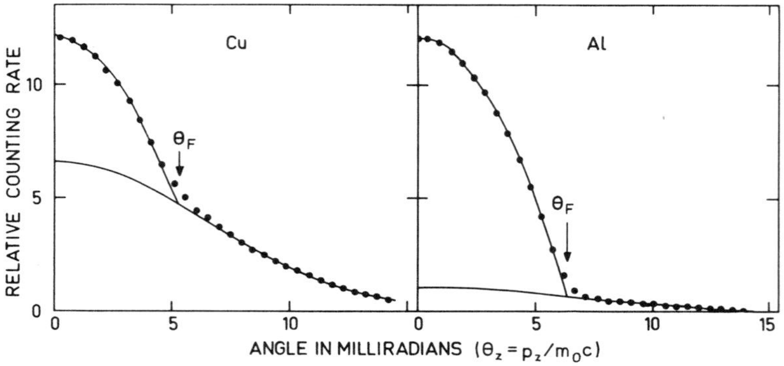

Fig.1.7. Typical angular correlation curves in metals (aluminum and copper). The inverted parabolas are due to free electrons and the Gaussian parts to core electrons

In contrast to lifetime values, the angular correlation curves are almost insensitive to many-body effects and they can be adequately interpreted using the independent-particle model. Therefore positrons are particularly suitable to momentum density and Fermi surface studies. For these purposes more sophisticated experimental systems than described in this section have been developed to measure two-dimensional momentum densities from single crystals. These topics are reviewed in Chap.2.

1.3.3 Line-Shape Measurements

As described in Sect.1.2 the motion of the electron-positron pair causes a Doppler shift on the energy of the annihilation radiation. As a consequence of (1.3) the line shape gives the distribution of the longitudinal momentum component of the annihilating pair, i.e., the distribution of p_y in the coordinate system of Fig.1.6. Thus the determination of the energy distribution of the 511 keV annihilation radiation is equivalent to the angular correlation measurement in the long-slit geometry, since the sample orientation with respect to the measurement system can naturally be chosen at will.

The only sensible method for line shape studies is to use a high-resolution solid-state detector. Figure 1.8 shows a typical installation. The source-sample sandwich is similar to that in the lifetime measurement. Annihilation radiation is detected with a lithium-drifted or intrinsic Ge crystal. The efficiency compared to the angular correlation system is roughly a hundred times larger, since the sample is placed close to the detector and no coincidence requirements exist. Typically a 1 h measurement with a 5 μCi positron source is enough for a sufficient statistical accuracy. The disadvantage, however, is the resolution of the system. The best Ge detectors have an energy resolution of about 1 keV at 511 keV corresponding to an equivalent angular resolution of about 4 mrad, which is an order of magnitude worse than the resolution of the angular correlation devices. Also electronic stability problems arise; a relatively small drift during the measurement can severely destroy the information obtained. The digital spectrum stabilizer shown in Fig.1.8 is essential, if reliable results are required.

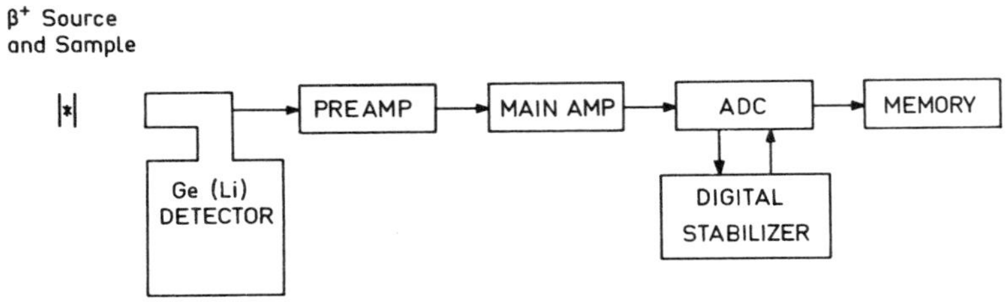

Fig.1.8. The system for measuring the shape of the Doppler-broadened annihilation line

Figure 1.9 shows the effect of Doppler broadening on the line shape of annihilation radiation. All the curves in the figure have been normalized to equal area. The high 514 keV peak from ^{85}Sr represents the response for monochromatic radiation. It is seen that the Doppler broadening is a remarkable effect. The line shape is different for different materials and it is also sensitive to positron trapping by lattice defects as is indicated in the figure for the case of deformed copper.

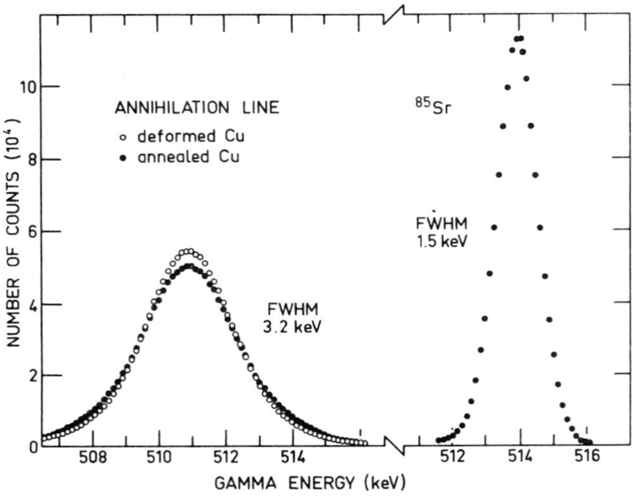

Fig.1.9. The effect of Doppler broadening on the annihilation line in annealed and deformed copper. The ^{85}Sr peak represents the energy resolution of the system. All the curves have been normalized to equal area

The resolution of the line shape measurement is, however, insufficient for the momentum density studies. Some deconvolution procedures have been developed but their reliability is always limited. As a consequence, various line shape parameters characterizing the width of the annihilation line are used [1.14]. The absolute values of these parameters do not include physical information but depend, e.g., on the resolution of the system. However, changes of these parameters can be used to follow various phenomena like the defect concentration of the sample (see Chap.3). Thus the line shape method is an effective tool in experiments where the underlying physical process is known and many measurements of the successive states of the sample are required.

1.3.4 Correlation Between Lifetime and Momentum

In principle, it is possible to correlate positron lifetime with the angle between the two gamma rays resulting from the annihilation. This means measurements of lifetime spectra at various angle deviations θ selected by an angular correlation device. Then a triple coincidence system is required, since in addition to the two annihilation gammas also the start gamma must be detected. Only a few experiments of this kind have been performed [1.15] because of the very low data accumulation rate.

The use of a Ge(Li) detector as a momentum-sensitive device improves the efficiency of the system but at the expense of momentum resolution. The solid-state detector can be used either as a third detector to measure the Doppler shift in coincidence with an ordinary lifetime spectrometer [1.16] or it can replace the stop detector given simultaneously both time and energy signals [1.17]. These systems are quite feasible to provide new information on the positron interaction with matter.

1.4 Positronium Formation and Annihilation

A positron in a molecular substance can capture an electron from the surrounding medium and a positronium atom (Ps), the bound state of the positron-electron pair, is formed. As the size of positronium is twice that of the hydrogen atom, positronium formation occurs mainly in molecular media which have relatively open structures.

The energetics of positronium formation is usually described by the so-called Ore gap model [1.18]. It states that positronium formation is most probable when the positron energy during its slowing down lies within a gap where no other electronic energy transfer process is possible. To capture an electron from a molecule of the medium with ionization energy E_i, the kinetic energy E of the positron must be greater than E_i-E_{Ps}, where E_{Ps} is the binding energy of positronium. In vacuum, E_{Ps} is 6.8 eV but may be smaller in the medium. When $E > E_i$ the positronium atom is formed with a kinetic energy greater than its binding energy and it will rapidly break up in collisions. Furthermore inelastic collisions will compete with positronium formation until the positron kinetic energy is less then E_{ex}, the lowest electronic excitation energy. Thus positronium formation is most probable with the energy in the range

$$E_i - E_{Ps} < E < E_{ex} \quad , \tag{1.5}$$

which is the Ore gap. Its width can be used to estimate the positronium yield, the fraction of positrons which have formed positronium.

The ground states of positronium are the singlet 1^1S_0 state (parapositronium) and the triplet 1^3S_1 state (orthopositronium). Their energy splitting is only 8.4×10^{-4} eV, the singlet state being the lower one. Accordingly, in the absence of ortho-para conversion, 1/4 of the positronium atoms are formed in the singlet state and 3/4 in the triplet state. The lifetime of parapositronium in self-annihilation into two photons is 125 ps, about the same as free positron lifetimes in dense metals. The 2γ-annihilation is, however, forbidden by selection rules in the spin-triplet case. Thus in vacuum, orthopositronium decays via three-photon emission with the lifetime of 142 ns, which is more than three orders of magnitude longer than the parapositronium lifetime. A competing mechanism, called "pick-off" annihilation, is always present when the orthopositronium atom exists in a medium. In the pick-off process the positron of orthopositronium suffers 2γ-annihilation in collision with a "foreign" electron having opposite spin. Consequently, the orthopositronium lifetime in condensed materials is reduced to a few nanoseconds.

As is quite evident, the annihilation characteristics are complex in the presence of positronium formation. Figure 1.10 shows a lifetime spectrum of positrons in quartz glass. The long lifetime of about 1.5 ns is due to pick-off annihilation of orthopositronium. In the angular correlation curve the self-annihilation of parapositronium produces a narrow component at small angles, because the center of mass of the positronium atom has a small momentum in the laboratory frame. Examples of some molecular materials are given in Fig.1.11 [1.19].

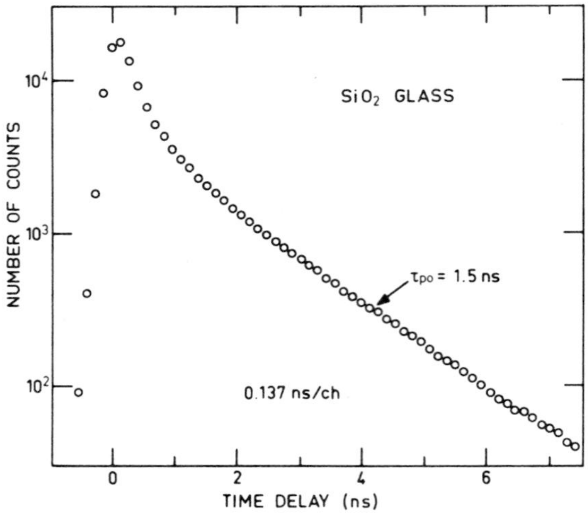

Fig.1.10. Positron lifetime spectrum in amorphous quartz, The 1.5 ns lifetime component represents the pick-off annihilation of orthopositronium

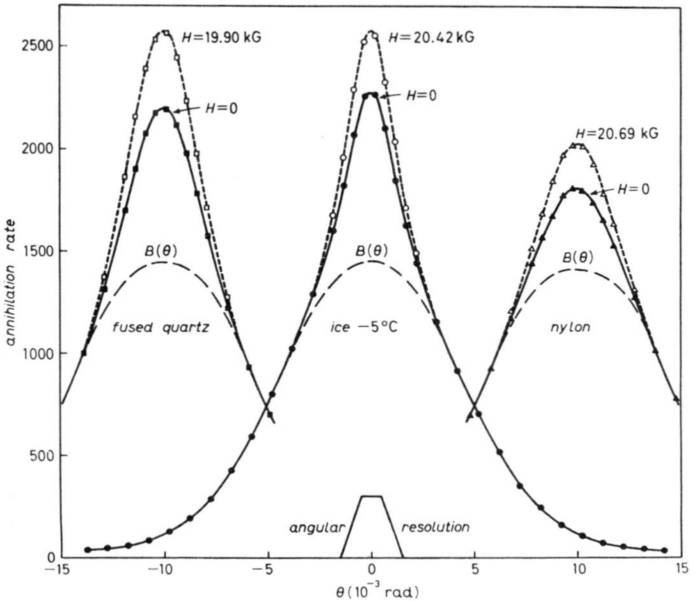

Fig.1.11. Angular correlation curves in some molecular materials. The narrow components are due to the self-annihilation of parapositroniums. The magnetic quenching of orthopositronium enhances the narrow component [1.19]

The pick-off lifetime of orthopositronium may be further reduced by other competing mechanisms. Such "quenching" processes are the ortho-para conversion, chemical reactions of positronium and magnetic quenching. The ortho-para conversion happens, e.g., in electron exchange between positronium and a paramagnetic molecule. In chemical reactions a complex molecule including the positron may be formed and then the positron is no longer correlated only with a single electron of parallel spin. In an external magnetic field the triplet $m = 0$ state is mixed with the singlet state, the mixing ratio being higher at higher fields. Through the singlet component the self-annihilation into two photons becomes possible. At sufficiently high field strengths this effect thus leads to a one-third reduction of the long-lived orthopositronium component in the lifetime spectrum and to a doubling of the parapositronium component in the angular correlation curve. Figure 1.11 shows also the enhancement of the narrow component due to applied magnetic field. The magnetic quenching is a convenient method to reveal the existence of positronium, and its applications to alkali halides are discussed in Chap.5.

1.5 Topics of Positron Studies

1.5.1 Metals

In metals positrons are effectively screened by conduction electrons and no positronium formation is possible. As a result, the annihilation rate, which is proportional to the electron density at the site of the positron, is enhanced from the value corresponding to the unperturbed density by a factor varying between 4 and 30. The enhancement also masks the electron-density dependence of the annihilation rate. In nearly-free-electron metals the positron lifetime varies only from 170 ps in aluminum to 420 ps in cesium, whereas the free-electron density changes by a factor of 20. This problem has been attacked by many-body calculations using the idealized model of a positron in a homogeneous electron gas. The form of the screening cloud, correlation energies and momentum-dependent enhancement factors are well understood in the case of an interacting electron gas (see Sects.2.1.2 and 4.1.4), although there still remain problems in the many-body computational methods [1.20]. However, the correlation effects due to more tightly bound core electrons or d electrons in transition metals pose a more difficult question, which still is quite open. Fortunately, the momentum distribution of annihilating pairs is quite insensitive to electron-positron correlation effects and therefore the angular correlation measurements of the 2γ annihilation radiation serve as a good tool for momentum density studies of metals, alloys and some dielectrics, as is described in Chap.2.

1.5.2 Metal Defects

At the end of the 1960s it was noticed that deformation or heating of the sample to elevated temperatures causes remarkable changes in the annihilation characteristics. Since these changes were too big to be attributed to bulk properties they were explained in terms of positron trapping by crystal imperfections. In defects, where atoms are missing or their density is locally reduced, the repulsion between the positron and ion cores is decreased. Also the redistribution of electrons causes a negative electrostatic potential at this type of defect. Thus positrons see defects like vacancies, voids and dislocations as strongly attractive centers in the crystal.

The localization of positrons at the defect sites has three important consequences:

I) The concentration of defects can be deduced from the ratio of trapped and free positrons. This has been successfully utilized, e.g., in the determination of vacancy formation energies.

II) The annihilation characteristics of trapped positrons reflect local properties of defects, giving thus unique information on their internal electronic structure.

III) The annihilation characteristics of trapped positrons are to some extent different for different defect configurations. For example, they can reveal vacancy agglomeration and give estimates on the size of microvoids which are too small to be detectable by any other methods.

The lattice defects of metals form perhaps the most rapidly developing field of positron studies. Positron annihilation has within a few years proved to be a new submicroscopic method which is significantly contributing to the physics of lattice defects. The related experimental and theoretical aspects are reviewed in Chapters 3 and 4, respectively.

1.5.3 Ionic Crystals

Positrons in alkali halides have always at least two lifetime components $\tau_1 \simeq 0.2$-0.3 ns and $\tau_2 \simeq 0.4$-0.7 ns with the intensity $I_2 \simeq 20$-60%. In addition, annihilation characteristics are found to be very sensitive to various types of defect centers. The shorter lifetime is evidently due to free positrons in the crystal. The problem on the origin of the intense second lifetime component has occupied positron physicists since the discovery of the complex structure of the positron lifetime spectra in 1963 by BISI et al. [1.21]. Several models have been presented. The second component might be due to positron trapping by lattice defects which are always present even in nominally perfect crystals. Also the coupling of positrons with lattice dilatations to a polaron like state has been suggested. As in molecular media the longer lifetime could be an indication of positronium formation with high pick-off annihilation rate because of the tightly packed crystal structure of alkali halides. The recent experimental and theoretical results seem to give an answer to this question. The problem on positrons in ionic crystals and their interactions with defect centers is reviewed in Chap.5.

1.5.4 Slow Positrons and Positronium

Some materials are observed to have a negative work function for positrons. After slowing down in the bulk a small fraction of positrons is emitted from the surface with a kinetic energy of the order of an electron volt. The handling and transportation of the low-energy positrons can be simply done with weak electrostatic and magnetic fields. Figure 1.12 shows an example of an apparatus to produce a slow-positron beam [1.22]. The low-energy positrons are obtained be moderating fast positrons from a 50 μCi ^{22}Na source in gold vanes coated with magnesium oxide. The positron energy upward of 1 eV can be adjusted by the accelerating potential on the vanes. Slow positrons are then guided by a magnetic field through a flight tube to a solid target, where they are detected by monitoring the annihilation radiation. Instead of radioactive isotopes also accelerators can be used to create

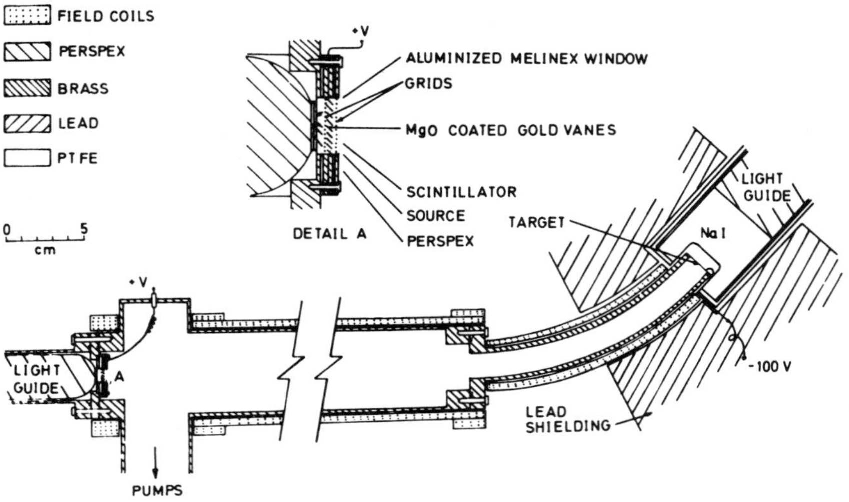

Fig.1.12. Schematic diagram of an apparatus for the low-energy positron beam. Detail A describes the moderator system in front of the positron source [1.22]

positrons via pair production of the bremsstrahlung [1.23] or through short-lived nuclear reaction products [1.24].

The main application of slow-positron beams has been the direct measurement of the total scattering cross sections for positron-atom or positron-molecule collisions. This is done simply by filling the flight tube with the target gas. Recently, a Ramsauer-Towsend minimum for positrons in argon has been observed by this method [1.25]. Evidently the slow-positron beam has a good potential as a tool for solid-state physics, too. Positrons and positronium atoms interact strongly with surfaces, as is described in Chaps.3 and 4, and also in BRANDT's review [1.26]. Therefore interesting applications of slow-positron beams to surface physics studies are to be expected in the near future.

As a bound state of electron and its antiparticle, the positronium atom is an ideal testing ground for quantum electrodynamics. Most recent calculations have been given by CASWELL et al. [1.27]. Since the discovery of positronium by DEUTSCH [1.28] the observation of its optical transitions has been a challenging problem. The earlier attempts failed mainly because of the high radiation background caused by energetic positrons. The first successful experiments by CANTER et al. [1.29-31] made use of a slow-positron beam. When striking a solid target, slow positrons are efficiently converted into positronium atoms far from the primary radioactive source. This makes it possible to study with high accuracy both the ground- and excited-state properties of the positronium atom. By combining optical, microwave, and gamma-spectroscopic methods, the authors were able to observe the Lyman-α (2P-1S) line and to measure the splitting $\Delta\nu$ of the 2^3S_1-2^3P_2 levels. Their result

$\Delta\nu$ = 8628.4 ± 2.8 MHz agrees within two standard deviations with the theoretical value, thus proving that quantum electrodynamics is correct at least to third-order terms in the fine-structure constant.

1.5.5 Gases and Low-Temperature Phenomena

The density of gases is so low that the slowing down and thermalization times are comparable to positron lifetimes. As a result, complex nonexponential features are seen in the lifetime spectra. Positronium formation is usually also present. The experimental annihilation rates, as well as the recent results obtained with slow-positron beams, give good reference data for computations of positron-atom or positronium-atom scattering. Because of the attractive electron-positron correlations, these computations form a stringent test to any approximate theories developed for the scattering of slow electrons by atoms and molecules. A general review on positrons in gases has been given by GRIFFITH and HEYLAND [1.32].

There are peculiar phenomena in low-temperature noble gases, seen especially in helium. Both positron and positronium may exist in localized states. The attractive polarization forces between a positron and helium atoms induce droplet formation around positrons when their kinetic energy is low enough [1.33,34]. The critical temperature of this local condensation is almost twice that of the bulk gas, the number of clustered atoms in the droplet is a few hundred and its density is close to that of liquid helium [1.34-36]. An essentially opposite effect occurs with positronium in a dense fluid. Because of the exchange repulsion between helium atoms and positronium, a cavity is formed at low temperatures and high densities [1.33,37]. The positronium bubble is analogous to the electron bubble [1.38]. Its radius is about 15 Å and can be roughly estimated from the balance between the outward pressure on the bubble walls due to zero-point motion of the trapped particle and the shrinking forces due to surface tension and external pressure. Since the overlap of the trapped particle with the bulk fluid is very small in the bubble state, the orthopositronium lifetime reaches about 100 ns, which is almost the vacuum value (142 ns) determined by 3γ annihilation.

Figure 1.13 [1.33] shows a typical lifetime spectrum in dense helium gas at low temperature, when both droplet and bubble formation are present. Note the very extended time scale; the horizontal axis covers almost 100 ns. The highest peak to the left determines the time zero and includes short lifetimes from parapositronium and from annihilations in the metallic chamber walls. The subsequent, almost linear, part is due to positrons which have not yet thermalized but reached an annihilation rate nearly independent of velocity. When the energy of the positron is low enough the clustering of atoms around the positron occurs. This leads to a sudden increase in the annihilation rate and is seen as a second peak followed by a steep decrease of the lifetime spectrum ("slow positrons" in the figure). The

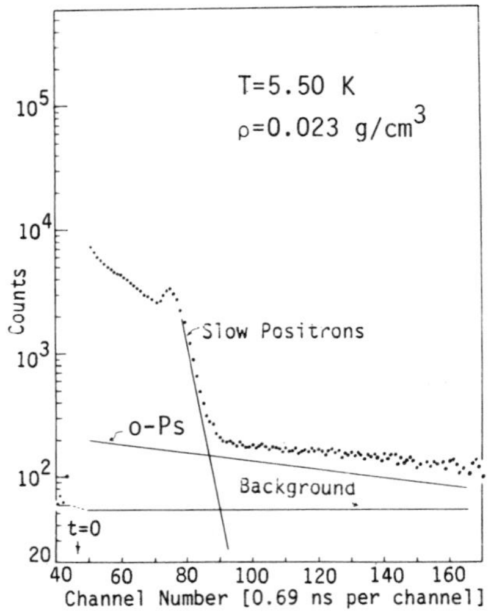

Fig.1.13. Positron lifetime spectrum in low-temperature helium gas showing both the cluster formation around slow positrons and bubble formation around positronium atoms [1.33]

lifetime of the orthopositronium in the bubble state is so long that the corresponding slope can hardly be distinguished from the constant background.

1.5.6 Molecular Solids

Positronium formation and interaction with the surrounding medium in molecular solids are still open questions to some extent. Positronium is mainly localized at open spaces present in the structure of these materials. The width of the narrow component in the angular correlation curves due to self-annihilation of parapositronium gives the degree of positronium localization, and the pick-off rate of orthopositronium is determined by the overlap with the surrounding molecules [1.39, 40].

However, it has been noticed that in some high-quality single crystals like quartz and CaF_2 [1.41] and ice [1.42], positronium may exist in a Bloch-type state. Because the center of mass is completely delocalized, the self-annihilation of the parapositronium produces a very narrow central peak together with discrete side peaks at the positions of the reciprocal lattice vectors. This is demonstrated in Fig.1.14 for the single crystal of ice [1.42]. Further, the Bloch-state positronium is sensitive to crystal imperfections which may localize it causing broadening and suppression of the central and side peaks. Thus there are possibilities to probe lattice defects of molecular crystals with the aid of positronium trapping in the same way as positron trapping is used for metal defects [1.43-45].

Polymers are molecular materials which have traditionally been under wide investigation. Positron lifetime spectra in polymers can normally be resolved into at least three different components with lifetimes $\tau_1 \simeq 0.3$ ns, $\tau_2 \simeq 0.7$ ns and $\tau_3 \simeq 3$ ns. The longest one is evidently due to pick-off annihilation of ortho-

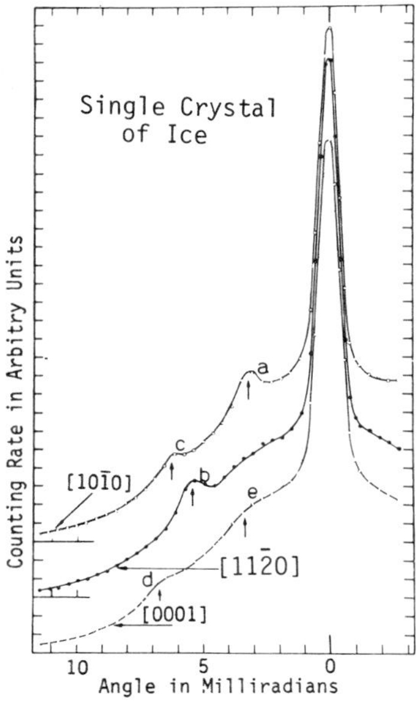

Fig.1.14. Angular correlation curves in single crystal of ice. The narrow central peak together with side peaks at the positions of reciprocal lattice vectors are due to Bloch-state parapositronium [1.42]

positronium. The shortest lifetime includes parapositronium and free positrons in the bulk. The problem has been the origin of the intermediate component, since it seems to be too long for free positrons. BRANDT [1.39] has attributed it to positrons which have not formed positronium but are bound to lower density amorphous domains. Recent studies on the electric field dependence of positron lifetime spectra [1.40] seem to support this idea. A weak electric field causes enhanced escape of positrons out of the Ore gap, thus decreasing the orthopositronium intensity I_3. This decrease levels off at strong electric fields, since free positrons below the Ore gap can be accelerated back into the gap. However, the intensity I_2 stays field independent, indicating that it is due to immobile positrons which have not formed positronium.

Increased attention has recently been paid to the study of glasses by positrons. Several problems have been surveyed: glass transition, impurity-induced ortho-para conversion, structure of binary glasses, and glass crystallization [1.46,47]. Perhaps the most potential application is the use of positrons to study dynamic phenomena connected with phase separation and crystallization processes. Lithium-silicate (Li_2O-SiO_2) glass has often been selected as the sample material because of its simple structure and the presence of all the basic processes. As a result of amorphous phase separation, segregation of nearly pure SiO_2 droplets from the bulk occurs. Heat treatment at moderate temperatures crystallizes the bulk glass but leaves the SiO_2 droplets amorphous. The longest 1.5 ns lifetime characteristic to amorphous quartz (see Fig.1.10) is due to pick-off annihilation inside the SiO_2 droplets, and its relative intensity is proportional to the volume fraction of the droplets. Crystallization, in turn, reduces the intermediate intensity corresponding to a lifetime of about 0.7 ns. Thus one can follow with positrons the kinetics of crystallization and phase-separation processes [1.48] and gain information on their interplay, which is an interesting and important problem in glass science.

1.5.7 Positronium Chemistry

The orthopositronium decay rate and its intensity depend strongly on the properties of the medium. Because of the hydrogenlike structure and relatively long lifetime, orthopositronium can take part in chemical reactions. As a result, the orthopositronium state is broken and the subsequent positron state (free positron, positron-molecule complex, etc.) has a lifetime comparable to that of free positrons. This phenomenon is called chemical quenching of orthopositronium and it is seen as a considerable decrease of the longer lifetime. Figure 1.15 shows an example of reactions with iodine molecules in cyclohexane solvent [1.49]. The measured decay rate of orthopositronium is

$$1/\tau_2 = \lambda_{po} + \lambda_{ch} \quad ,$$

where λ_{po} and λ_{ch} are the pick-off rate and chemical reaction rate of orthopositronium, respectively. At low reagent concentrations C a linear relationship holds and $\lambda_{ch} = kC$. Here the constant k is the reaction rate constant in complete analogy with chemical reactions between ordinary atoms. An advantage of positron lifetime measurements is evident: the chemical reaction rate constant k is obtained in an absolute time scale. Thus problems having importance in chemistry can be attacked by positron annihilation and this field is traditionally called positronium chemistry. For special reviews see [1.8,50,51].

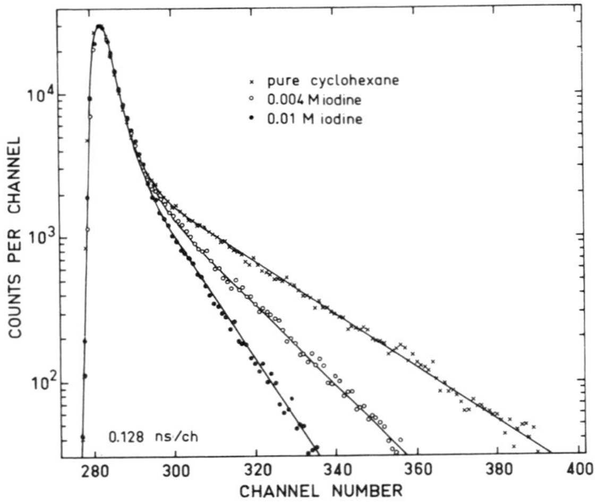

Fig.1.15. Positron lifetime spectrum in cyclohexane, where the orthopositronium lifetime is seen as the long tail. Reaction with added iodine molecules is seen to reduce the lifetime value [1.49]

The bubble model developed for positronium in low-temperature noble gases has been adopted to explain the pick-off rate also in pure organic liquids [1.52]. Positronium is localized in a hole with a radius of about 5 Å determined by the balance between the surface tension and pressure due to the zero-point motion of the trapped positronium. Experimentally, good correlation between surface tension and pick-off rate has been established [1.53]. A detailed model of how the chemical reaction occurs in the bubble state of the positronium has been presented by GOLDANSKII and SHANTAROVICH [1:51].

The suggestion by MOGENSEN [1.54] that positronium formation is a spur reaction has stimulated new activity in the field. The positron spur is a group of electrons and ions produced in the last ionization collisions during the slowing down of the positron. According to the model, positronium is formed when the positron captures an electron in its own spur. This means that positronium formation must then compete with electron-ion recombination, with electron and positron reactions with solvent molecules and impurities, with diffusion of electrons and the positron out of the spur, etc. To test the predictions of the spur model, the results of positronium yield measurements have been successfully correlated with the properties of excess electrons studied in radiation chemistry (for references see JANSEN and MOGENSEN [1.55]), but the model is still subject to discussion.

1.6 Summary

Initially fast positrons annihilate after coming to thermal equilibrium with the surrounding matter. The annihilation characteristics give information on the electron-positron system which can be directly related to the electronic structure of the medium. There are three conventionally measurable quantities: the positron lifetime, the angular correlation of 2γ annihilation radiation, and the Doppler-broadened annihilation line shape. The positron lifetime measurement is effective in revealing different positron states in matter. The lifetime values themselves are strongly affected by electron-positron correlation effects. The lifetime of free positrons in solids is 0.1-0.4 ns. The angular correlation measurements are widely used to investigate electron momentum densities in metals and alloys. The width of an angular correlation curve is typically about 10 mrad. The line shape of the annihilation radiation also reflects the momentum distribution of annihilated electrons, but with a poor resolution. However, it provides a fast method to label the state of a sample, e.g., with respect to defect concentration.

Positronium formation is often present especially in molecular materials and it brings about considerable changes in the annihilation characteristics. Self-annihilation of parapositronium produces a narrow component on the angular correlation curve. The annihilation rate of the orthopositronium through the pick-off processes is relatively slow leading to lifetimes of about 1-4 ns.

Positrons are widely used to study momentum density of solids. The extreme sensitivity of positrons to crystal imperfections has found a variety of applications in the physics of crystal defects. These topics, together with applications in ionic crystals, are reviewed in this book. Other fields, such as low-energy positron beams, cluster and bubble formation in noble gases, molecular solids, and positronium chemistry, which are also under active investigation, have been only very brielfy discussed in this chapter.

References

1.1 R.M. Lambrecht: "Antimatter-Matter Interactions, 1. Positron and Positronium, a Bibliography 1930-1974"; Tech. Rpt. BNL 50510, Brookhaven National Laboratory (1975)

1.2 A.T. Stewart, L.O. Roellig (eds.): *Positron Annihilation* (Academic Press, New York 1967)

1.3 Proc. of 2nd Int. Conf. on Positron Annihilation, Kingston, Ontario 1971 (unpublished)

1.4 P. Hautojärvi, A. Seeger (eds.): Proc. of 3rd Int. Conf. on Positron Annihilation, Offprints from Appl. Phys. *3-5*

1.5 Proc. of 4th Int. Conf. on Positron Annihilation, Helsingør, Denmark 1976 (unpublished)

1.6 R.N. West: Adv. Phys. *22*, 263 (1973)

1.7 I.Ya. Dekhtyar: Phys. Rpt. C *9*, 243 (1974)

1.8 V.I. Goldanskii: At. Energy Rev. *6*, 3 (1968)

1.9 R.M. Singru, K.B. Lal, S.J. Tao: Atomic Data and Nuclear Data Tables *17*, 271 (1976)

1.10 P.A.M. Dirac: Proc. Camb. Phil. Soc. Math. Phys. Sci. *26*, 361 (1930)

1.11 R. Myllylä: Nucl. Instr. Methods *148*, 267 (1978)

1.12 H. Weisberg, S. Berko: Phys. Rev. *154*, 249 (1967)

1.13 V.H.C. Crisp, I.K. MacKenzie, R.N. West: J. Phys. E *6*, 1191 (1973)

1.14 J.L. Campbell: Appl. Phys. *13*, 365 (1977)

1.15 J.D. McGervey, V.F. Walters: Phys. Rev. B *2*, 2421 (1970)

1.16 I.K. MacKenzie, B.T.A. McKee: Appl. Phys. *10*, 245 (1976)

1.17 O. Brümmer, G. Brauer, V. Andrejtscheff, L. Käubler: Phys. Status Solidi (b) *71*, 59 (1975)

1.18 A. Ore: Naturvidenskap Rekke No. 9, Univ. i Bergen Årbok (1949)

1.19 P. Colombino, B. Fiscella: Nuovo Cimento B *3*, 1 (1971)

1.20 J. Arponen, E. Pajanne: J. Phys. C *8*, L152 (1975)

1.21 A. Bisi, A. Fiorentini, L. Zappa: Phys. Rev. *131*, 1023 (1963)

1.22 P.G. Coleman, T.C. Griffith, G.R. Heyland: Appl. Phys. *4*, 89 (1974)

1.23 D.G. Costello, D.E. Groce, D.F. Herring, J.Wm. McGowan: Phys. Rev. B *5*, 1433 (1972)

1.24 W.E. Kauppila, T.S. Stein, G. Jesion, M.S. Sababneh, V. Pol: Rev. Sci. Instrum. *48*, 822 (1977)

1.25 W.E. Kauppila, T.S. Stein, G. Jesion: Phys. Rev. Lett. *36*, 580 (1976)

1.26 W. Brandt: In *Radiation Effects on Solid Surfaces*, ed. by M. Kaminsky, Advantages in Chemistry Series, No. 158 (American Chemical Society, Washington D.C. 1976) Chap.9, pp.219-244

1.27 W.E. Caswell, G.P. Lepage, J. Sapirstein: Phys. Rev. Lett. *38*, 488 (1977)

1.28 M. Deutsch: Phys. Rev. *82*, 455 (1951)
1.29 K.F. Canter, A.P. Mills Jr., S. Berko: Phys. Rev. Lett. *33*, 7 (1974)
1.30 K.F. Canter, A.P. Mills Jr., S. Berko: Phys. Rev. Lett. *34*, 177 (1975)
1.31 A.P. Mills Jr., S. Berko, K.F. Canter: Phys. Rev. Lett. *34*, 1541 (1975)
1.32 T.C. Griffith, G.R. Heyland: Phys. Rpt. C *39*, 169 (1978)
1.33 K.F. Canter, J.D. McNutt, L.O. Roellig: Phys. Rev. A *12*, 375 (1975)
1.34 P. Hautojärvi, K. Rytsölä, P. Tuovinen, A. Vehanen, P. Jauho: Phys. Rev. Lett. *38*, 842 (1977)
1.35 M.J. Stott, E. Zaremba: Phys. Rev. Lett. *38*, 1493 (1977)
1.36 M. Manninen, P. Hautojärvi: Phys. Rev. B *17*, 2129 (1978)
1.37 P. Hautojärvi, K. Rytsölä, P. Tuovinen, P. Jauho: Phys. Lett. A *57*, 175 (1976)
1.38 A.L. Fetter: In *The Physics of Liquid and Solid Helium*, ed. by K.H. Benneman, J.B. Ketterson (Whiley and Sons, New York 1976) pp.207-305
1.39 W. Brandt: Ref. 1.2, pp.155-182
1.40 W. Brandt, J. Wilkenfeld: Phys. Rev. B *12*, 2579 (1975)
1.41 W. Brandt, G. Coussot, R. Paulin: Phys. Rev. Lett. *23*, 522 (1969)
1.42 O.E. Mogensen, G. Kvajič, M. Eldrup, M. Miloševič-Kvajič: Phys. Rev. B *4*, 71 (1971)
1.43 G. Coussot, R. Paulin: J. Appl. Phys. *43*, 1325 (1972)
1.44 M. Eldrup, O. Mogensen, G. Trumpy: J. Chem. Phys. *57*, 495 (1972)
1.45 O.E. Mogensen, M. Eldrup: "Positronium Bloch Function, and Trapping of Positronium in Vacancies, in Ice", Risø Rpt. No. 366, Risø National Laboratory, Denmark (1977)
1.46 P. Hautojärvi, I. Lehmusoksa, V. Komppa, E. Pajanne: J. Non-Cryst. Solids *18*, 395 (1975)
1.47 K.P. Singh, R.N. West, A. Paul: J. Phys. C *9*, 305 (1976)
1.48 P. Hautojärvi, A. Vehanen, V. Komppa, E. Pajanne: J. Non-Cryst. Solids *29*, 365 (1978)
1.49 B. Lévay, P. Hautojärvi: J. Phys. Chem. *76*, 1951 (1972)
1.50 H.J. Ache: Angew. Chem. Int. Edit. Engl. *11*, 179 (1972)
1.51 V.I. Goldanskii, V.P. Shantarovich: Appl. Phys. *3*, 335 (1974)
1.52 A.P. Buchikhin, V.I. Goldanskii, A.O. Tatur, V.P. Shantarovich: Zh. Eksp. Teor. Fiz. *60*, 1136 (1971) [Engl. transl.: Sov. Phys.-JETP *33*, 615 (1971)]
1.53 S.J. Tao: J. Chem. Phys. *56*, 5499 (1972)
1.54 O.E. Mogensen: J. Chem. Phys. *60*, 998 (1974)
1.55 P. Jansen, O.E. Mogensen: Chem. Phys. *25*, 75 (1977)

2. Electron Momentum Densities in Metals and Alloys

P.E. Mijnarends

With 15 Figures

Among the methods available for the study of the electronic structure of solids, the investigation of the electron momentum density with the aid of low-energy positron annihilation takes a place of increasing importance. During the 35 years that have elapsed since the first observation of the angular correlation between the two annihilation quanta by BERINGER and MONTGOMERY [2.1] the positron annihilation technique has developed into a method capable of yielding relevant information about various aspects of the electronic structure such as Fermi surface dimensions and wave functions. So far, the technique has been applied mainly to pure metals which have served as a practicing ground. For metals, however, the dimensions of the Fermi surface may be obtained by other methods such as the de Haas-van Alphen (dHvA) effect, the magnetoresistance effect, the rf size effect, etc., that are often capable of a greater precision than positron annihilation. Lately, nondilute disordered alloys and high-temperature phases of ordered materials have been gaining interest. In these systems excessive electron scattering due to the short mean free path of the electrons precludes the use of the above-mentioned methods at solute concentrations $\gtrsim$1%. At higher concentrations, other methods like the measurement of Kohn anomalies in X-ray diffuse scattering [2.2] or inelastic neutron scattering [2.3], and the Faraday effect [2.4,5] can still be used for investigations of the Fermi surface geometry. The Compton effect [2.6] is able, in principle, to provide direct information about the entire electron momentum distribution, but cannot compete in resolution with positron annihilation. Thus, positron annihilation is one of the very few useful techniques for the study of nondilute alloys, although it is not entirely free of problems. Positrons in crystalline solids have a high affinity to low-density lattice defects, but this problem can be overcome in many cases by a careful preparation of the specimens. Moreover, the recent development of two-dimensional detector geometries has greatly increased the resolution with which the momentum distribution may be studied. Hence a considerable increase in activity in the field of momentum density studies with positrons may be anticipated in the near future. Much of the work done with the conventional long-slit geometry will sooner or later be repeated with two-dimensional instruments, and this may well result in a revision of some of the earlier conclusions. The present moment therefore may be well suited to survey the field and to give a review of the work done up to now.

In this chapter we shall focus on the work done in the last five to ten years, since earlier work is suitably covered in previous reviews by STEWART and ROELLIG [2.7], WEST [2.8], DEKHTYAR [2.9], BERKO and MADER [2.10], and BERKO [2.11]. We shall also limit ourselves to "perfect" metals and alloys, as the defect solid state is extensively discussed in other chapters of this book. This also implies that only perfectly ordered and perfectly disordered substitutional alloys will be considered and that the complex effects of clustering will be ignored. The chapter divides into two parts. In the first part the present state of the art will be reviewed. In Sect.2.1 the independent-particle theory of the two-photon momentum density in metals and alloys will be treated, followed by a discussion of the various interactions between the positron and its environment in the form of positron-electron and positron-phonon interactions. The relation between the momentum density, the electron and positron wave functions, and the electronic band structure will be the subject of Sect.2.2. The following section is dedicated to a discussion of recent developments in the experimental techniques, both in the measurement of the angular correlation or Doppler broadening curves and in the subsequent retrieval of the momentum distribution from these curves. Sections 2.4 and 2.5 together form the second part of this chapter in which the recent work in metals and alloys is reviewed.

2.1 Theory

The theory of the two-photon momentum density as it will be discussed in Sect. 2.1.1 is an independent-particle theory that ignores the correlations of the two annihilating particles with each other and with the surrounding electrons. At first sight it seems strange that such a coarse approximation can lead to useful results, because it is known that calculations of positron lifetimes based on this model fail to reproduce the experimental values by an order of magnitude. As known, the positron lifetime measures the electron density at the site of the positron. Many-body theory shows that correlation effects are indeed essential in explaining positron lifetimes and positron thermalization, but that the momentum dependence of these effects is relatively weak. Consequently, independent-particle theory forms a very good first approximation, to which many-body corrections can be applied at a later stage. The present section will therefore first deal with the two-photon momentum density in the independent-particle model (IPM) before addressing itself to the many-body corrections to the momentum density (Sect.2.1.2), the problem of positron thermalization, and various thermal effects (Sect.2.1.3).

2.1.1 Momentum Density

A thermalized positron in a metal annihilates with an electron under emission of two or three gamma quanta. The two-photon process, in which two 511 keV γ quanta are emitted practically collinearly in opposite directions, has a probability which in a metal with random spin states is 372 times larger than that of the 3γ decay mode (in ferromagnetic media, however, this ratio may be different). The 3γ decay mode can therefore be ignored in most cases.

In the IPM, the probability $\Gamma(\underline{p})$ per unit of time of annihilation under emission of a photon pair carrying away a total momentum $\hbar p$ is[1] [2.12]

$$\Gamma(\underline{p})d\underline{p} = (2\pi)^{-3}\pi r_0^2 c\rho(\underline{p})d\underline{p} \quad , \tag{2.1}$$

with

$$\rho(\underline{p}) = \sum_{\text{occ.}} \left| \int \exp(-i\underline{p} \cdot \underline{r})\psi(\underline{r})\psi_+^*(\underline{r})d\underline{r} \right|^2 \tag{2.2}$$

being called the "photon-pair momentum density" or briefly "momentum density". Here $r_0 = e^2/mc^2$ is the classical electron radius, $\psi(\underline{r})$ and $\psi_+(\underline{r})$ represent the electron and positron wave functions respectively, and the summation extends over all occupied electron and positron states.

At temperature T the thermalized positron is near the bottom of the positron conduction band (the number of positrons present in the sample at any one time is of the order one). Its momentum distribution has the Boltzmann form

$$f_+(p_+) = (\pi^{\frac{1}{2}}x)^{-3} \exp(-p_+^2/x^2) \tag{2.3}$$

where $x = (2m^*k_BT)^{\frac{1}{2}}$, m^* is the positron effective mass and k_B the Boltzmann constant. The distribution of the electrons over the available states labelled by the wave vector $\underline{k}$ is governed by the Fermi-Dirac distribution function

$$f[E(\underline{k})] = [\exp\{[E(\underline{k}) - E_F]/k_BT\} + 1]^{-1} \quad , \tag{2.4}$$

where E_F is the Fermi energy. Since the annihilation photons carry away the combined momenta of the electron and the positron, the momentum distribution $\Gamma(\underline{p},T)$ at finite temperatures is given by the convolution of (2.3) and (2.1). At T = 0, the positron is in its ground state $\underline{k}_+ = 0$ and $f_+(p)$ is a delta function at p = 0, while $f[E(\underline{k})]$ is unity for occupied states $\underline{k}$ and zero if the state is empty. Hence

---

[1] Atomic units $\hbar = m = e = 1$ will be used throughout with momenta expressed in inverse Bohr radii a_0^{-1} (where $a_0=\hbar^2/me^2$). Energies, however, are expressed in Rydbergs.

the momentum distribution of a free-electron gas at T = 0 with Fermi momentum p_F is given by a step function $\rho(\underline{p}) = \theta(p_F - p)$. In a periodic crystal of volume V the electrons are in Bloch states $\psi_{\underline{k},j}(\underline{r}) = V^{-\frac{1}{2}} u_{\underline{k},j}(\underline{r}) \exp(i\underline{k} \cdot \underline{r})$ where j labels the energy band. Equation (2.2) can then be written

$$\rho(\underline{p}) = 2 \sum_{\underline{k},j} f[E_j(\underline{k})] |A_j(\underline{p},\underline{k})|^2 \quad , \tag{2.5}$$

where

$$A_j(\underline{p},\underline{k}) = \int \exp(-i\underline{p} \cdot \underline{r}) \psi_{\underline{k},j}(\underline{r}) \psi_+^*(\underline{r}) d\underline{r} \tag{2.6}$$

$$= V^{-\frac{1}{2}} \int \exp[-i(\underline{p} - \underline{k}) \cdot \underline{r}] u_{\underline{k},j}(\underline{r}) \psi_+^*(\underline{r}) d\underline{r} \quad . \tag{2.7}$$

The factor of two takes account of spin degeneracy in nonmagnetic systems. By making use of the periodicity of $u_{\underline{k},j}$ and ψ_+ one can reduce the integration in (2.6) to one over the unit cell

$$A_j(\underline{p},\underline{k}) = (N/\Omega)^{\frac{1}{2}} \delta(\underline{p} - \underline{k} - \underline{G}) \int_{\text{cell}} \exp[-i(\underline{p} - \underline{k}) \cdot \underline{r}] u_{\underline{k},j}(\underline{r}) \psi_+^*(\underline{r}) d\underline{r} \quad . \tag{2.8}$$

Here $\underline{G}$ is a vector of the reciprocal lattice and use has been made of

$$V^{-\frac{1}{2}} \sum_n \exp[-i(\underline{p} - \underline{k}) \cdot \underline{R}_n] = (N/\Omega)^{\frac{1}{2}} \delta(\underline{p} - \underline{k} - \underline{G}) \quad , \tag{2.9}$$

valid for a rigid lattice of N atoms (one per unit cell of volume Ω) situated at the positions $\underline{R}_n$. A generalization to lattices with more than one atom per unit cell has been given by BRANDT et al. [2.13]. From (2.5-8) it follows that an occupied Bloch state $\underline{k}$ in the jth band not only contributes to the momentum density $\rho(\underline{p})$ at the photon-pair momentum $\underline{p} = \underline{k}$ ("Normal" contribution), but also at $\underline{p} = \underline{k} + \underline{G}$ ("Umklapp" contribution). The magnitude of each contribution is given by $|A_j(\underline{p},\underline{k})|^2$ and depends on the overlap of the electron and positron wave functions. Thus, as BERKO and PLASKETT [2.14] (see also [Ref.2.11, p.287]) have shown, full bands give a momentum density that is continuous in $\underline{p}$ space, while partly filled bands give rise to discontinuities in $\rho(\underline{p})$ at those momenta $\underline{p}$ for which $\underline{k} = \underline{p} - \underline{G}$ is situated on the Fermi surface in the first Brillouin zone [provided that the factor $|A_j(\underline{p},k)|^2$ multiplying f in (2.5) is not zero]. The result is that the total momentum density $\rho(\underline{p})$ consists of a continuous underground with superimposed on it a series of Fermi surfaces centered at the reciprocal lattice points and each modulated by $|A_j(\underline{p},\underline{k})|^2$. Thus, although the Fermi surface is a periodic function of $\underline{k}$ in the extended zone scheme, $\rho(\underline{p})$ is aperiodic in $\underline{p}$ space. It does, however, possess the full point symmetry of the crystal lattice.

Sometimes it is useful to display the Bloch character of the electron and positron wave functions explicitly by writing

$$\psi_{\underline{k},j}(\underline{r}) = V^{-\frac{1}{2}} \sum_{\underline{G}} a_{\underline{G}}^{(j)}(\underline{k}) \exp[i(\underline{k} + \underline{G}) \cdot \underline{r}] \quad . \tag{2.10}$$

Adoption of a similar formula with coefficients $b_{\underline{G}''}$ for the positron in its ground state $\underline{k}_+ = 0$ and substitution into (2.8) yields for the momentum density $\rho_{\underline{k}}(\underline{p})$ contributed by this electron state

$$\rho_{\underline{k},j}(\underline{p}) = 2f[E_j(\underline{k})] \sum_{\underline{G},\underline{G}'} \left|a_{\underline{G}}^{(j)}(\underline{k}) b_{\underline{G}+\underline{G}'}^{*}\right|^2 \delta(\underline{p} - \underline{k} - \underline{G}') \quad , \tag{2.11}$$

where $\underline{G}' = \underline{G}'' - \underline{G}$. The coefficient b_0 is by far the largest of the positron Fourier coefficients, and for nearly-free conduction electrons with $\underline{k}$ not too close to the Brillouin zone boundary $|a_{\underline{G}=0}(\underline{k})| > |a_{\underline{G}\neq 0}(\underline{k})|$. Hence it follows from (2.11) that in this case the weight of the Fermi surface in the first zone will be much larger than the weights of the Fermi surfaces in the higher zones in the extended zone scheme.

So far we have concentrated on periodic systems. In systems with substitutional disorder $\underline{k}$ is not a good quantum number owing to the lack of periodicity of the lattice potential, and the concept of wave function loses its usefulness. Yet, various experiments have shown that the ensuing blurring of the Fermi surface in many cases is sufficiently small to warrant the continued use of concepts like wave vector and Fermi surface. An alternative formulation of $\rho(\underline{p})$ in terms of single-particle Green's functions is then

$$\rho(\underline{p}) = (2/\pi^2 V) \int d\underline{x} \int d\underline{y} \exp[-i\underline{p} \cdot (\underline{x} - \underline{y})] \int dEf[E(\underline{k})] \int dE_+ f_+(E_+)$$
$$\cdot \operatorname{Im}\{G_e^a(\underline{x}\underline{y},E)\} \operatorname{Im}\{G_+^a(\underline{x}\underline{y},E_+)\} \quad . \tag{2.12}$$

Here f and f_+ are again the electron and positron distribution functions and G_e^a and G_+^a the respective advanced Green's functions. HONG and CARBOTTE [2.15] have calculated the configurational average $\langle\rho(\underline{p})\rangle$ in the coherent-potential approximation (CPA). Care must be excercised in taking the configurational average of the product of Green's functions in (2.12). The electrons and positrons see the same impurity configuration at the same time but react differently to it, so the averages may not be taken separately. Model calculations of the smearing of the Fermi break due to the disorder show that the positron and the electron make comparable contributions. This is in contrast to thermal smearing which is caused practically entirely by the positron (Sect.2.1.3).

From the above discussion it will have become clear that the information obtainable from a study of $\Gamma(\underline{p})$ or $\rho(\underline{p})$ in metals and alloys is twofold. The Fermi breaks in $\Gamma(\underline{p})$ provide information on the Fermi-surface geometry through the distribution function $f[E_j(\underline{k})]$, while the overall shape of $\Gamma(\underline{p})$ gives information on the wave functions of the investigated system. Most of the remainder of this chapter will be concerned with the relationship between the momentum density and these two aspects of the electron band structure of solids: Fermi surface and wave functions. But first we shall discuss various many-body and thermal effects.

2.1.2 Many-Body Effects

In the previous section the theory of positron annihilation has been discussed on the basis of the IPM, i.e., without taking account of positron-electron ($e^+ - e^-$) and electron-electron ($e^- - e^-$) correlation effects. In reality a positron in an electron gas attracts a cloud of electrons around it that effectively screens its positive charge. Hence the electron density at the position of the positron is higher than predicted by the IPM and the annihilation rate will consequently be enhanced. This effect is obvious in positron lifetime studies where measured lifetimes are an order of magnitude shorter than expected on the basis of the IPM. Measured angular correlations, on the other hand, are reasonably well described by band-structure calculations ignoring correlation effects. The successful quantitative explanation of these two facts is one of the great successes of many-body theory.

The enhancement has been intensively studied with the help of Green's function techniques by a number of authors. An extensive review of the work up to 1973 has been given by WEST [2.8]. As known, the momentum distribution of a free-electron gas in the absence of interactions has the form of a unit step function: $\rho_0(\underline{p}) = \theta(1 - \gamma)$, with $\gamma = p/p_F$. When $e^- - e^-$ correlations are taken into account the discontinuity remains but becomes somewhat smaller, and tails form in the momentum range $1.0 < \gamma < 1.6$ [2.16]. By summing the set of all ladder graphs in the perturbation series for the positron-electron Green's function KAHANA [2.17,18] has shown that the enhancement arising from the inclusion of the $e^+ - e^-$ interactions has the effect of multiplying the IPM distribution $\rho_0(\underline{p})$ by a factor $\varepsilon(\underline{p}) = \rho_{enh.}(\underline{p})/\rho_0(\underline{p})$ given by

$$\varepsilon(\underline{p}) = a + b\gamma^2 + c\gamma^4 \quad , \quad \gamma = p/p_F \leqq 1 \quad . \tag{2.13}$$

The constants a, b and c depend on the electron density represented by $r_s = [4\pi n a_0^3/3]^{-1/3}$ and are listed in Table 2.1. Here n is the number of electrons per unit of volume. Table 2.2 shows some values of ε for typical values of γ. Annihilation rates obtained with this theory agree reasonably well with experimental lifetimes of WEISBERG and BERKO [2.19], except at low densities ($r_s > 5$ a.u.) where the

Table 2.1. Coefficients a, b and c in the expression for the momentum-dependent enhancement $\varepsilon(\gamma) = a + b\gamma^2 + c\gamma^4$ in metals of different densities [2.18]

r_s	a	b	c
2	3.480	0.600	0.387
3	6.172	1.292	0.967
4	11.225	2.940	2.617

Table 2.2. Enhancement factors for different momenta $\gamma = p/p_F$ in metals of different densities[a]

	$\gamma = p/p_F$				
r_s	0	0.2	0.5	0.7	1.0
2	3.480 *1.000*	3.505 *1.007*	3.654 *1.050*	3.867 *1.111*	4.467 *1.284*
3	6.172 *1.000*	6.225 *1.009*	6.555 *1.062*	7.037 *1.140*	8.431 *1.366*
4	11.225 *1.000*	11.347 *1.011*	12.124 *1.080*	13.294 *1.184*	16.782 *1.495*

[a]The entries in italics give the enhancement normalized to unity at zero momentum

ladder approximation diverges [2.20,21]. The momentum dependence of $\varepsilon(\underline{p})$ is significant but not strong. For momenta $p > p_F$ the $e^+ - e^-$ interactions cancel the tails resulting from the $e^- - e^-$ correlations to a great extent [2.22]. MAJUMDAR [2.23] has shown formally that the position of the discontinuity at the Fermi surface is not affected by the $e^+ - e^-$ interaction. Hence the net effect of this interaction is to increase the discontinuity at the Fermi surface. Figure 2.1 shows the effects of the $e^- - e^-$ and $e^+ - e^-$ correlations on $\rho(\underline{p})$ for an electron gas with $r_s = 4$ a.u. A similar graph by BERKO [2.11] for $r_s = 2$ a.u. shows smaller correlation effects. The angular correlation curves derived from $\rho(\underline{p})$ display a noticeable bulge when compared with the free-electron parabola which can easily be observed experimentally in metals like sodium, where lattice effects are weak. This is illustrated by Fig.2.2 which shows the angular correlation of Na measured by DONAGHY and STEWART [2.24]. A Gaussian core distribution has been subtracted from the data.

Later work concerns various refinements to this basic theory. The unphysical pileup of electron charge around the positron in the original formulation of Kahana, arising from summing only over the ladder graphs, can be avoided by also including other Feynman diagrams [2.25,26]. This procedure has little effect on the annihilation rate because the latter is determined by the electron density at the position of the positron, whereas the total displaced charge is sensitive to long-range

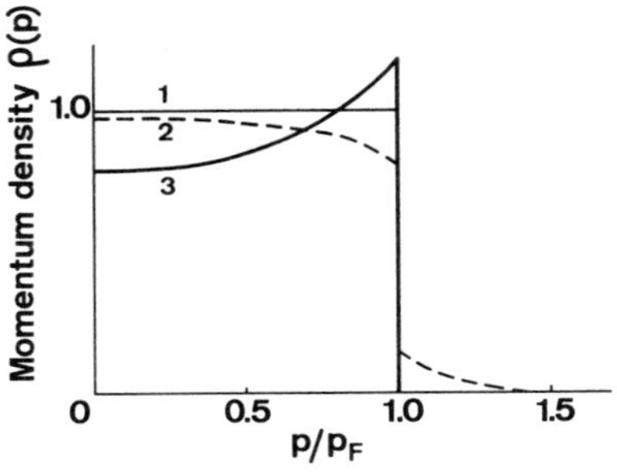

Fig.2.1. Momentum density $\rho(\underline{p})$ for r_s = 4 a.u. 1: noninteracting free electrons, 2: interacting electron gas [2.16], 3: with $e^+ - e^-$ interactions. Curves normalized to same total density. In modified form from BERKO [2.11]. Copyright 1977 McGraw-Hill, Inc. Used with permission of McGraw-Hill Book Company

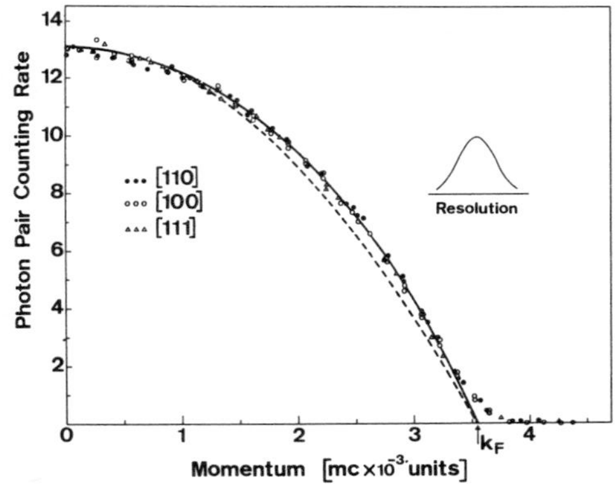

Fig.2.2. Angular correlation in sodium (r_s = 3.93 a.u.) measured by DONAGHY and STEWART. Dashed curve: free-electron parabola. Solid curve: theoretical prediction by Kahana. A Gaussian core contribution has been subtracted [2.24]

$e^+ - e^-$ correlations. Modifications to the screened interaction potential [2.20, 27-29] and corrections to the ladder sums due to particle-hole interactions [2.20, 22,28,30] likewise do not significantly alter the rate of annihilation. BHATTACHARYYA and SINGWI [2.21] have shown that the divergence at low densities [2.31] can be prevented by explicitly introducing three-particle correlations into a theory of dielectric screening ([2.32] and papers cited therein). After a small correction due to core annihilation these authors obtained good agreement with the experimental lifetimes of [2.19], also for the low-density metals Rb and Cs. The extreme low-density regime ($r_s \gg 5$ a.u.) in which the electrons eventually form a Wigner crystal has been investigated by ARPONEN [2.33], and MAJUMDAR and co-workers [2.34,35]. The calculated annihilation rate approaches that of spin-averaged positronium, i.e., $2 \times 10^9\ s^{-1}$.

All studies discussed so far concern an interacting electron gas with a uniform background of positive charge ("jellium" model). Inclusion of the lattice introduces tails in the momentum distribution due to Umklapp and core annihilations. HEDE and CARBOTTE [2.36] have studied the enhancement of the Umklapp contributions in the presence of a weak lattice potential. The enhancement factor at $p > p_F$ shows a very similar momentum dependence to that for $p < p_F$, increasing somewhat near the edge of the Fermi spheres. FUJIWARA and co-workers [2.37-39] have considered the effect of electronic intraband and interband transitions under the influence of the $e^+ - e^-$ interaction in a two-band nearly free-electron model. Although there is a net enhancement of the annihilation rate they find the Umklapp components dehanced as a result of interband transitions. The dehancement is expected to be the largest in solids with full bands [2.40]. A calculation of the ladder approximation to core annihilation by CARBOTTE and SALVADORI [2.41,42] yields core enhancement factors that are smaller than those for the conduction electrons but again very little momentum dependent, thus preserving the predictions of the IPM. The theory, when applied to crystalline argon where all electrons are tightly bound, predicts the lifetime and the angular correlation rather well [2.43], but it seems to be less successful for the heavier inert-gas solids [2.44]. The implicit assumption of coherence between the IPM wave function and the correlation correction valid for Bloch electrons is, however, not correct for strongly localized electrons, as pointed out by CHIBA et al. [2.45]. Their predicted annihilation rates based on incoherence show satisfactory agreement with experimental data for a large number of insulators. BRANDT and REINHEIMER [2.46] have discussed the enhancement in semiconductors.

It is generally felt that the enhancement of the conduction electrons in simple metals with a weak lattice potential is now well understood, although there is still some controversy about the enhancement or dehancement of the Umklapp components. Accurate numerical calculations of core enhancement factors are still lacking. The least satisfactory is the situation in d metals where a relatively wide d band overlaps and hybridizes with the conduction band. Persistent discrepancies between experimental and calculated momentum distributions (Sect.2.4.5) suggest the presence of significant enhancement effects in these metals but it is unclear how they could be corrected for at present.

2.1.3 Positron Thermalization, Effective Mass and Other Thermal Effects

a) Thermalization

The problem of positron thermalization in solids has received considerable attention. In brief, a positron emitted by a radioactive source with an energy of several 100 keV is slowed down to energies of the order of 0.01 eV in a time that is a small fraction of its lifetime of a few times 10^{-10} s. At first, energy dissipation takes place through ionization, excitation of atoms and plasmon excitation, and later by creation of electron-hole pairs and phonons. LEE-WHITING [2.47] predicted a time of 3×10^{-12} s required for thermalization down to 0.025 eV by assuming a screened Coulomb interaction between the positron and a conduction electron and ignoring all other interactions. CARBOTTE and ARORA [2.48] treated the thermalization by electron-hole pair creation with Green's function techniques and found that at temperatures of the order of 100 K and high electron densities thermalization may be incomplete. The relaxation of the distribution has been studied by WOLL and Carbotte [2.49] using a Boltzmann-equation approach. DEBENEDETTI et al. [2.12] were the first to consider thermalization by positron-phonon scattering, but the subject was nor pursued further until MIKESKA [2.50] pointed out the importance of this interaction. At low energies positron-phonon scattering is more effective in dissipating the positron energy than electron-hole pair creation, and hence considerably shortens the time required for thermalization [2.51,52]. The theoretical work on positron thermalization has been reviewed by BERGERSEN and PAJANNE [2.53]. The consensus of opinion among theorists is that positrons in a solid should reach thermal equilibrium down to temperatures of a few tens of degrees before being annihilated.

Experiment has long been in disagreement with this view. As discussed earlier, the momentum density in metals with partially filled bands at zero temperature displays a discontinuity at the Fermi surface, even in the presence of $e^- - e^-$ and $e^+ - e^-$ interactions. The thermal broadening of the Fermi-Dirac distribution at nonzero temperatures is beyond present limits of detection, but the smearing of this discontinuity due to positron thermal motion can be measured. The positron momentum distribution is given by (2.3). A measurement of the smearing at various temperatures yields the effective positron temperature $T_{eff} = (m^*/m)T$, but corrections have to be made for thermal expansion and possibly for the effect of finite mean free path [2.54]. STEWART and co-workers [2.55-58] have measured the smearing for the alkali metals. They found that at temperatures down to about 200 K, T_{eff} equalled the specimen temperature T if m^*/m was assumed to range from 1.8 ± 0.3 for Li and Na to 2.3 ± 0.3 for Rb. At lower temperatures the curve showing T_{eff} vs T levelled off, indicating that the positron had reached a minimum energy and did not attain thermal equilibrium. Measured minimum values for T_{eff} ranged from 60 ± 50 K for Rb to 200 ± 80 K for Li. These relatively high minimum temperatures would rather severely limit the momentum resolution attainable with the positron annihilation technique. However, recent, more accurate measurements by KUBICA and STEWART [2.59] have produced evidence for thermalization down to 10-30 K and hence the limitation on the resolution due to positron thermal motion is less serious than once feared.

Little is known about positron thermalization in semiconductors and insulators. Estimates of the thermalization time from measured positron mobilities in Ge and Si [2.60] yield values considerably smaller than the positron lifetimes in these materials and suggest complete thermalization even at a few degrees Kelvin.

b) Effective Mass

The high effective mass m^* of the positron still constitutes a problem. Contributions to this quantity come from the positron band structure [2.56,61], positron-electron [2.62,63], and positron-phonon [2.50-52] interactions. The first two processes each give rise to an effective mass of about (1.10-1.15) × m, i.e., much smaller than the experimental value. The positron-phonon interaction gives a negligible contribution (< 1%) to the effective mass, but accompanying quasiparticle lifetime effects cause a deformation of the Boltzmann distribution that leads to a larger thermal smearing of the Fermi break than expected from the small effective mass. If this deformed distribution is fitted with an effective Boltzmann distribution, the width of the latter is greater than that of the original distribution. This can be described by an apparent effective mass m^{**} that approaches the experimental value [2.51] but after more detailed calculations cannot fully account for it [2.64-66]. Since the experimental uncertainties are rather large (± 0.3) more accurate measurements of the effective mass would be valuable.

In semiconductors one would expect somewhat lower values of m^* than in metals because of the less efficient screening of the positron by the electrons due to the energy gap. SHULMAN et al. [2.67] have compared angular correlation curves for Ge at T = 300 K and 5 K and have used a somewhat different type of analysis. A remaining small discrepancy between the data and the model used by these authors may be due to neglect of the Debye-Waller factor (see below). The analysis resulted in $m^* = (1.23 \pm 0.17)m$, indeed significantly smaller than the value found for the alkali's. An even smaller value might be obtained if positron-phonon coupling in semiconductors would lead to a similar broadening of the Boltzmann distribution as in metals.

c) Other Thermal Effects

Besides the temperature-dependent smearing of Fermi breaks due to positron thermal motion, other thermal effects must be considered. First, thermally generated lattice defects can trap positrons. This effect is discussed elsewhere in this book. Secondly, the thermal expansion of the crystal lattice causes a corresponding contraction of the reciprocal lattice and an increase of the momentum density in order to preserve normalization. Thirdly, the temperature also affects the intensity of the high-momentum components (Umklapp contributions) of $\rho(\underline{p})$. In deriving (2.8) it was assumed that the ions are situated on a rigid lattice. The thermal vibrations of the ions about their equilibrium positions reduce the Umklapp contributions. This can be taken into account by the introduction of a Debye-Waller factor exp(-2W) in (2.8) [2.13]

$$|A_j(\underline{p},\underline{k})|^2 = N\delta(\underline{p} - \underline{k} - \underline{G}) \exp[-2W(\underline{G},T)]|\Phi_{\underline{k},j}(\underline{p})|^2 \quad , \qquad (2.14)$$

where

$$\Phi_{\underline{k},j}(\underline{p}) = \Omega^{-\frac{1}{2}} \int_{cell} \exp[-i(\underline{p} - \underline{k}) \cdot \underline{r}] u_{\underline{k},j}(\underline{r}) \psi_+^*(\underline{r}) d\underline{r} \qquad (2.15)$$

represents the Fourier transform of the wave-function product and it can be shown that for cubic materials (see, e.g., [2.68])

$$W = \frac{3G}{2Mk_B\theta_D}\left[\frac{1}{4} + (T/\theta_D)\phi(\theta_D/T)\right] \quad . \qquad (2.16)$$

Here M is the ionic mass in a.u., $k_B = 0.31686 \times 10^{-5}$ is the Boltzmann constant in a.u./K, θ_D is the Debye temperature, and $\phi(x)$ represents the Debye function. Contrary to X-ray diffraction, positron annihilation samples only the first few reciprocal lattice vectors and, except in light materials with a low θ_D, incorporation of the Debye-Waller factor will usually (but not always) mean a small correction. Table 2.3 shows values of this factor for a few materials.

Table 2.3. Debye-Waller factor $\exp[-2W(\underline{G},T)]$ for some materials[a]

	Structure	θ_D [K]	$(a/2\pi)\underline{G}$	$\exp[-2W(\underline{G},T)]$	
				T = 78 K	T = 300 K
Li	bcc	363	110 002	0.89 0.78	0.72 0.52
Na	bcc	150	110 002	0.90 0.81	0.69 0.47
K	bcc	100	110 002	0.94 0.89	0.73 0.53
Rb	fcc	58	111 002	0.86 0.82	0.58 0.48
Cs	fcc	42	111 002	0.86 0.81	0.56 0.46
Al	fcc	390	111 002	0.97 0.96	0.92 0.90
Cu	fcc	330	111 002	0.98 0.97	0.94 0.93
Ge	diamond	290	111 220	0.99 0.98	0.97 0.93
Pb	fcc	88	111 002	0.97 0.95	0.88 0.84

[a]Debye temperatures θ_D from [2.54]

2.2 Wave Functions

2.2.1 Positron Wave Function

If positron annihilation momentum density measurements are made in order to obtain information about the electronic structure of solids, some previous knowledge of the spatial distribution of the positron in the solid is essential. As a result of its positive charge, the positron is strongly repelled from the regions of the ion cores and hence annihilates preferentially with the outer electrons. Its spatial distribution in the lattice is determined by the lattice potential [2.69]. This potential differs from the one seen by the electrons mainly in its sign and in the absence of exchange terms. An additional potential term comes from $e^+ - e^-$ correlation. In metals this term can be neglected since the correlation energy is a slow function of electron density [2.70]. The corresponding potential will therefore be approximately constant over the interstitial region. In the high-electron-density region inside the ion cores the correlation potential is swamped by the much larger Hartree potential; moreover the wave function in that region is small. In substitutional binary alloys, however, there will be a considerable nonuniformity of the valence-electron density if the valence of the two constituents is not the same, and hence there will be a contribution to the positron potential from $e^+ - e^-$ correlation. In addition there is a potential term due to charge transfer effects.

The problem of positron band structure in a periodic lattice is similar to its electron counterpart. On account of its previous thermalization the positron annihilates from a state very close to $\underline{k}_+ = 0$ in its lowest energy band. Early wave-function calculations [2.14,71] were therefore based on the Wigner-Seitz method that yields accurate results in the vicinity of the nucleus but is unable to reproduce the correct lattice symmetry in the outer parts of the atomic polyhedron where the wave function is largest. The latter problem was solved [2.72,73] by expansion of the wave function in symmetrized plane waves. The wave function so obtained has the proper crystal symmetry but for an accurate description of its strong curvature in the core many plane waves are required. It also does not have a cusp at r = 0 as it should in order to satisfy the Schrödinger equation with a Coulomb potential. KUBICA and STOTT [2.74] combined the qualities of both methods by writing the wave function as a product of a slightly modified Wigner-Seitz function $u_{WS}(\underline{r})$, centered at $\underline{R}$, that has the correct behavior in the core region, and a smoothly varying pseudowave function ψ_{ps} that displays the correct angular behavior:

$$\psi_+(\underline{r}) = u_{WS}(\underline{r} - \underline{R})\psi_{ps}(\underline{r}) \quad . \tag{2.17}$$

ψ_{ps} satisfies a Schrödinger equation with a pseudopotential weak enough to justify the use of first-order perturbation theory, and can be represented by a rapidly converging plane-wave expansion. The method is not limited to crystal potentials of the muffin type[2] and can also be readily adapted to the calculation of positron wave functions in vacancies and alloys. A slight modification [2.75] seems to yield even weaker pseudopotentials.

Other methods used to calculate positron wave functions in pure metals are the Korringa-Kohn-Rostoker (KKR) and the augmented-plane-wave (APW) methods well known from electron band theory [2.76]. The fast approximation scheme of HUBBARD [2.77,78], based on the former, was used by MIJNARENDS [2.79] in early calculations on copper and iron. WAKOH et al. [2.80] employed the APW method in niobium and vanadium, while the MAPW method, which differs from the conventional APW method in that not only the wave function but also its first derivative is made continuous everywhere, was applied by BROSS and STÖHR [2.81] to the calculation of the positron wave function in Cu. The latter authors also calculated the energies of positron states at $\underline{k}_+ = 0$ belonging to representations other than Γ_1 and found that the next-higher state ($\Gamma_{2'}$) lies more than 2.7 Ry higher in energy.

The spatial distribution of positrons in a crystal is very sensitive to the presence of vacancies, voids, and other lattice defects. A vacancy represents a site with a potential sufficiently attractive to bind a positron. Annihilation from such a trapping site will be characterized by a long positron lifetime and a high valence to core electron ratio in the angular correlation curve. Atomic vacancy concentrations down to $\sim 10^{-7}$ can be measured by positron annihilation [2.82]. Other lattice defects will also affect the observed annihilation characteristics. The interaction between positrons and defects will be extensively discussed in Chaps.3 and 4.

In disordered alloys the problem of the spatial positron distribution is considerably more complex. Not only are there additional terms in the lattice potential coming from $e^+ - e^-$ correlation and charge transfer effects, but the disorder also prevents a description in terms of Bloch waves. Moreover, one constituent of the alloy may have a larger positron affinity than the other, resulting in preferential annihilation with electrons of the first. The difference in pseudopotential between the two kinds of atoms is a measure of their relative positron affinity [2.69] that is reflected in the slowly varying pseudowave function, whereas the Wigner-Seitz parts of the wave function are little affected by the alloying. Since the core-electron wave functions are also insensitive to the atomic environment one expects that the total core annihilation rate in an alloy would be given by a weighted sum of the core rates of the two pure metals, the weights being the positron pseudo-density at the two constituents. Thus it is possible to obtain information about

---

[2]As known, a muffin-tin potential is a potential that is locally spherically symmetric within spheres centered about the atoms in the unit cell and constant in the interstitial region.

positron affinities in alloys by studying the core annihilation rate vs. concentration from combined lifetime and angular correlation measurements [2.83,84]. The positron spatial distribution will also affect the shape of the angular correlation. The so-called h parameter for a polycrystal, defined by

$$h = N(p_z = 0)\Delta p_z / \int_{-\infty}^{\infty} N(p_z)dp_z \quad , \qquad (2.18)$$

contains the same information and may be a more sensitive indicator in cases where the core rates of the pure constituents and of the alloy are very similar [2.83]. KOENIG [2.85] has discussed the effects of charge transfer in the virtual-crystal and coherent-potential approximations.

Although the enhancement of the positron wave function at isolated impurities may be sizable, the pseudopotential differences are not large enough to result in a positron-impurity bound state, and the positron wave function extends over the entire crystal [2.69]. Clusters of atoms, however, may be sufficiently attractive to cause positron localization that may manifest itself in an additional smearing of the Fermi break. KUBICA et al. [2.84] have found evidence for positron localization in Li clusters in Li-Mg alloys from an analysis of the core rates and the smearing of the Fermi break (Sect.4.2.4). It is uncertain, however, whether this smearing is attributable to positron localization alone. Disorder scattering of the electrons will also contribute (Sect.2.1.1) and hence the positron localization effect may have been overestimated.

2.2.2 Electron Band Structure and Wave Functions

The technique of band-structure calculation in pure metals has undergone many years of development since 1933 when WIGNER and SEITZ [2.86] first applied the cellular method to sodium. It would be beyond the scope of this article to give a full review of the calculational methods that have been developed during that period. The numerous aspects of the subject are covered by a number of excellent review articles and books [2.76,87-89]. For a long time the efforts have been concentrated on an accurate determination of the energy bands $E(\underline{k})$. The calculation of wave functions for the subsequent computation of various quantities of interest in solid-state physics is much more recent. We shall therefore give some attention to the form of the wave functions and to those aspects of wave-function calculation that are of interest for momentum-density calculations. A comparison of the calculated distributions with experimental results will be made in Sect.2.4.

The band-structure methods most commonly employed in momentum-density calculations are the augmented-plane-wave (APW), the Korringa-Kohn-Rostoker (KKR), and the orthogonalized-plane-wave (OPW) methods. Of these, the OPW method is conceptually the most transparent and will therefore be dealt with first.

a) OPW Method

In many crystals the conduction electrons behave as if they are nearly free. Yet, a straightforward expansion of their wave function ψ in a Bloch sum of plane waves (2.10) converges only slowly because the rapid oscillations in the wave function near the nuclei require the presence of high momenta in (2.10). The convergence is greatly improved by expanding ψ in orthogonalized plane waves (OPW's) $\chi_{\underline{k}+\underline{G}}(\underline{r})$ obtained by making each plane wave orthogonal to the core orbitals according to

$$\chi_{\underline{k}+\underline{G}}(\underline{r}) = V^{-\frac{1}{2}}\{\exp[i(\underline{k} + \underline{G}) \cdot \underline{r}] - \mu_{\underline{k}+\underline{G}}\phi_{\underline{k}}(\underline{r})\} \quad . \tag{2.19}$$

Here $\phi_{\underline{k}}(\underline{r})$ is a Bloch sum of tightly bound nonoverlapping core orbitals $\phi_c(\underline{r})$ centered about the atom at $\underline{R}$

$$\phi_{\underline{k}}(\underline{r}) = \sum_{\underline{R}} \exp(i\underline{k} \cdot \underline{R})\phi_c(\underline{r} - \underline{R}) \quad , \tag{2.20}$$

while

$$\mu_{\underline{k}+\underline{G}} = \int \exp[i(\underline{k} + \underline{G}) \cdot \underline{r}]\phi^*_{\underline{k}}(\underline{r})d\underline{r} \tag{2.21}$$

causes $\chi_{\underline{k}+\underline{G}}$ to be orthogonal to $\phi_{\underline{k}}$. By the orthogonalization process a certain amount of high-momentum components is mixed into each plane wave which speeds up the rate of convergence of the expansion, but even so the number of OPW's required may run into several hundreds. The coefficients $a_{\underline{G}}(\underline{k})$ in the expansion of ψ into OPW's are found variationally by minimizing the expectation value of the one-electron Hamiltonian. This gives rise to a determinantal equation, the zeros of which are the energy eigenvalues.

The momentum density can now be calculated by substitution of (2.19) into (2.6), together with an appropriately chosen positron wave function. Expanding the latter into plane waves with coefficients $b_{\underline{G}''}$ one readily obtains

$$\rho(\underline{p}) = 2 \sum_{\underline{k}} f[E(\underline{k})] \left| \sum_{\underline{G},\underline{G}'} a_{\underline{G}}(\underline{k})b^*_{\underline{G}+\underline{G}'}\delta(\underline{p} - \underline{k} + \underline{G}') \right.$$

$$\left. - (1/V) \sum_{\underline{G},\underline{G}'} a_{\underline{G}}(\underline{k})b^*_{\underline{G}+\underline{G}'}\mu_{\underline{k}+\underline{G}} \int_V \exp[-i(\underline{p} + \underline{G} + \underline{G}') \cdot \underline{r}]\phi_{\underline{k}}(\underline{r})d\underline{r} \right|^2 \quad . \tag{2.22}$$

The integral in the second term extends over the entire crystal. By making use of the Bloch character of $\phi_{\underline{k}}$ it can be reduced to an integration over the atomic cell. Together with (2.9) one obtains

$$\rho(\underline{p}) = 2 \sum_{\underline{k}} f[E(\underline{k})] \Big| \sum_{\underline{G},\underline{G}'} a_{\underline{G}}(\underline{k}) b^{*}_{\underline{G}+\underline{G}'} [\delta(\underline{p} - \underline{k} + \underline{G}')$$

$$- (1/\Omega) \mu_{\underline{k}+\underline{G}} \mu^{*}_{\underline{p}+\underline{G}+\underline{G}'} \sum_{\underline{H}} \delta(\underline{p} - \underline{k} + \underline{H})] \Big|^{2} \quad . \qquad (2.23)$$

The delta function in the first term within the brackets limits $\underline{G}'$ to that reciprocal lattice vector that reduces $\underline{p}$ to a vector $\underline{k}$ in the first zone. In the second term there is no such limitation on the choice of $\underline{G}'$; the delta function in that term only expresses that $\underline{p}$ and $\underline{k}$ should differ by an arbitrary vector $\underline{H}$ of the reciprocal lattice.

It is of interest to compare the OPW expression (2.23), the results of which are shown schematically in Fig.2.3, with the plane-wave result (2.11). The first term in (2.23) represents the plane-wave contribution, divided into Normal ($\underline{G}' = 0$) and Umklapp ($\underline{G}' \neq 0$) terms. The second term represents the contribution of the orthogonalization terms. Since the μ's are Fourier transforms of the tight-binding core functions this contribution blends with the flat tail in the momentum distribution arising from annihilations with the core electrons. However, the presence of Fermi breaks in the orthogonalization term (Fig.2.3) distinguishes it from the true core contribution and shows that this term originates from electrons in the conduction band.

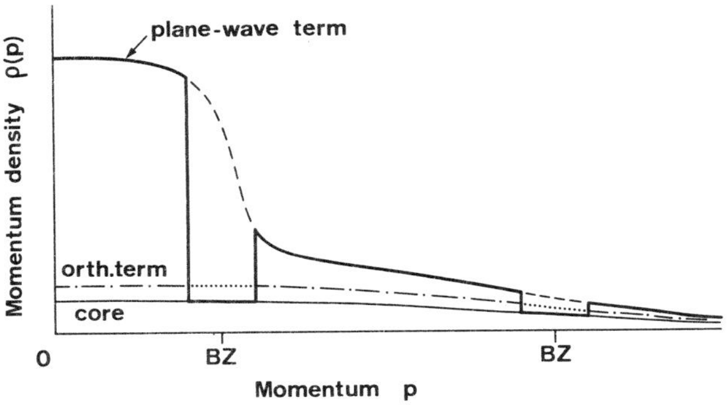

Fig.2.3. Momentum density for a metal calculated by the OPW method. The partially filled conduction band causes Fermi breaks in the plane-wave and orthogonalization contributions

The advantage of the OPW method is its ability to use potentials that need not have the muffin-tin form. The method can therefore be applied to materials with strongly directional covalent bonds. It is less well suited for solids with resonant d states like the transition and noble metals, although a modified version of the method has been applied to calculations of the band structure of niobium [2.90] and copper [2.91].

b) APW Method

The APW method is widely used in band-structure and momentum-density calculations and, like the OPW method, is based on a (modified) plane-wave expansion of the electron wave functions. It employs a muffin-tin potential (although methods have been devised to treat deviations from the muffin-tin form [2.92]), which makes it particularly suited for application to closely packed structures. The wave functions are expanded in linear combinations of "augmented" plane waves $\chi_{\underline{k}+\underline{G}}(\underline{r})$ with expansion coefficients $a_{\underline{G}}(\underline{k})$. In the interstitial region where the potential is assumed constant the natural form of the APW is a plane wave

$$\chi_{\underline{k}}(\underline{r}) = V^{-\frac{1}{2}} \exp(i\underline{k} \cdot \underline{r}) \quad , \quad r > r_i \quad . \tag{2.24}$$

Inside the muffin-tin sphere with radius r_i around the atom at $\underline{r}_n$ (there may be more than one atom per unit cell) the potential is spherically symmetric and the APW is written as

$$\chi_{\underline{k}}(\underline{r}) = 4\pi V^{-\frac{1}{2}} \exp(i\underline{k} \cdot \underline{r}_n)$$

$$\cdot \sum_L i^\ell j_\ell(kr_i) \frac{R_\ell(|\underline{r}'|,E)}{R_\ell(r_i,E)} Y_L^*(\hat{k}) Y_L(\hat{r}') \quad , \quad r < r_i \quad , \tag{2.25}$$

where $\underline{r}' = \underline{r} - \underline{r}_n$, the index L stands for the combination ℓ,m and $R_\ell(r,E)$ is the solution of the radial Schrödinger equation for the energy E that is regular at r = 0. The other factors in (2.25) ensure that the two parts of the APW match at the surface $|\underline{r}'| = r_i$ of the muffin-tin spheres, although its derivative is discontinuous. As in the OPW method, the expansion coefficients $a_{\underline{G}}(\underline{k})$ are determined variationally.

LOUCKS [2.93] has derived a formula for the momentum density based on the APW method. He assumed a positron wave function that is constant in the interstitial region and satisfies the radial Schrödinger equation with appropriate boundary conditions inside the muffing-tin spheres. We shall extend his derivation slightly by using a more realistic plane-wave positron wave function.

Since the APW's defined by (2.24,25) have the Bloch form the integration in the overlap integral (2.6,7) can be reduced following (2.8) to an integral over the unit cell. This integral is then expressed as the sum of two contributions, one from the interstitial region and the other from inside the spheres. The first contribution is obtained by integrating over the entire unit cell and then subtracting the contribution from the regions inside the spheres

$$\int_{\text{interst.}} = (1/\Omega) \sum_{\underline{G},\underline{G}''} a_{\underline{G}}(\underline{k}) b^*_{\underline{G}''}[\Omega\delta(\underline{p} - \underline{k} + \underline{G}'' - \underline{G})$$

$$- \sum_{\underline{r}_n} \exp(i\underline{K} \cdot \underline{r}_n) \cdot 4\pi r_i^2 j_1(Kr_i)/K] \quad , \tag{2.26}$$

where $\underline{K} = \underline{k} - \underline{p} + \underline{G} - \underline{G}''$. Inside the spheres the calculation is simplified considerably if one replaces the plane-wave expansion of $\psi^*_+(\underline{r})$ by its spherical average $\psi^*_0(r)$. Substitution of (2.25) into (2.6) followed by some straightforward analysis then yields

$$\int_{\text{sphere at } \underline{r}_n} = \Omega^{-1} \sum_{\underline{G}} a_{\underline{G}}(\underline{k})\delta(\underline{p} - \underline{k} - \underline{G} - \underline{K}) \exp[i(\underline{k} + \underline{G} - \underline{p}) \cdot \underline{r}_n]$$

$$\cdot \sum_{\ell} (2\ell + 1) j_\ell(|\underline{k} + \underline{G}|r_i) P_\ell(\cos\theta_{\underline{G}}) T_\ell(p,E_j) \quad , \tag{2.27}$$

where $\theta_{\underline{G}}$ is the angle between $\underline{k} + \underline{G}$ and $\underline{p}$, and

$$T_\ell(p,E_j) = 4\pi V^{\frac{1}{2}} \int_0^{r_i} j_\ell(pr)[R_\ell(r,E_j)/R_\ell(r_i,E_j)]\psi^*_0(r) r^2 dr \quad . \tag{2.28}$$

Hence, upon summation over all spheres in the unit cell, one obtains for the overlap integral

$$A_j(\underline{p},\underline{k}) = \Omega^{-1} \sum_{\underline{G}} a_{\underline{G}}(\underline{k}) \Bigg\{ \sum_{\underline{G}''} b^*_{\underline{G}''}[\Omega\delta(\underline{p} - \underline{k} + \underline{G}'' - \underline{G})$$

$$- \sum_{\underline{K}} S(\underline{K})(4\pi r_i^2/K) j_1(Kr_i)\delta(\underline{p} - \underline{k} - \underline{G} + \underline{K} + \underline{G}'')]$$

$$+ \sum_{\underline{K}} S(\underline{K})\delta(\underline{p} - \underline{k} - \underline{G} - \underline{K}) \sum_{\ell} (2\ell + 1) j_\ell(|\underline{k} + \underline{G}|r_i) P_\ell(\cos\theta_{\underline{G}}) T_\ell(p,E_j) \Bigg\} \quad , \tag{2.29}$$

after which the momentum density follows from (2.5). In (2.29)

$$S(\underline{K}) = \sum_{\underline{r}_n} \exp(i\underline{K} \cdot \underline{r}_n) \tag{2.30}$$

represents the structure factor and the summation extends over all atoms in the unit cell.

At first sight the computational effort involved in the numerical evaluation of (2.29) over all of momentum space seems formidable, but when suitably organized the calculations become more tractable. The most time-consuming parts of the calculation (apart from finding the eigenvalues and the $a_{\underline{G}}$'s) are the computation of $T_\ell(p,E)$ and the multiple sums in the first two terms of (2.29) (because of the large number of plane waves (100-200) necessary for an accurate description of the positron wave function). It is possible, however, to perform these parts of the computations once and for all for a specific material and store the results in the form of tables. Let $\underline{H}$ be the reciprocal lattice vector by which $\underline{p}$ and $\underline{k}$ differ: $\underline{p} = \underline{k} + \underline{H}$. A rather limited number of $\underline{H}$'s suffices to compute $\rho(\underline{p})$ up to the fifth or sixth zone in $\underline{p}$ space. Given a specific positron wave function it is then possible to perform the summation over $\underline{G}''$ and $\underline{K}$ in the first two terms of (2.29) and store the results in an array with indices corresponding to $\underline{G}$ and $\underline{H}$. The array moreover is symmetric and can be stored in the symmetric storage mode. The quantities $T_\ell(p,E)/p^\ell$ are smooth functions of both E and p and can be obtained by interpolation in a previously prepared two-dimensional table. The remaining quantities in (2.29) depend explicitly on $\underline{k}$, but their computation is not particularly time consuming. Applications of the APW method to the computation of two-photon momentum distributions will be discussed in Sect.2.4.

c) KKR and Related Methods

In the KKR or Green's function method, as developed by KORRINGA [2.94], and by KOHN and ROSTOKER [2.95], the wave function is expanded in spherical waves throughout the entire atomic cell,

$$\psi_{\underline{k}}(E,\underline{r}) = \sum_L C_L(\underline{k})R_\ell(r,E)Y_L(\hat{r}) \quad , \tag{2.31}$$

where, as before, inside the muffin-tin sphere $R_\ell(r,E)$ is the solution of the radial Schrödinger equation, while outside the muffin tin it is given in terms of the scattering phase shifts η_ℓ (with $\kappa = E^{\frac{1}{2}}$) by

$$R_\ell(r,E) = \exp(i\eta_\ell)[j_\ell(\kappa r)\cos\eta_\ell - n_\ell(\kappa r)\sin\eta_\ell] \ , \quad r > r_i \quad . \tag{2.32}$$

The $C_L(\underline{k})$ in (2.31) are determined so as to make the inverse T-matrix of the crystal stationary. As before, this gives rise to a determinantal equation, the zero's of which are the energy eigenvalues $E_j(\underline{k})$. However, since experience shows that in most metals the phase shifts become negligibly small for $\ell \geq 3$, the order of this determinantal equation is nine, i.e., much smaller than in any of the plane-wave methods. For an accurate representation of the wave functions in the interstitial region, however, terms with higher ℓ may be required. The $C_L(\underline{k})$'s needed for this are easily determined once the eigenvalues are known [2.96].

Owing to the irregular form of the atomic polyhedron it may be more convenient to expand $\psi_{\underline{k}}$ in plane waves in the interstitial region [2.97] before evaluating the Fourier transform (2.6). In this latter form the KKR method has been used by WAKOH and YAMASHITA [2.98] to compute the momentum density for Compton scattering in vanadium. A pure Green's function formulation of the momentum density for Compton scattering that avoids the explicit calculation of the $C_L(\underline{k})$'s and is also applicable to disordered alloys has been presented by MIJNARENDS and BANSIL [2.99]. No applications of the KKR method in one of these forms to problems in positron annihilation are known. However, a pseudopotential plane-wave formulation of the KKR method due to ZIMAN [2.100] (called the KKRZ method) has provided a basis for an approximation scheme developed by HUBBARD [2.77] for the rapid calculation of the electronic band structure of d metals. The Hubbard scheme is related to earlier interpolation schemes, but whereas in those schemes the matrix elements are expressed in a number of adjustable parameters to be obtained from a least-squares fit to an existing band structure, Hubbard's scheme allows a first-principle calculation of the matrix elements, starting from a muffin-tin potential. The scheme was adapted by MIJNARENDS [2.79] to calculate momentum densities in positron annihilation. As the wave functions are expanded in plane waves outside and according to (2.31) inside the muffin-tin spheres, the overlap integral $A_j(\underline{p},\underline{k})$ can be written in the form (2.29). The scheme has been applied by several authors to a variety of metals (Sect.2.4).

d) Other Methods

In insulating compounds the electrons are well localized and their wave functions may be described by linear combinations of atomic orbitals (LCAO). Along with suitably chosen positron wave functions this method has found application in the computation of the momentum density in certain insulating compounds [2.101,102]. Since the subject is somewhat outside the scope of this chapter the interested reader is referred to the original publications. The renormalized-free-atom (RFA) method [2.103], originated by CHODOROW [2.104] and extended by SEGALL [2.105], has been applied successfully to the calculation of the electron momentum density for Compton scattering in various transition metals [2.106]. However, this method seems less suited for positron annihilation, because the wave functions derived by it are least accurate in the interstitial regions of the unit cell that are preferentially sampled by the positron [2.107].

2.2.3 Symmetry Properties of $A_j(\underline{p},\underline{k})$

In the previous subsection much attention has been given to the accurate numerical computation of the momentum distribution by band-structure methods. It is possible, however, to make certain qualitative predictions concerning the momentum distribution solely on the basis of the symmetry properties of the wave functions, and hence of the overlap integral $A_j(\underline{p},\underline{k})$, without a need for numerical calculations. For example, the size of the discontinuity in $\rho(\underline{p})$ at a Fermi break is determined by the value of the corresponding $|A_j(\underline{p},\underline{k})|^2$. It is therefore important to know for which momenta $\underline{p}$ a small or zero value of $A_j(\underline{p},\underline{k})$ can be expected due to symmetry alone before embarking on an experimental search for such a break. A thorough study of the symmetry properties can save many hours of data taking that otherwise might be wasted. It can also elucidate the origin of certain features of the momentum distribution that without a consideration of the symmetry aspects of the problem would remain obscure. We shall first consider the radial behavior of $A_j(\underline{p},\underline{k})$ in $\underline{p}$ space, followed by its directional symmetry properties.

a) Radial Behavior

The radial behavior of $A_j(\underline{p},\underline{k})$ is best studied by considering a tight-binding Bloch function

$$\psi_{\underline{k}}(\underline{r}) = \sum_{\underline{R}} \exp(i\underline{k} \cdot \underline{R})\phi(\underline{r} - \underline{R}) \quad , \tag{2.33}$$

where the $\phi(\underline{r})$ are normalized, nonoverlapping atomic wave functions. By inserting this into the overlap integral (2.6) one obtains (dropping the band index j)

$$A(\underline{p},\underline{k}) = N\delta(\underline{p} - \underline{k} - \underline{G}) \int_{\Omega} \exp(-i\underline{p} \cdot \underline{r})\phi(\underline{r})\psi_+^*(\underline{r})d\underline{r} \quad . \tag{2.34}$$

Noting that the leading term ψ_0^* in the expansion of the positron wave function is spherically symmetric, writing $\phi(\underline{r}) = R_\ell(\underline{r})Y_L(\hat{r})$, where R_ℓ is the radial part of the wave function, expanding $\exp(-i\underline{p} \cdot \underline{r})$ into partial waves and integrating over $\hat{r}$, one finds

$$A(\underline{p},\underline{k}) = 4\pi N\delta(\underline{p} - \underline{k} - \underline{G})i^\ell Y_L(\hat{p})$$
$$\cdot \int_0^\infty j_\ell(pr)R_\ell(r)\psi_0^*(r)r^2dr + \text{higher terms} \quad . \tag{2.35}$$

Hence, the radial behavior of $A(\underline{p},\underline{k})$ is determined by the factor $j_\ell(pr)$, which for small p behaves as p^ℓ. In particular, a d-band contribution to $\rho(\underline{p})$ increases like p^4 at low momenta and thus one should not expect to find d-band Fermi breaks at low momenta, for instance due to a small Fermi surface sheet centered at the origin $\underline{p} = 0$. However, such a break may be visible through an Umklapp process at an equivalent point in a higher zone. On the other hand, Fermi surface sheets on which the electron wave functions have s symmetry will be easily visible at low p since $j_0(pr) \to 1$ as $p \to 0$.

b) Directional Symmetry

It follows from (2.35) that the directional symmetry of $A(\underline{p},\underline{k})$ in $\underline{p}$ space is the same in principle as that of $\psi_{\underline{k}}(\underline{r})$ in $\underline{r}$ space. In a crystal lattice the symmetry properties can be conveniently described with the aid of group theory. The symmetry of $\psi_k(\underline{r})$ is determined by its transformation properties under the group $G_0(\underline{k})$ of $\underline{k}$, which is that group of rotations and reflections that transform $\underline{k}$ into itself or an equivalent vector $\underline{k}' = \underline{k} + \underline{G}$. Group theory shows that each energy band and its corresponding wave functions belong to an irreducible representation of $G_0(\underline{k})$ labelled by the index j. An integral of the form (2.6), in which $\psi_{\underline{k},j}(\underline{r})$ belongs to the jth irreducible representation of $G_0(\underline{k})$, can only be nonzero if $\exp(-i\underline{p} \cdot \underline{r})$ contains a part that belongs to that same representation [2.79]. HARTHOORN and MIJNARENDS [2.108] have given two methods to determine whether this is so and have collected the results of this selection rule for the simple cubic (O_h^1), fcc (O_h^5), and bcc (O_h^9) structures in a set of tables. From these tables one can immediately locate the planes and lines in $\underline{p}$ space where symmetry causes $A_j(\underline{p},\underline{k})$ of a particular band j to have a node and thus avoid these when searching for a sheet of the Fermi surface connected with that band.

2.3 Experimental Techniques

In this section the experimental techniques used in the determination of momentum distributions will be discussed. Besides a description of the setups used to measure one or more components of the photon-pair momentum it includes a discussion of various corrections that must be applied to the raw data and of procedures in use for the analysis of the measurements.

2.3.1 2γ Angular Correlation Measurements

The two quanta created in the annihilation process carry off a momentum $\underline{p}$ and a total energy $E = 2mc^2 - E_b$, where E_b is the binding energy of the positron and the electron to the solid, m the electron mass and c the velocity of light. In the center-of-mass system the quanta are emitted collinearly. In the laboratory system the momentum components (p_y and p_z) perpendicular to the direction of the annihilation radiation cause a deviation from the 180° angle given by $p_\perp/mc$ and of the order of 5-10 mrad (1 mrad corresponds to a momentum of $mc \times 10^{-3} = 0.137$ a.u.). The parallel component p_x gives rise to an energy difference between the two quanta $\Delta E \approx cp_x$, typically 2-5 keV.

An accurate measurement of the angle between the quanta by their detection in coincidence requires a high angular resolution which is achieved by the use of small detector openings and a specimen-detector distance of several meters. The resulting coincidence rates are low. The first angular correlation measurements were performed with Geiger counters [2.1], but these were soon replaced by more efficient NaI(Tl) scintillation detectors. By the use of detectors subtending a large angle (100-250 mrad) in the y direction the opening angle of the counters is increased while a high resolution (0.2-1.0 mrad) along z is maintained. Since the Doppler shift is small compared to the energy resolution of the detectors and the opening angle in the horizontal plane is now much wider than the angular correlation, all information concerning p_x and p_y is lost and the coincidence rate $N(p_z)$ is given by

$$N(p_z)\Delta p_z = \text{const.}\Delta p_z \int_{-\infty}^{\infty} dp_x \int_{-\infty}^{\infty} dp_y \Gamma(\underline{p}) \quad . \tag{2.36}$$

Here $p_z = mc\theta_z$ and $p_y = mc\theta_y$, where θ_y and θ_z are the projected angles of deviation. This is the so-called long-slit geometry commonly employed in angular correlation work. Its coincidence rate is proportional to $\Gamma(\underline{p})$ integrated over a slice in momentum space perpendicular to z. Figure 2.4 shows a schematic diagram. The radioactive source (^{22}Na, up to 100 mCi; ^{64}Cu, reactor produced, up to several Curies; ^{58}Co, several 100 mCi) is placed close to the specimen and is shielded from direct view by the counters. The coincidence rate may be increased by focussing the positrons onto the sample with a strong magnetic field (1.0-1.5 T). In copper-based materials positron activity may be induced in situ by irradiation of the specimen itself with thermal neutrons [2.109], but great care has to be taken to avoid complications due to radiation-induced lattice defects. The background due to accidental coincidences can easily be kept below 0.5% of N(0) with the aid of conventional fast-slow coincidence circuitry (resolving time τ = 10-30 ns). The angular correlation curve is scanned many times to minimize errors from electronic drift and source decay. The consistency of the results of subsequent scans may be checked with the aid of a χ^2 test [2.110]. This provides a sensitive test on electronic drift and other possible sources of error.

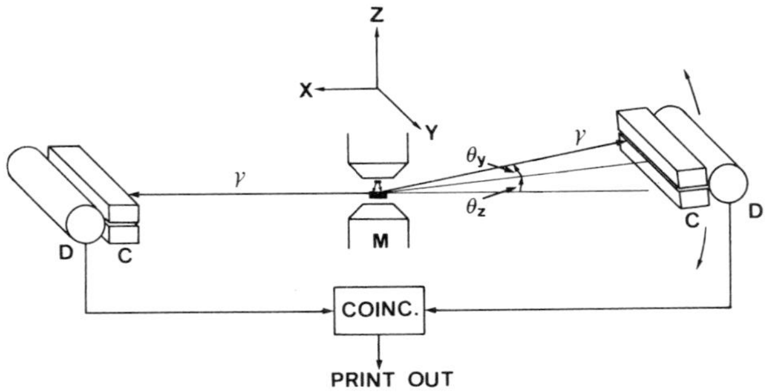

Fig.2.4. Schematic diagram of long-slit geometry. C = collimator, D = detector, M = magnet

The double integration inherent in the long-slit geometry causes a considerable loss of information. More detailed information can be obtained using point detectors that resolve two of the three momentum components according to

$$N(p_y,p_z)\Delta p_y\Delta p_z = \text{const.}\Delta p_y\Delta p_z \int_{-\infty}^{\infty} dp_x\Gamma(\underline{p}) \quad . \qquad (2.37)$$

The first point-detector setup in which p_y was restricted to zero was built by COLOMBINO et al. [2.111]. The counting rate was very low but the data showed a considerable amount of detail not obtainable with the long-slit geometry. A compromise between counting rate and resolution was sought by the use of so-called crossed slits or short slits that limit the integration over p_y to values of the order of the Fermi momentum. This partly suppresses the core contribution and thereby gives a relative amplification of the effects of Fermi surface geometry. This technique has been used in combination with in situ sources to study the Fermi surface of copper and copper-based alloys by several authors [2.109,112]. McGERVEY [2.113], however, has made a comparison of the relative merits of crossed vs. long slits and concluded that at equal source strength and counting time the long-slit geometry, owing to its higher counting rate, has advantages for the observation of the necks in copper. The complicated geometry of short slits precludes the use of reconstruction methods (Sect.2.3.6) so that a comparison between theory and experiment is only possible by folding theory with the instrument response function according to

$$N(p_z)\Delta p_z = \text{const.}\Delta p_z \int_{-mc\Delta}^{mc\Delta} dp_y R(p_y) \int_{-\infty}^{\infty} dp_x\Gamma(\underline{p}) \quad . \qquad (2.38)$$

Here Δ is the angle subtended by each of the slits in the xy plane, while the resolution function $R(p_y)$ describes the effect of the short slits. For small specimens $R(p_y) = 1 - |p_y|/mc\Delta$ has the form of a triangle [2.114].

The considerable loss in counting rate inherent in the use of point detectors or crossed slits can be compensated for by the use of multiple detector systems that allow truly two-dimensional measurements. A few years ago BERKO and his group [2.10] developed a machine with 11 pairs of small NaI detectors on a line having a resolution $\Delta p_z \times \Delta p_y = 0.5 \times 2.0\ \text{mrad}^2$. Data could be taken at a preset value of p_y that could be changed from run to run. This machine was the prototype of a much larger setup (sample-detector distance L = 10 meters) that has recently been completed by the same group. It employs 2 × 32 NaI counters arranged in two square arrays which provide a resolution of typically $0.5 \times 1.5\ \text{mrad}^2$ [2.115,116]. Figure 2.5 illustrates the capabilities of this setup. The material is an oriented single crystal of quartz. One can clearly see the sharp peak at p = 0 and the Umklapps at momenta corresponding to the reciprocal lattice, caused by the annihilation of positronium from a Bloch state in the crystal [2.117,118]. WEST and co-workers [2.119] have constructed a two-dimensional machine using a pair of gamma-ray or Anger cameras in coincidence. Each camera consists of a 50.8 cm diameter NaI crystal optically coupled to a closely packed array of thirty-seven 7.6 cm photomultiplier tubes. Analogue processing of the tube outputs yields a time-energy signal (τ = 200 ns) and y and z position signals. Preliminary setting up at L = 14 m provides a resolution of 0.6 × 0.6 mrad^2 and a coincidence rate of 100 Hz with a 20 mCi ^{22}Na source. A multiple detector system employing solid-state detectors is being developed by TRIFTSHÄUSER and co-workers [2.120,121].

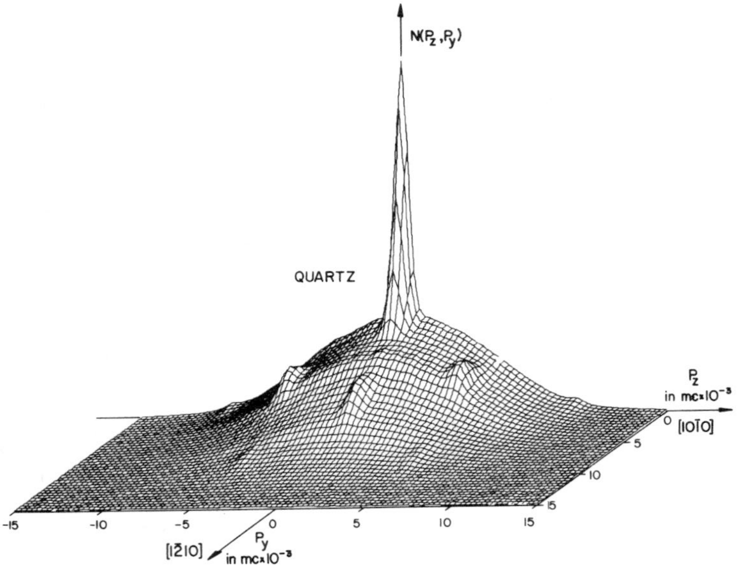

Fig.2.5. Two-dimensional angular correlation in oriented quartz. Each crossing of lines corresponds to a measured coincidence rate [2.116]

A very interesting method consists of the use of spark chambers or multiwire proportional chambers. HOWELLS and OSMON [2.122-124] originally built a spark chamber in combination with a NaI point detector. Read-out took place through two television cameras and a resolution of 0.75 × 0.75 mrad^2 was attained, but the efficiency was low (2%). A better resolution in combination with a higher detection efficiency (15%) is possible with the multiwire proportional chambers developed at CERN and incorporated in a setup at the University of Geneva [2.125]. In these chambers the annihilation photon is converted into a photo-electron in a high-Z converter, followed by detection of the electron in the chamber [2.126,127]. Spatial resolutions of 1 × 1 mm^2 have been achieved implying an angular resolution of 0.1 to a few tenths of a milliradian in either direction. At Queen's University, Kingston, Ontario, a similar machine with larger chambers is being developed [2.128]. This technique holds great promise for future momentum density work.

2.3.2 Rotating Specimen Method

A slightly different use of the point-geometry has been made by SUEOKA [2.129], and WILLIAMS et al. [2.130]. By rotating the sample while keeping both counters fixed at an angle of exactly 180° these authors measure line integrals along radial directions in momentum space. In principle this would yield diameters of the Fermi surface. The results are not affected by smearing due to positron motion, but possible anisotropies of the core contribution and Umklapp processes complicate the interpretation of the results. The method has also been used with long slits [2.131].

2.3.3 Doppler Broadening

As pointed out earlier, the Doppler shift of the photon energy is proportional to p_x. The subsequent broadening of the annihilation line can be measured with a Ge(Li) solid-state detector [2.132,133] and the information obtained is similar to that in the long-slit geometry. The experiment is easy to perform, the data acquisition rate is high, and a small source strength ($\lesssim$ 10 μCi) suffices. CAMPBELL et al. [2.134] have shown that surface effects can be reduced by the use of high-energy positron sources like ^{68}Ge. The resolution (~ 1.0-1.5 keV, equivalent to 4-6 mrad), however, is more than an order of magnitude lower than in an angular correlation measurement, which makes the Doppler broadening technique less suited for accurate momentum density work, at least in the valence electron region. Recently, high-purity Ge detectors have become available. While offering a comparable resolution these have the advantage over Ge(Li) detectors of being storable at room temperature without damage.

By a combination of the Ge(Li) detector with a NaI counter [2.135] or an other Ge(Li) detector [2.136] in coincidence the background can be significantly reduced to a fraction of ~ 10^{-5} of the peak counting rate so that the core electron contribution can be observed up to very high momenta. This may be of interest for the determination of spatial positron distributions in solids, in particular in alloys (Sect.2.2.1), and for the investigation of core enhancement. It also provides a solution to the low-energy tailing of Doppler curves that is mainly due to incomplete charge collection of solid-state detectors and causes uncertainties in the subtraction of the background [2.137]. A combination of the Doppler broadening technique and the point-slit angular correlation method would allow the simultaneous determination of all three momentum components. SINGRU [2.138] has demonstrated the feasibility of such an experiment, but there are obvious counting rate problems.

2.3.4 Specimen Preparation

Irrespective of which technique is chosen to determine the momentum density, great care in specimen preparation is of utmost importance. As mentioned in Sect.2.2.1 the repulsive interaction between the positrons and the ion cores causes lattice defects (vacancies, dislocations, voids, etc.) to act as traps which can bind a positron. The angular correlation of a specimen with defects represents the momentum density in the traps rather than in the bulk of the material. Atomic vacancy concentrations as low as $\sim 10^{-7}$ can have a noticeable effect on the experimental results, so that great care in sample preparation is needed. Cutting of metallic samples by erosion with a fine spark in combination with prolonged annealing and electro-polishing or repeated etching is usually effective in producing specimens with low defect concentrations. In alloys cluster formation can be a problem, especially if the positron affinity of the constituents of the alloy is different. Since the positron lifetime is often a more sensitive indicator of the presence of multiple decay modes than the angular correlation, lifetime measurements can be useful in characterizing the specimen.

2.3.5 Corrections

Various corrections must be applied to the raw data, depending on whether the angular correlation or the Doppler broadening technique is used. Some of these like the subtraction of the background due to accidental coincidences are obvious, others deserve a brief discussion.

a) "Beam Profile" Correction

The noncoincident counting rate in the movable detector may depend somewhat on the position of the detector as a result of direction-dependent γ-ray absorption in the specimen, especially in high-Z materials. This can be avoided to a large extent by tilting the specimen by an angle of $\sim$ 2-3 degrees towards the stationary counter [2.139]. The residual effect can be measured and corrected for. Only pulses in the 511 keV photopeak should be accepted in this measurement, even when the angular correlation curves to be corrected have been measured with a wider setting of the energy window. If not, low-energy annihilation quanta that have suffered Compton scattering in the specimen are included in the measurement of the correction factor with erroneous results.

b) Diffraction Effect

When a single crystal is oriented such that the annihilation quanta leave the specimen at an angle $< 1^{\circ}$ with a set of lattice planes the angular correlation curve may be distorted by Bragg reflection of the photons from these lattice planes [2.139]. Since the wavelength of 511 keV γ radiation is 0.00243 nm, the Bragg angle is $\sim 0.5^{\circ}$. The effect has been carefully investigated by HYODO et al. [2.140,141]. A reliable correction is difficult to give but one can avoid the effect altogether by either orienting the crystal exactly (i.e., within 1 mrad) with its lattice planes along the direction to the stationary counter, or by deliberately missetting the crystal by an angle of 2-3°.

c) Angular Resolution and Positron Thermal Motion

The angular resolution function $R(p_z)$ consists of a convolution of a number of contributions. First, there is the optical resolution formed by the small but finite heights of the collimator slits in front of the two detectors and the effective specimen thickness. The latter consists of the positron implantation profile [2.142,143] and the apparent thickness of the sample which results from its small tilt. After convolution these factors produce a triangular resolution curve with rounded corners (the exponential positron implantation is usually not deep enough to cause any appreciable asymmetry, except possibly for ^{68}Ge). This curve must then be convoluted with the Maxwell-Boltzmann curve (2.3) describing positron thermal motion. The result can be considered approximately Gaussian. In Doppler broadening experiments the resolution function is given by the detector response. Finally, the experimental angular correlation or Doppler broadening curve is deconvoluted. Various procedures for this have been described [2.137,144,145]. Obviously the deconvolution problem is most important in the analysis of Doppler broadening experiments where the width of the resolution function is comparable to that of the curve to be measured. The presence of noise in the data and uncertainties in the resolution function preclude "perfect" deconvolution and lead to a "residual instrument function", introduced by PAATERO et al. [2.146]. Theory should be convoluted with this function before it is compared with the deconvoluted experimental curves [2.147]. Another possibility is to convolute theory with the full resolution curve instead of attempting to deconvolute the experimental data.

d) Finite Slit Length

In (2.36) it was assumed that the long-slit detectors are of sufficient length to warrant integration over all p_y. Strictly taken this is not always the case, and a correction for the finite detector length has to be made. This correction has been discussed first by ARIFOV et al. [2.148], and later by MIJNARENDS [2.149] for the case of both anisotropic and isotropic $\Gamma(\underline{p})$. MOGENSEN [2.150] has pointed out that the finite-slit effect may be reduced by the use of specimens of several centimeters wide, while others [2.131,151] use detectors with different horizontal opening angles to minimize this correction. A realistic numerical example may illustrate the size of the correction. Consider an angular correlation, consisting of an inverted parabola with a cutoff angle of 5 mrad resting on a Gaussian core contribution of the same amplitude at $\theta = 0$ and a full width at half maximum (FWHM) of $2\theta_{\frac{1}{2}} = 18$ mrad; values typical of noble metals. When this angular correlation is measured with long-slit detectors each subtending an angle $\Delta = 100$ mrad in the horizontal plane, the Gaussian part is attenuated by $(\pi \ln 2)^{-\frac{1}{2}}(\theta_{\frac{1}{2}}/\Delta) = 6\%$ with respect to the infinite-slit case, while the parabolic part suffers an attenuation of only 2%. Moreover, the latter is slightly deformed by the admixture of a correction term. For $\Delta = 50$ mrad these figures are doubled [2.149]. It is clear that the resultant overall distortion of the angular correlation curve is large enough to deserve serious consideration.

2.3.6 Analysis

When the angular correlation data have been collected, one is faced with the problem of extracting information about the underlying momentum distribution from them. Basically there are two methods for this. The first one assumes a model distribution with a number of parameters and calculates the angular correlations following from the model. The parameters may then be obtained by fitting the calculated curves to the observed ones. This method is used by many workers and is particularly suitable if the general shape of the momentum distribution is known and easy to parametrize. However, it is not always clearly realized that in general a number of different models may yield fits of a comparable quality, i.e., the solution found may not be unique. This situation is reminiscent of the sampling theorem in Fourier analysis [2.152]. The number of possible models consistent with the data can be reduced by increasing the number of single crystal orientations for which the angular correlation is measured and by improving the statistical accuracy of the data. Yet, any finite amount of data will inevitably result in a lack of uniqueness of the momentum distribution derived from them. This is obvious in the second method of analysis: reconstruction of the momentum distribution $\Gamma(\underline{p})$ from a number of angular correlation curves measured for different orientations of a single crystal [2.153, 154]. This method is useful if no parametrized model of the momentum distribution is available. Since $\Gamma(\underline{p})$ possesses the full point symmetry of the crystal lattice, it is expanded into a (formally infinite) set of symmetry-adapted basis functions $\xi_i(\underline{p})$

$$\Gamma(\underline{p}) = \sum_i a_i \xi_i(\underline{p}) \quad . \tag{2.39}$$

The use of symmetry-adapted basis functions enables one to exploit the crystal symmetry to the fullest extent by reducing the number of terms occurring in the expansion (2.39). The basis functions can be functions of both length and direction of $\underline{p}$, e.g., symmetrized combinations of plane waves

$$\Gamma(\underline{p}) = \Gamma_0(p) + \sum_{\underline{R}} a_{\underline{R}} \exp(-i\underline{p} \cdot \underline{R}) \quad , \tag{2.40}$$

as introduced by MUELLER [2.155], or they may depend only on the direction $\hat{p}$ of $\underline{p}$ [2.153]

$$\Gamma(\underline{p}) = \sum_{\ell} a_{\ell}(p) F_{\ell}(\hat{p}) \quad . \tag{2.41}$$

In the latter case the expansion coefficients become functions of $|\underline{p}|$; the F_{ℓ} are lattice harmonics. In either case the expansion coefficients are determined from appropriate integrals containing the experimental data [2.154]. In the process the set of ξ_i has to be truncated since only a finite amount of data has been collected. This introduces series termination errors that represent another aspect of the nonuniqueness encountered earlier. The effect of these errors is to limit the resolution of the reconstruction procedure and to distort the momentum distribution one wishes to measure. Their effect can be described in terms of a convolution of the true distribution with a resolution function S. The latter represents the resolution of the reconstruction method and should not be confused with the resolution of the experimental setup. An estimate of the resolution of the reconstruction method may be obtained from the first neglected basis function. It is well known from diffraction theory [2.156] that in the case of (2.40) the resolution function is spherically symmetric and is given by $S(\underline{p}' - \underline{p}) = j_1(x)/x$, where $x = |\underline{p}' - \underline{p}|R_{max}$, j_1 is a spherical Bessel function and R_{max} the length of the last $\underline{R}$ vector before truncation. The FWHM of $S(\underline{p}' - \underline{p})$ is found from FWHM = $5.0/R_{max}$ with both FWHM and R_{max} in a.u. Expansion into lattice harmonics and truncation at $\ell = \ell_{max}$ in principle yields an infinitely high radial resolution, but the tangential resolution is given by [2.157]

$$S(\hat{p}' - \hat{p}) = (4\pi)^{-1} \sum_{\ell}{}' (2\ell + 1) P_{\ell}(\cos\omega) \quad , \tag{2.42}$$

where ω is the angle between the directions $\hat{p}$ and $\hat{p}'$ and the summation runs only over those ℓ values present in (2.41). The FWHM of (2.42) is approximately $180/\ell_{max}$ degrees. Figure 2.6 shows the difference between resolution functions typical for the two cases and compares them with the neck of the copper Fermi surface. It is clear that the choice of reconstruction method will be strongly influenced by the nature of the problem. The reconstruction of isolated pockets of electrons or holes around points in higher Brillouin zones would require lattice harmonics of a rather high order, and the use of a plane-wave expansion may be preferable in that case. At low momenta the lattice harmonic method provides a better resolution. The resolution of the reconstruction procedure also affects the angular resolution with which the data should be taken. If the latter would be very much higher than the former it may be a better policy to widen the collimator slits and use the resulting higher counting rate to improve the statistics and increase the number of crystal orientations. In that way a larger number of basis functions may be included before truncation, thus improving the resolution of the reconstruction procedure to a value more comparable with the angular resolution in data taking[3].

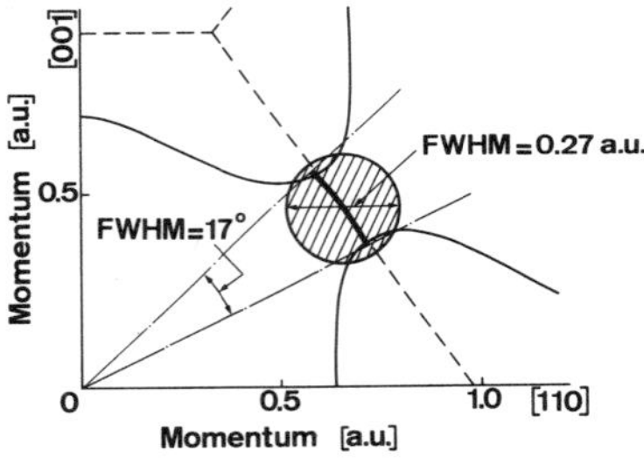

Fig.2.6. Typical resolution functions for the lattice harmonic (heavily drawn segment) and plane-wave (hatched circle) methods of reconstruction. The former corresponds to $\ell_{max} = 12$, the latter is representative for MUELLER's reconstruction [2.155] of Compton profile data for Si. The resolution functions are compared with the Fermi surface of copper

The propagation of statistical errors in reconstruction methods is another point that deserves attention. The reconstruction process results in a strong correlation between the errors in neighboring points of the final $\Gamma(\underline{p})$ distribution. Consequently, random statistical errors in the angular correlations may appear after reconstruction as fluctuations in $\Gamma(\underline{p})$ that extend over several neighboring points and hence may not be recognized for what they really are: correlated noise. The interpretation of small features in the reconstructed $\Gamma(\underline{p})$ in terms of details of Fermi surface geometry is therefore dangerous and may well be fortuitous. Experimental data to be used for a reconstruction should be of a high statistical accuracy ($> 10^5$ counts at $\theta = 0$) and be taken in more than just a few crystal directions.

---

[3]In his reconstruction of REED and EISENBERGER's [2.158] Compton profile data for Si, MUELLER [2.155] uses 65 stars of plane waves. Together with his choice of "lattice constant" of the $\underline{R}$ lattice this yields a resolution function with a FWHM of 0.27 a.u. = 2.0 mrad, as shown in Fig.2.6.

Besides the number the choice of the crystal orientations is important. Conventionally, data are taken with p_z along directions of high symmetry. This is the most convenient choice if the measurements are subsequently analyzed by comparison with some (parametrized) model. If on the other hand the analysis is performed by reconstruction, this choice is not necessarily optimal and measurements along a few special directions of low symmetry may give more accurate results [2.159-161].

The reconstruction of long-slit measurements following (2.41) has been applied to a large number of materials, among which Cu [2.162], for which the results are shown in Fig.2.7. Terms up to $\ell = 12$ have been included in the reconstruction. The smearing of the neck of the Fermi surface due to the tangential resolution function of Fig.2.6 is clearly visible. Other materials to which the method has been applied are Fe [2.163], Li [2.164], Ni [2.165], Fe_3O_4 [2.166], etc. (Sect.2.4). HOWELLS and OSMON [2.123], and MAJUMDAR [2.167] have discussed the reconstruction of a three-dimensional $\Gamma(\underline{p})$ from point-slit measurements, while CORMACK [2.168] has formulated the method in two dimensions. MUELLER [2.155] has applied the plane-wave method to Compton profiles in Si.

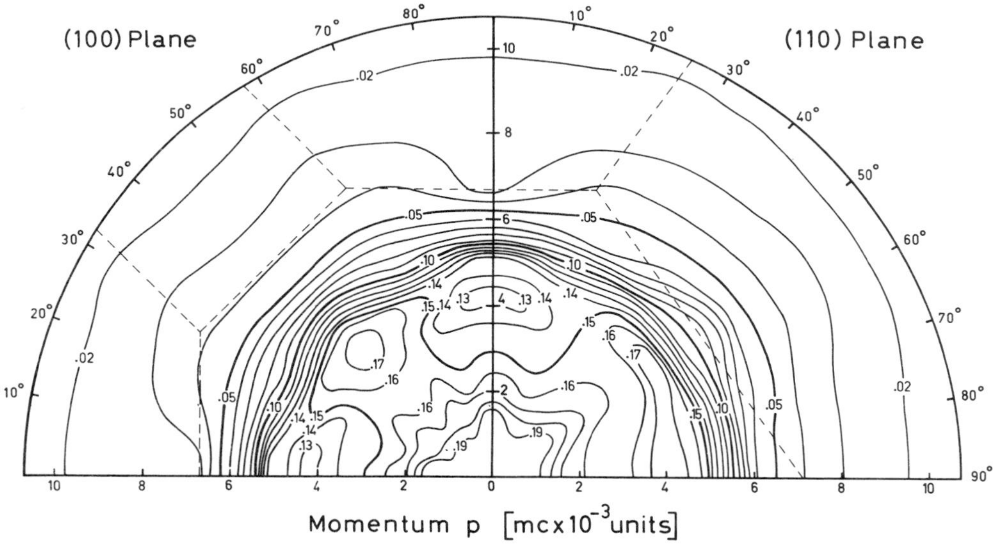

Fig.2.7. Reconstructed momentum distribution $\Gamma(\underline{p})$ in copper (100) and (110) planes

LOCK, CRISP, and WEST (LCW) [2.169] have developed an interesting method to analyze long-slit angular correlations in terms of Fermi-surface cross sections. These authors superimpose identical profiles at intervals equal to the distance between reciprocal lattice planes perpendicular to the p_z direction. In this way they obtain a (periodic) curve that reflects the dependence of the areas of cross-sectional cuts through the Fermi surface on p_z. Full bands give a constant contribution. Within the IPM this "LCW theorem" is exact for Compton profiles. In positron annihilation the theorem is only approximately valid. Experiments in Cu and Ge have tested its validity and limitations [2.170,171].

It appears that the one-dimensional Fourier transform (FT) of directional Compton profiles (and therefore to a somewhat lesser extent also the transform of angular correlation curves) constitutes a source of information that so far has been insufficiently tapped. PATTISON and WILLIAMS [2.172] have pointed out that the Fermi momentum of a free-electron gas can be obtained simply from the zero's of the FT. Resolution plays no role because it does not affect the position of the zero's. The authors have derived the Fermi momentum of Al with a precision of 0.07 mrad, i.e., 1%, from Compton profiles taken with resolutions equivalent to 2.7 and 5.1 mrad! It is also felt that the Fourier transformed profiles can yield information about ionic bonding [2.173]. SCHÜLKE [2.174] has formulated an interesting set of theorems applicable to the one-dimensional FT's of directional Compton profiles. Some of these can be applied directly to positron annihilation, others are valid subject to the same restrictions as the LCW theorem. An example of the second kind is the theorem which states that the one-dimensional FT of an angular correlation (Compton profile) $N(p_z)$ in an insulator or semiconductor must be zero at points corresponding to the lengths of those lattice vectors that lie along the z axis. A thorough investigation of the properties of the Fourier transformed curves seems justified.

2.4 Momentum Density Work in Metals

In the remaining sections of this chapter the momentum density work, first in metals, and in Sect.2.5 in alloys, will be reviewed. As already mentioned in the introduction, the earlier momentum density work starting with the pioneering experiments of DEBENEDETTI et al. [2.12], and STEWART [2.175], and the oriented crystal measurements of BERKO et al. [2.14,139] has been extensively covered in other reviews. We shall therefore only briefly mention the earlier work and refer to it whenever necessary and rather focus our attention on the work of the last five to ten years.

2.4.1 Alkali Metals

The momentum density of the alkali metals has been studied from two different points of view. In the first place, the electrons in sodium form the closest approximation of a free-electron gas of any metal and its Fermi surface is very nearly spherical.

This makes it a suitable system for checking the predictions of many-body theory and for investigating the smearing of Fermi surface breaks due to thermal effects. These topics have been discussed in Sect.2.1.2,3 respectively. Secondly, the angular correlation technique can be used to measure the anisotropy of the Fermi surface. A large amount of data concerning Fermi surface dimensions in the alkali metals has been gathered through the use of conventional techniques like the measurement of the dHvA effect. This work has been reviewed by LEE [2.176]. These techniques are in general capable of yielding a higher precision than positron annihilation, but depend on mean free paths that are considerably larger than the electron orbits in the presence of an externally applied magnetic field. Low temperatures and highly perfect single crystals are therefore required. Sodium and lithium, however, both show a partial martensitic transformation to the hcp structure on cooling (at 36 and 78 K, respectively [2.177]) and the incompletely transformed material contains a high density of defects. The application of these conventional techniques is therefore difficult in Na [2.178,179] and practically impossible in Li. The positron annihilation technique, on the other hand, not being limited to low temperatures, can provide valuable information on the anisotropy of the Fermi surface with relatively little effort.

A certain amount of experimental data on Li has been collected by the measurement of various transport properties, optical absorption, and soft X-ray emission spectra. However, these data can only yield information on the Fermi surface in a rather indirect way. Band-structure calculations result in practically spherical Fermi surfaces for Na and, to a lesser extent, also for K and Rb. In Li most calculations yield a Fermi surface that is extended along the <110> directions with an anisotropy ranging from 4.2 to 7.2% between various authors [2.180]. COHEN and HEINE [2.181] proposed a simple model of the band structure of the alkali's, the noble metals and their α-phase alloys. This model could correlate a large amount of experimental data on these materials but predicted that the Fermi surface of bcc Li should actually contact the Brillouin zone faces in the <110> directions. This would imply an anisotropy $\geq$ 14% and a multiply connected Fermi surface linked by necks. Some of the experimental data, in particular the anomalous negative pressure dependence of the resistivity [2.182], seemed to confirm this prediction, but measurements with the positron annihilation technique present a different picture. The first measurements with positrons on oriented single crystals of Na and Li were done by DONAGHY and STEWART [2.24,183]. They found the Fermi surface of Na to be spherical within the experimental accuracy of 1.5% and the annihilation probability to follow Kahana's predictions concerning enhancement (Fig.2.2). An analysis of their Li data led the authors to conclude that $k_{110}/k_{100} \sim 1.05$ and hence that the Fermi surface does not make contact with the zone faces in the <110> directions. MELNGAILIS and DEBENEDETTI [2.184] compared the same data with a more refined OPW calculation and came to substantially similar conclusions. Inclusion of the Kahana enhancement factor was essential to obtain quantitative agreement between calculation and experiment. STACHOWIAK [2.164], using the three profiles of Donaghy to reconstruct the momentum density, came to the conclusion that the existence of necks cannot be excluded, but measurements by PACIGA and WILLIAMS [2.185] with the rotating-specimen technique gave a most probable value of 2.9% and an upper limit of 4.8% for the anisotropy. An estimate of the Fourier component V_{110} of the potential yielded a value of $\sim$ 0.10 Ry. Both results are in good agreement with a self-consistent band-structure calculation by RUDGE [2.186]. Recent information on the shape of the Fermi surface in Li comes from measurements of the dHvA effect in dispersions of small Li particles in paraffin-wax by RANDLES and SPRINGFORD [2.180]. By taking the grain size small enough the martensitic transformation can be suppressed and the bcc structure retained down to liquid helium temperatures. Although the effective polycrystallinity of the sample makes the interpretation of the data less straightforward than in the case of large single crystals, the authors derive an anisotropy of 2.6 ± 0.9%, in agreement with [2.185] but significantly less than given by most band-structure calculations using local exchange-correlation potentials. This result would lend support, however, to recent calculations [2.187,188] which consider the effects of nonlocal exchange and correlation on the Fermi surface of metals. These calculations yield Fermi surface distortions that are substantially reduced in comparison with those obtained from local calculations. NICKERSON and VOSKO [2.189] have performed model calculations in an alkali-like system. They show that the Fermi surface distortions resulting from the use of local and nonlocal approximations to exchange and correlation bracket the distortion found with exact Hartree-Fock theory, and that the nonlocal calculation is not necessarily closer to the exact result than the local one. These authors also point out that the Fermi surface is more sensitive than other quantities like electron densities, chemical potentials and bandwidths to the approximations used to treat exchange and correlation. These results suggest that new, high-precision measurements with positrons of the Fermi radii in Li could be of great use in helping to decide the best way of treating exchange and correlation effects in band-structure theory.

Since the remaining alkali metals do not undergo low-temperature phase transitions, conventional techniques of Fermi surface determination can be applied. The momentum density and Fermi surface in these metals have not been investigated with positron annihilation, apart from a study of the solid-liquid transition in polycrystalline Rb and Cs [2.190] and a search for the existence of a spin density wave in K [2.191].

The alkali metals are highly compressible and hence lend themselves to studies of the effect of electron density on the annihilation rate. MACKENZIE et al. [2.192] measured the effect of pressure on the positron annihilation rate in sodium and other alkali metals, and found that the annihilation rate was both higher and varied more rapidly than predicted for annihilation with conduction electrons alone. This suggests a significant contribution from the core electrons. Compression of the material reduces the interstitial volume and hence the fraction of core annihilations should increase. In this way KUBICA and STOTT [2.193] could explain the observed annihilation rates, especially for the lighter alkali metals. Initial Doppler broadening measurements by MACKENZIE et al. [2.192] did not substantiate this increasing core fraction, but subsequent more accurate angular correlation measurements showed a clear increase in the high-momentum region. GUSTAFSON and WILLENBERG [2.194] measured the angular correlation in Na over the entire range from 4 to 68 kbar. The valence electron contribution shows a free-electron behavior up to ~ 55 kbar, but above this pressure deviations that suggest nonfree-electron-like behavior set in. This could be consistent with the behavior of the resistivity of Na under high pressure. These authors also studied the behavior of enhancement under pressure. Although their results agree well in absolute magnitude with the theoretical predictions by KAHANA [2.18], and CARBOTTE [2.26], the pressure dependence of the enhancement cannot be explained.

2.4.2 Other Simple Metals

Besides the alkali metals there are a number of polyvalent metals for which the nearly free-electron model provides a reasonably good description of their electron structure. Their Fermi surfaces overlap into the second and third zones and may be distorted from sphericity due to zone boundary effects. Among these, beryllium shows a very large anisotropy. The angular correlation curves for this metal have been measured by STEWART et al. [2.195], and BERKO [2.196]. SHAND and STEWART [Ref.2.7, p.291] compared the data of [2.195] taken with p_z along the c axis with the results of a nearly free-electron calculation and derived a value of ~0.34 ± 0.16 Ry for the band gap at the (0002) zone boundary. However, the band structure of Be deviates too much from the free-electron model to allow a good fit near the zone faces with only two plane waves. In a more extensive calculation SHAND [2.197] therefore employed 23 plane waves and achieved a rather good fit to the measured curves despite the use of a constant positron wave function. Berko [2.11] has compared the positron data with results of Compton profile measurements. In view of the relative simplicity of the electronic structure of Be, its small core contribution, the large anisotropy of its long-slit angular correlations, the availability of Compton profile data for this material, and finally the great progress in band-theoretical methods during the last 15 years, a new and thorough study of Be would be of interest.

Magnesium, zinc and cadmium have a nearly free-electron-like band structure with an approximately spherical Fermi surface. Early measurements on Mg [2.196] are in agreement with this picture. BECKER et al. [2.198] applied the point-slit geometry to Mg and Zn and compared their curves with a free-electron model in which the Umklapp contributions were only caused by the high-momentum components in the positron wave function. After enhancement was included they obtained a fair agreement with experiment for Mg, but in the case of Zn this model is apparently too simple. However, when the Fermi surface is distorted so that it follows the boundaries of the second Brillouin zone the fits improve [2.199]. KUBICA and STEWART [2.59] measured the angular correlation along the c axis of Mg with a high resolution, obtaining the Fermi radius in that direction with an accuracy of 0.1% and a band gap at the [0002] zone face of 0.05 Ry, in good agreement with a local pseudopotential calculation [2.200]. Other measurements on Zn have been performed by various authors [2.201-203]. KONTRYM-SZNAJD and STACHOWIAK [2.204] analyzed the data of [2.203] by postulating certain deviations from the free-electron model, guided by the results of a reconstruction. However, a reconstruction based on only three directions cannot be expected to yield reliable results, in particular for symmetries lower than cubic. The postulated model therefore need not be the best solution. More detailed experimental data are needed. Similar studies have been performed for Zn, Mg and Cd by other authors [2.205,206].

Of the group of trivalent metals aluminum has been studied most extensively. BERKO and PLASKETT [2.14] found the Fermi surface to be isotropic within ± 1% and calculated the angular correlation with a Wigner-Seitz wave function for the positron and free conduction electrons. An interesting approach was followed by STROUD and EHRENREICH [2.72]. These authors made use of the fact that the Fourier coefficients of the lattice potential seen by the positron are closely related to the X-ray form factors. Their calculated angular correlation curves were practically isotropic in spite of the anisotropy of the positron wave function. Also under pressure (up to 100 kbar) Al continues to display free-electron behavior [2.207], in agreement with band theory [2.208]. Recently, OKADA et al. [2.209] have performed accurate measurements of the anisotropy in Al with fairly short slits. Although the angular correlations look parabolic, the slopes of these curves show significant structure. The authors also plot the directional differences $N_{110} - N_{100}$, $N_{111} - N_{110}$ and $N_{111} - N_{100}$, which show that a large part of the anisotropy is due to the [111] direction. KUBO et al. [2.210] have performed APW calculations of the directional Compton profiles and positron angular correlation curves and have compared these with various experimental Compton profiles and with the positron data of [2.209]. The calculations explain both the observed structure in the slopes and the directional differences in terms of zone boundary effects on the Fermi surface. These same effects show up in a more pronounced way in the work of MADER et al. [2.211]. This study convincingly illustrates the great value of the two-dimensional geometry. The authors studied the momentum distribution for several single-crystal orientations and compared their results with an OPW calculation (Fig.2.8). A new feature, of which the long-slit geometry is incapable, was the observation of struc-

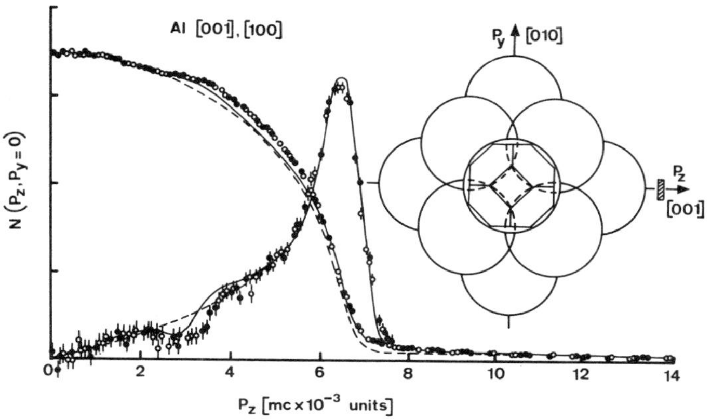

Fig.2.8. Angular correlation in aluminum with $\{p_z, p_x\}$ = {[001],[100]} and p_y = 0, together with its first derivative by MADER et al. [2.211]. Full curve is OPW prediction, dashed curve corresponds to simple sphere model shown in inset. The shaded rectangle indicates the resolution along p_y and p_z

ture in the Umklapp components without the disturbing presence of the main Fermi surface. This shows unequivocally that a large fraction of the tail of the distribution in Al is due to Umklapp contributions. The results were also accurate enough to distinguish between the use of OPW wave functions and the corresponding pseudo-wave functions obtained from the former by leaving out the core orthogonalization terms discussed in Sect.2.2.2. Finally, it appeared necessary to enhance the core contribution less than the high-momentum conduction electron part by a factor of ~ 2. Since the enhancement factor for valence electrons is ~ 4 (cf. Table 2.2), these figures imply a core enhancement of ~ 2, somewhat less than the theoretical figure of 2.7 from CARBOTTE and SALVADORI [2.42]. Similar results were obtained earlier for the alkali metals by TERRELL et al. [2.212] who found $\varepsilon_{cond.}/\varepsilon_{core} \sim 4$. LYNN et al. [2.136] found little core enhancement, but their measurements cover a much larger momentum range. These authors studied the Doppler broadening in Al with two Ge(Li) detectors in coincidence. A peak-to-background ratio > 10^5 enabled them to observe the $(1s)^2$ core contribution, as may be seen in Fig.2.9. The authors put their results on an absolute scale with the aid of a lifetime measurement and found by comparison with IPM calculations that the core enhancement at large momenta could not be more than a few tens of percent. All these results reflect the lower polarizability of the core electrons, especially close to the nucleus, compared to the conduction electrons. The technique of Lynn et al. offers great possibilities for further study of the core enhancement problem.

The next column in the periodic system contains carbon in the form of graphite and diamond and the covalently bonded semiconductors silicon and germanium that will all be discussed in Sect.2.4.3. Furthermore the metals tin and lead are found

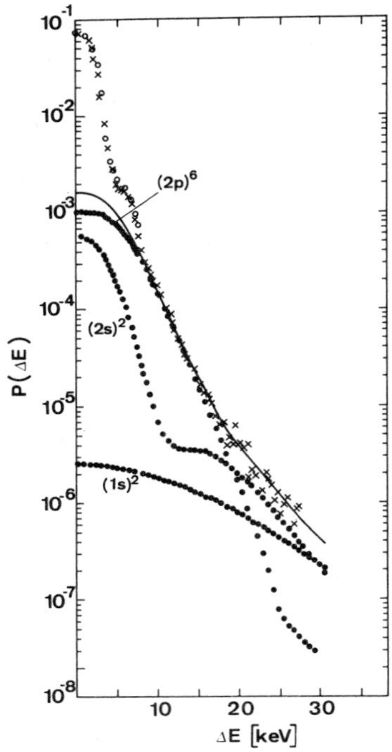

Fig.2.9. Experimental deconvoluted (crosses) and calculated Doppler broadening curves in aluminum by LYNN et al. [2.136]. The black dots represent the calculated contributions of the individual shells, the solid curve the sum of these contributions. The open circles denote angular correlation data by HAUTOJÄRVI [2.213]. ΔE = 1.0 keV corresponds to $\Delta\theta = 1.96 \times 10$-3

in this column. The angular correlations for tetragonal β-tin have been measured by MOGENSEN and TRUMPY [2.214] and further analyzed by KONTRYM-SZNAJD [2.215]. There are signs of important Umklapp components. Lead has not been studied in detail [2.202]. An attempt to observe a change in the Fermi momentum due to superconductivity yielded a negative result [2.216]. Arsenic, antimony and bismuth in the neighboring column are semimetals. Their low electric conductivity is a result of the fact that the Jones zone, corresponding to five electrons per atom, is practically full. There are hole and electron pockets in the 5th and 6th bands. The number of electrons (holes) in these pockets is of the order 10^{-5} per atom in Bi and 10^{-3} per atom in Sb. There exists a vast body of experimental information obtained by different methods about the electronic structure of Bi and, to a lesser extent, Sb and As [2.217]. Positron data on oriented single crystals [2.214,218, 219] are relatively scarce and partly in contradiction with results obtained by other methods.

2.4.3 Oriented Graphite, Diamond, Silicon, and Germanium

The first measurements on oriented single crystals were done by BERKO et al. [2.139] on graphite. Since the positron preferentially annihilates with the π electrons the angular correlations are strongly anisotropic, which is illustrated even more clearly by measurements with the point-slit geometry [2.111]. By contrast, Compton scattering samples the π and σ electrons uniformly and yields a momentum density that is only slightly extended along the c axis [2.220]. The work of SCHRADER and KIM [2.221], who analyzed the role of symmetry in positron annihilation and Compton scattering in a benzene molecule, is interesting in this connection. If the positron is in a bound molecular orbital it may give rise in certain cases to a strongly anisotropic momentum distribution, whereas Compton scattering results are much more isotropic. Experiments on oriented single crystals of benzene or similar substances would be of interest to test these ideas (however, see also [2.222]).

Silicon and germanium, both with the diamond structure, have been studied by a number of workers. COLOMBINO et al. [2.223] investigated these materials with both the long-slit and the point-slit geometry and reported significant structure in the tails of the distribution. Later measurements by ERSKINE and McGERVEY [2.224], however, did not confirm this structure. The data of the latter authors displayed a large anisotropy which they could correlate with the shape of the Jones zone. MOTT and JONES [2.225] have shown that in the extended-zone scheme no energy gap occurs at those Bragg reflection planes for which the structure factor is zero (e.g., the {200} and {222} planes in the diamond structure). The volume bounded by the set of {220} planes (the Jones zone) can just hold four electrons and the structure factor for these planes is particularly large. It would then be energetically favorable for the electrons to first fill up all states inside the Jones zone before spilling over into higher zones. Simple calculations based on this model could accurately reproduce the observed structure along the [110] direction, but for other directions more sophisticated computations that take account of the higher momentum components are required. (Since there are no partially filled bands, $\rho(\underline{p})$ is continuous across the zone boundary [2.14]). Recent two-dimensional measurements [2.10] have shown up a large anisotropy, particularly at small angles, indicating important deviations from the simple Jones zone interpretation. The Jones zone model has also been invoked to explain Compton scattering measurements in diamond (a singularly unpopular material among positron workers!) [2.226]. The work of STROUD and EHRENREICH [2.72] has already been mentioned in connection with aluminum. Their pseudopotential computation for Si, in which a positron wave function obtained from experimental X-ray structure factors is used, gives excellent agreement with the measured anisotropy $[N_{100} - N_{110}]/N_{100}$. The use of a constant positron wave function worsens the fit to the positron data and reduces the anisotropy. A part of the anisotropy is thus due to the manner in which the positron weighs the various regions of momentum space. The results with constant ψ_+ are in good agreement with recently measured Compton profiles by REED and EISENBERGER [2.158]. These authors point out that the N_{100}-N_{110} directional difference is strongly correlated with the shape of the Jones zone, whereas $N_{100} - N_{111}$ reflects more the crystal potential. This explains the insensitivity of the anisotropy to changes in the pseudopotential, observed by STROUD and EHRENREICH. The influence of the positron wave function in Si was also studied by DLUBEK and BRÜMMER [2.227], who compared angular correlations measured along the three symmetry directions with Compton profiles of SCHOLKE [2.228]. An analysis of both sets of data by reconstruction shows significant differences in the respective momentum distributions, in accordance with IPM computations. An interesting approach to the study of momentum densities in solids is the work of SETH and ELLIS [2.229]. They calculated momentum densities and Compton profiles for the isostructural series diamond, 3C-silicon carbide and Si using approximate Hartree-Fock-Slater crystal wave functions and analyzed the anisotropies by means of expansions in cubic harmonics. The spherical components ($\ell = 0$) are very similar when scaled according to the lattice momentum $q_0 = 2\pi/a$, but the anisotropic components are not. The $\ell = 6$ components are qualitatively similar, but do not scale in a simple way, while the $\ell = 4$ and $\ell = 8$ components are quite different. The $\ell = 6$ component is dominant in the $N_{100} - N_{110}$ directional difference (although the $\ell = 4$ component cannot be neglected). Combination of these observations with those of REED and EISENBERGER [2.158] described above leads to the conclusion that the $\ell = 6$ term, in spite of its failure to scale, is strongly correlated with the geometry of the Jones zone, while the term with $\ell = 4$ that dominates the $N_{100} - N_{111}$ difference is mainly determined by the crystal potential. It would be interesting to perform similar studies with Compton scattering and positrons on other series of closely related materials.

The momentum distribution in Ge has been the subject of a combined Compton and positron study by SHULMAN et al. [2.67]. There is substantial agreement between this work and previous work on Si and Ge [2.72,158,224]. New was the observation in the high-momentum tail of Umklapp images of the Jones zone. At high momenta the Compton data lie consistently above the positron curve. A similar reduction of the high-momentum components in the positron data below the predictions of the IPM had already been observed in Si by FUJIWARA and HYODO [2.40], who interpreted their observations as evidence for the dehancement of the Umklapp component due to interband transitions of the electrons (Sect.2.1.2). Another study of Ge concerns the applicability of the LCW theorem [2.170].

2.4.4 Noble Metals

This section and the following one will be dedicated to the group of d metals. In this group, consisting of the transition metals and the noble metals, the conduction band overlaps and hybridizes with the d bands. The noble metals include copper, of which the band structure and Fermi surface are accurately known by other methods. It has therefore often been used as a test case for new experimental techniques and computational procedures. A discussion of the positron work on Cu up to 1974 has been given by SINGRU [2.230]. The present section will deal only with the noble metals while the transition metals, that are of interest because of their complex

Fermi surface topology and magnetic properties, will be discussed in Sect.2.4.5 together with the rare earths.

After the classical work of BERKO and PLASKETT [2.14] on oriented single crystals of Cu had shown up a sizable anisotropy in the angular correlations, much effort has been spent on a further investigation of this anisotropy and its correlation with the electronic structure. Measurements of the anisotropy with long slits at both low and high momenta [2.231,232] can largely be explained in terms of cross-sectional cuts through a series of Fermi surfaces of the HALSE [2.233] model. These surfaces, each carrying a weight $|A_G(\underline{p},\underline{k})|^2$ that depends only on $\underline{G} = \underline{p} - \underline{k}$, are centered on the reciprocal lattice points in the extended zone scheme. Small residual discrepancies can be ascribed to a slight $\underline{k}$ dependence of the $|A_{\underline{G}}|^2$'s, core anisotropy and possibly (anisotropic) enhancement. The relative independence of the weights on $\underline{k}$, which is reminiscent of a nearly free-electron system, was clarified by band-structure calculations by MIJNARENDS [2.79] using HUBBARD's method [2.77]. In spite of the complexity of the momentum density contributions of the individual (s-d hybridized) bands, the sum over all occupied bands is approximately constant over the interior of the Fermi surface at $\underline{G} = 0$, while displaying some variation with $\underline{k}$ over the Fermi spheres centered at $\underline{G} \neq 0$. In the interstitial space between the Fermi surfaces the momentum density has the form of a slightly anisotropic core distribution. A similar model to that of BERKO et al. [2.231] was used by SENICKI et al. [2.234] to explain the anisotropy observed in a point-slit measurement. A prediction of the directional profiles $N_{ijk}(\theta)$ themselves, rather than differences between them, appears to be more difficult [2.235]. The problems lie mainly in the size of the calculated core contribution, which is too small (see also [2.73]). The use of HERMAN-SKILLMAN [2.236] free-atom wave functions for the 3d electrons ignores the variation of the wave functions over the energy range between top and bottom of the d band first observed by WOOD [2.237]. This neglect results in a positron-electron wave-function overlap which is too small and too concentrated in real space. Band-structure calculations that properly account for this, produce a higher and more narrow core contribution and achieve a better agreement with experiment, although some slight discrepancy remains [2.238, 239]. It is interesting to see, however, that even if the *size* of the core contribution is quite incorrect, its *shape* is predicted with some accuracy. This follows from a mutual comparison of the experimental core distributions of copper, silver, and gold which is in reasonable agreement with simple Herman-Skillman theory [2.240]. Apparently the core distribution is strongly correlated with the ionic size and the lattice parameter.

In all these and similar [2.241,242] studies computations based on some model are compared with experiment. The opposite approach, viz. reconstruction of $\Gamma(\underline{p})$ from a set of eight directional profiles N_{ijk} has been followed by MIJNARENDS [2.162]. The results are shown in Fig.2.7. This analysis alleviates the problem of separating core- and conduction-electron contribution to some extent and allows determination of the shape of the Fermi surface from the measurements. In the neighborhood of the <100> directions the results were in good agreement with dHvA measurements, but close to <110> the Fermi momentum was found to be 5% too large. Although this was first ascribed to $e^+ - e^-$ correlation effects, later checks showed that a systematic alignment error, affecting in particular the measurements in the <110> region, may have been the cause.

In view of the interest in Fermi surfaces in Cu-based alloys, precise measurements of the belly radii and neck dimensions in Cu are of value. FUJIWARA and SUEOKA were the first to observe the necks using the crossed-slit technique [2.109]. Application of extra pairs of slits restricts the integration with respect to p_y (2.38) to an angle of only 5.5 mrad (FWHM), thus suppressing a large part of the core contribution (but see also [2.113]). The accompanying loss of counting rate was compensated for by the use of an in situ source. The necks are also clearly visible with the rotating specimen method (Sect.2.3.2) developed by SUEOKA [2.129, 243], and WILLIAMS et al. [2.130]. The latter found some core anisotropy in accordance with band theory [2.79]. MORINAGA [2.244], and AKAHANE and FUJIWARA [2.245] have performed measurements following this method. The latter authors compared their results with pseudopotential calculations of $\rho(\underline{p})$ which take account of the Umklapp components. A following rotating-specimen study with long slits [2.131] yielded the neck radius to within 0.5°. In all these investigations remaining discrepancies in the <111> directions were interpreted as indicative of an increased enhancement in the neck region [2.38]. This conclusion was also reached by MELNGAILIS [2.246] from a comparison of measurements with different techniques. Belly radii can often be measured very well with the long-slit geometry but for some orientations, notably k_{110}, the necks interfere and point-slits are indicated [2.10].

A number of other techniques have been tested by applying them to copper. SINGRU has obtained the equivalent of a crossed-slit geometry by combining a Ge(Li) detector with one long-slit detector [2.247]. Replacement of the latter by a point detector allows the simultaneous measurement of all three momentum components [2.138], but the counting rate is low. HOWELLS and OSMON [2.123] were the first to measure a two-dimensional angular correlation with their spark chamber setup (Sect.2.3.1). More detailed two-dimensional data have been obtained by BERKO and MADER [2.10]. Figure 2.10 shows a set of two-dimensional angular correlations as a function of p_y and p_z. The necks can be seen twice within the first zone, at $p_y = 0$, $p_z = 4.75$ mrad and also at $p_y = \frac{1}{2}\underline{G}_{111} = 5.81$ mrad, $p_z = 0$.

LOCK et al. [2.169] have tested the LCW theorem named after them on copper (Sect. 2.3.6). Application to measurements with p_z along [110] gives excellent agreement with calculated Fermi surface cross sections, but there are substantial discrepancies along the [100] direction. Further studies [2.170,171] show that these are most probably due to high-momentum components in the positron wave function.

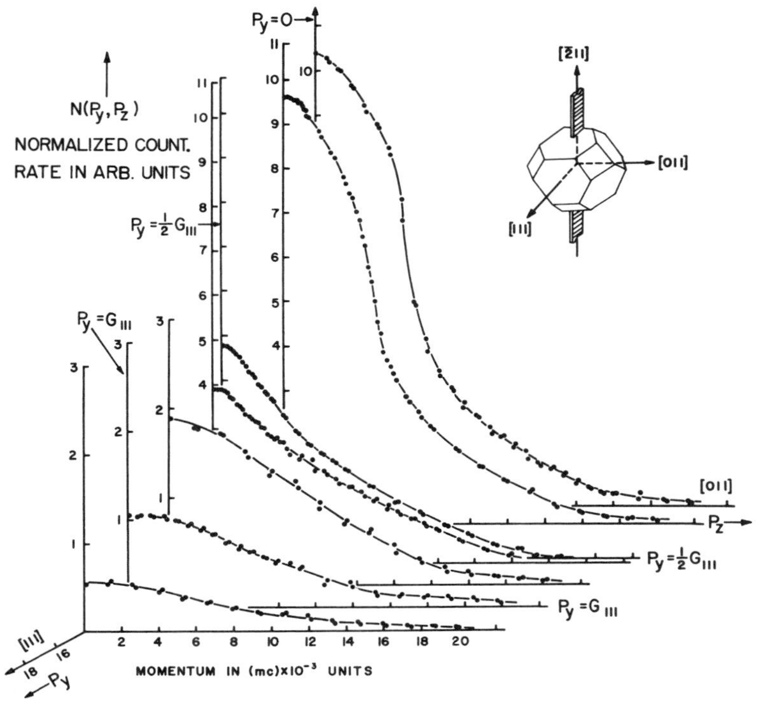

Fig.2.10. Two-dimensional angular correlations in copper by BERKO and MADER. The crystal orientation is shown in the inset. The hatched figure represents the resolution function [2.10]

As concerns theory, some of the band-structure calculations of $\rho(\underline{p})$ have already been mentioned [2.79,238]. BROSS and STÖHR [2.81] have reported results of a modified APW calculation, while KANHERE and SINGRU [2.107] have tested the applicability of the renormalized free-atom model [2.103] to Cu and Ni. This model yields reasonably good Compton profiles in transition metals but is less successful in predicting positron angular correlations.

In contrast to copper, silver and gold have hardly been studied. PAJAK and ROZENFELD [2.205,248] compared experimental angular correlations, after subtraction of a Gaussian core, with cross sections of the HALSE [2.233] model. An earlier study concerned the core contributions [2.249]. HARTHOORN [2.250] performed measurements along five directions and analyzed his data by reconstruction. A fit of a Fourier series of the type used in [2.233,251] to the Fermi momentum obtained from his reconstruction enabled him to extend his results into the neck region. The Fermi surface thus obtained is in reasonably good agreement with that of Halse, but some discrepancy remains.

2.4.5 Transition Metals and Rare Earths

The fact that the Fermi level intersects the d bands in the transition metals leads to a complex Fermi surface topology. The strong $\underline{p}$ dependence of the momentum density contributions corresponding to the various Fermi surface sheets, both radially and directionally (Sect.2.2.3), combined with the existence of important Umklapp components, severely complicates the interpretation of the experimental data. The problem has become more tractable since detailed band-structure calculations of the momentum density have become available. These calculations enable one to relate observed structure in the angular correlations or in the momentum distribution derived from these to details of the band structure, and to elucidate the role played by many-body effects. Moreover, since the momentum distribution is more sensitive to the form of the valence electron wave functions than for instance X-ray form factors [2.106], a confrontation of theory and experiment can also give important information about the accuracy of wave functions obtained by various methods of band-structure calculation. The rare-earth metals have been included in this section because their electronic structure strongly resembles that of the transition metals. This is due to the fact that 5d and 6s states in the rare earths overlap and hybridize in a way similar to that found in the transition metals, while the 4f electrons are strongly localized and form narrow deep-lying bands. The band structure of the rare earths has been reviewed extensively by DIMMOCK [2.89].

Among the 3d transition metals, iron, cobalt and nickel stand out because of their ferromagnetic properties. By magnetization of the specimen with the aid of an external magnetic field $\underline{B}$ the spin degeneracy of the electrons can be lifted and an excess of electron spins created in a direction opposite to $\underline{B}$. Furthermore, parity nonconservation in beta decay results in a partial polarization of the positron beam. A selection rule obeyed by the annihilation process states that two-photon decay can only take place if the annihilating particles have opposite spin (singlet state) and allows one in principle to study the momentum distributions of the two electron-spin populations separately. Since the average positron polarization is usually not accurately known, it is in practice more convenient to consider the sum and difference of the measured momentum distributions. The latter can be considered to represent the "spin density" in momentum space (as sampled by the positron). BERKO [Ref.2.7, pp.61-79] has discussed the theory of polarized positron annihilation and reviewed early studies in this field. A full discussion of the work up to date may be found in two recent reviews [2.11,252]. We shall therefore only briefly mention the most important recent studies in the field.

The momentum density in magnetized iron has been studied by MIJNARENDS [2.79, 163]. A reconstruction of his angular correlation data resulted in a broad shell of negative spin density between 4.5 and about 10 mrad and a positive density at higher angles, in agreement with earlier measurements [2.253]. In the analysis of the data, use was made of a small magnetic field dependence of the three-photon decay rate measured by BERKO and MILLS [2.254]. Earlier, the negative part of the spin density was thought to imply a negative polarization of the conduction electrons, in the sense that there would be an excess of conduction electrons with a polarization opposed to the net polarization of the 3d electrons. This would confirm the interpretation of early measurements with polarized neutrons [2.255]. Recent band-structure calculations [2.256], however, show that the conduction electrons are positively polarized. The negative magnetization found in the polarized neutron measurements originates in the spin dependence of the d electron wave functions. The positron results can be explained along the same lines. A band-structure calculation of the momentum density [2.79] indicates that the negative spin density in momentum space is caused by the spin dependence of the wave functions belonging to low-lying *fully* occupied d bands, while the positive part of the spin density originates in Umklapp processes involving s-d hybridized electrons with unpaired spin. A complicating factor in the interpretation of the positron data is the unequal way in which the positron samples the various groups of electrons. A spin dependence of the screened e^+ - e^- Coulomb interaction [2.257,258] may result in spin-dependent enhancement factors and render the problem still more complex. Compton scattering is not beset with these complications and it is therefore of interest to measure the Compton profile of the magnetic electrons in iron by observing the spin-dependent Compton scattering of circularly polarized γ rays, as done by SAKAI and ÔNO [2.259] (Fig.2.11). They find a positive spin density coming from the 3d electrons with a dip at low momenta. APW calculations by WAKOH and KUBO [2.260] are in fair agreement with both positron and Compton results, but there are some discrepancies.

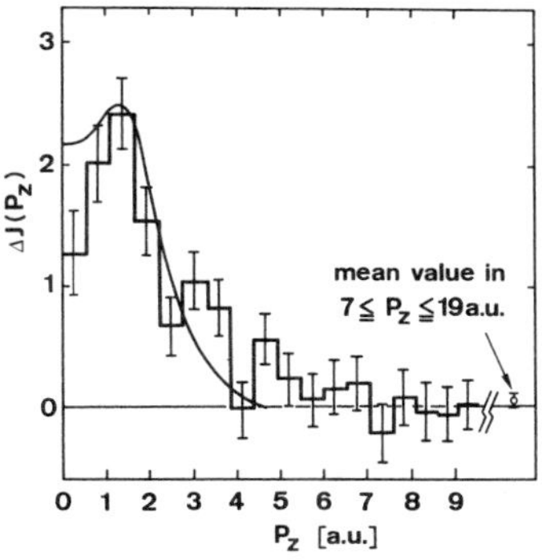

Fig.2.11. Magnetic Compton profile in iron, measured by SAKAI and ÔNO [2.259]. The solid curve represents an APW calculation by WAKOH and KUBO [2.260], convoluted with the instrumental resolution function and normalized to the area of the experimental histogram

Of the three ferromagnets cobalt has been investigated the least. ROZENFELD and co-workers [2.261,262] have measured the angular correlations in five oriented single crystals of hcp Co. The momentum distribution appears to be flattened along the c axis. Polarized-positron measurements in magnetized nickel have been done by several authors [2.254,263]. Extensive band-structure calculations of $\rho(\underline{p})$ by SINGRU and MIJNARENDS [2.239] show the complicated behavior of the momentum density of d bands and the large extent to which that density is affected by crystal symmetry [2.108]. These calculations also illustrate why attempts to explain angular correlations for transition metals in terms of Fermi surface cross sections with $\underline{k}$-dependent weight factors are not very successful, as has been demonstrated already by SHIOTANI et al. [2.264]. KONTRYM-SZNAJD et al. [2.165] have studied the momentum density in nickel at temperatures just below and above the Curie point. In view of the limited statistical accuracy and the violent oscillations displayed by their reconstructed curves, some of the conclusions of the authors regarding changes in the electronic structure at the Curie point may be affected by noise. The directional Compton profiles in Ni have been studied experimentally by EISENBERGER and REED [2.265] and have been calculated by RATH et al. [2.266] with the aid of the LCAO method. The agreement, both with regard to the full profiles and the structure in the anisotropy, is excellent, as shown by WANG and CALLAWAY [2.267]. According to these authors, this structure can be understood qualitatively in terms of Fermi surface cross sections, but in view of the discussion at the beginning of this subsection and the results of SHIOTANI et al. [2.264] it is to be expected that important deviations will show up in a more quantitative comparison.

Chromium can exist in three magnetic phases. Above the Nêel temperature (T_N = 410 K) it is paramagnetic, below T_N antiferromagnetic. The latter state is caused by a transverse spin-density wave, that changes its polarization to longitudinal at 120 K. Angular correlation measurements at temperatures above and below 120 K yield identical results [2.268,269] so that any changes in the electronic band structure accompanying this phase transition are not clearly reflected in the momentum density.

Of the remaining 3d transition metals titanium, vanadium, and manganese, the first and the last have not been investigated in any detail. The anisotropy of the angular correlations in V and Cr (and also in Nb and Mo) has been measured by SHIOTANI et al. [2.268,270] and calculated via APW by WAKOH et al. [2.80]. As far as the anisotropy is concerned, the agreement is rather good in the case of V, Cr, and Mo and excellent for Nb, as shown in Fig.2.12. The relation between the observed structure and the topology of the Fermi surface is clarified in an earlier paper on Compton profiles in V [2.98], where the authors also show the contributions of each separate band to the anisotropy. This allows a better insight into the relationship between momentum density profiles and band structure. The success of this work and of related studies [2.271,272] shows that the anisotropy of the momentum density in transition metals can be fully understood in terms of their band structure. The same cannot be said, however, of the overall shape of the momentum distribution or the profiles derived from it. Most band-structure calculations consistently overestimate the high-momentum content of the profiles [2.80,239,268].

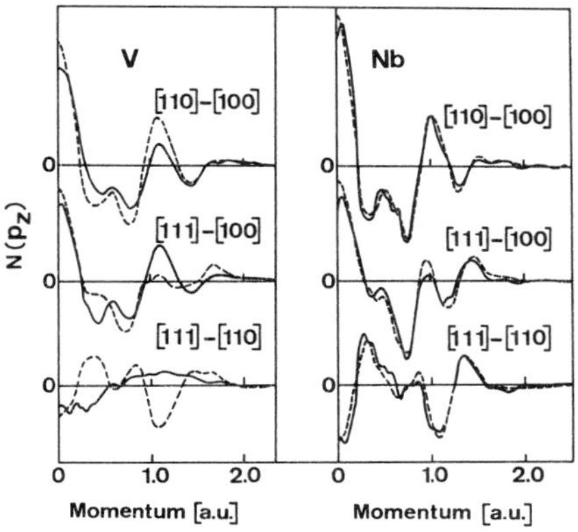

Fig.2.12. Experimental anisotropy curves in V and Nb by SHIOTANI et al. [2.270] compared with the results of APW calculations (dashed curves) by WAKOH et al. [2.80]

The possibility that this signals deficiencies in the implementation of the computational methods cannot be excluded, but these discrepancies are more likely due to $e^+ - e^-$ correlation effects. The excellent agreement between the measured and calculated anisotropies would then suggest that also in transition metals the enhancement is largely isotropic.

The 4d transition metals yttrium and zirconium have the hcp structure. Since their energy band structure is very similar to that of the rare earths they have been studied together with Gd, Tb, Dy, Ho, and Er [2.273-276]. Except in the case of Gd [2.277 and references cited therein], the low purity of these materials, together with magnetic ordering at low temperatures, has prohibited the use of the dHvA technique so far. These positron investigations therefore represent practically the only direct experimental information concerning the Fermi surface geometry of these metals. Moreover, extremely high neutron absorption cross sections in some of the rare earths preclude the measurement of Kohn anomalies with thermal neutrons. Other experimental evidence, for instance relating to electrical resistivity, magnetic susceptibility, Hall effect, etc., is of a rather indirect nature. The two-photon angular correlations with p_z along the c axis display a characteristic bump for these materials. Band-structure calculations [2.93,278] show that this bump is due to flat electron and hole sheets of the Fermi surface normal to the c axis. The observed structure in the angular correlation curves is not quite as pronounced as expected from a folding of theory with the optical resolution and positron motion. The authors ascribe this to many-body effects. IYAKUTTI et al. [2.279] find that inclusion of $e^- - e^-$ correlation can cause significant changes in the energy bands. HOHENEMSER et al. [2.280] have reported a study of the spin distribution in Gd at different temperatures with polarized positrons. The spin density is highly anisotropic and there is a relatively large 3γ effect [2.254]. It is an interesting speculation whether the spin polarization of the 4f electrons in Gd would be observable at all in such an experiment. Experiments under pressure and at different temperatures have been done in an attempt to clarify the role of the f electron in the γ, α, and α' phases of cerium, but the free-electron model used to interpret the data appears inadequate [2.273,281,282]. Now that more is known about positron effective masses and the influence of impurities, and also band-structure calculations for these materials have become more refined and purer samples are becoming available, it would be interesting to repeat some of these investigations in order to establish the band structure of these interesting materials more firmly and to elucidate the role played by correlation effects.

Relatively little positron work has been done on the 4d and 5d transition metals. The band structure of molybdenum has first been studied by KIM and BUYERS [2.283] with point slits. They compared their data with a calculation employing an interpolation scheme and energy levels of tungsten, which has a very similar band structure. The recent studies of SHIOTANI et al. on Mo and Nb have already been mentioned. KANHERE and SINGRU [2.284] calculated the momentum density of palladium with the aid of the Hubbard scheme. HARTHOORN [2.250] recently measured the angular correlations for several crystal orientations and obtained the momentum distribution by reconstruction.

To conclude this section, mention should be made of some recent systematic investigations of the Doppler-broadened line shape of annihilation radiation in metals [2.285,286]. After deconvolution of the instrumental resolution functions the line shapes can be described in terms of an inverted parabola resting on a Gaussian. The standard deviation of, and the fraction of annihilations under, the Gaussian showed a characteristic and repetitive behavior when plotted against atomic number.

2.5 Disordered Alloys and Ordered Metallic Compounds

2.5.1 Disordered Alloys

A detailed knowledge of the electronic structure of alloys is of great importance for a better understanding of their properties. This is obvious for properties like specific heat, magnetic susceptibility, etc., that are directly related to the density of states at the Fermi energy. The electronic structure also plays an important role, however, in determining the phase stability in alloys. When the concentration x of a binary substitutional alloy $A_{1-x}B_x$ is varied from 0 to 1, the alloy passes through a number of different crystal structures. HUME-ROTHERY [2.287] has pointed out in 1931 that in alloys of noble metals each of these structures tends to be stable around a particular value of the number of valence electrons per atom e/a, independent of other properties of the alloy. For instance, it has been shown experimentally that the limit of the α-phase solubility of many B-group elements in the noble metals is reached at e/a $\sim$ 1.36. According to the explanation forwarded by JONES [2.225,288] this would be so because at that e/a ratio the (free-electron)

Fermi sphere makes contact with the octahedral faces of the Brillouin zone. This would result in a peak in the density-of-states curve N(E) followed by a rapid decrease. Additional electrons can then only be accommodated at the cost of a relatively large increase in energy, and a different crystal structure may energetically be more favorable. Later theories, put forward after it had been found that the Fermi surfaces of the noble metals contact the zone face even at e/a = 1, explain phase transformation by Fermi-surface contact with other zone faces [2.289] or in terms of the rapid variation of the $\underline{k}$-dependent dielectric function $\varepsilon_{\underline{k}}$ in the vicinity of $\underline{k} = 2\underline{k}_F$ [2.290], while d-d overlap also seems to play a rolē [2.291]. It is clear that in order to clarify this complex problem more information about the band structure and Fermi surfaces in nondilute alloys is needed. For reasons mentioned in the introduction to this chapter the positron annihilation technique is particularly suited to provide this information in spite of some inherent difficulties such as preferential annihilation, influence of clusters, etc.

In the last few years great progress has been made in our understanding of the electronic structure of alloys. The application of multiple-scattering theory to this problem by a number of authors (see [2.292] for a full bibliography) has resulted in the development by BANSIL et al. [2.292] of a scheme for the calculation of the electronic band structure in disordered alloys using muffin-tin potentials. This approach, based on the average t-matrix approximation (ATA) and related to the KKR method for pure metals, yields complex eigenvalues, the imaginary parts of which give the finite lifetimes of the $\underline{k}$ states. As a result of this damping the energy bands $E(\underline{k})$ and the Fermi surface are blurred, the latter typically by a few percent of k_F. The damping can be observed in dilute ($\leq$ 1 at.%) alloys by measuring the Dingle temperature in the dHvA effect [2.293]. Recently, STOCKS et al. [2.294,295] have presented a CPA formulation of this theory. MIJNARENDS and BANSIL [2.99,296] have used the ATA approach to calculate the electron momentum density for Compton scattering in alloys. Their theory awaits extension into a CPA treatment of positron annihilation. The CPA model calculation by HONG and CARBOTTE [2.15] that shows the effect of inclusion of the positron is a first step in this direction.

The positron work in alloys up to 1974 has been extensively reviewed by BERKO and MADER [2.10] and will therefore be mentioned only briefly. Most measurements to date have been done on Cu alloys in view of the fact that the Fermi surface of pure Cu is well known and because of the possibility of using in situ sources. When Cu is alloyed with a polyvalent metal its Fermi surface swells and the neck size increases in order to accommodate the extra electrons. BERKO and MADER have collected the measurements of neck sizes for the α phases of CuZn [2.244,297-299], CuGe and CuGa [2.300-302], and CuAl [2.303-305], by various authors and compared them with theoretical curves based on the rigid-band model and the averaged t-matrix calculations by BANSIL et al. On the whole there is good agreement between the experimental data taken with different geometries and the band-structure calculations.

However, there are strong indications that the rotating specimen technique underestimates the neck radius. The results of MORINAGA [2.244], and of WILLIAMS [2.298] obtained with this technique are in good agreement with each other, but lie consistently below the data obtained with the various geometries in which the specimen is kept stationary. BECKER et al. [2.299] have discussed some of the problems inherent in the rotating specimen method.

Table 2.4 Abbreviations used in Figs.2.13-15

Abbreviation	Authors	References
AK	Akahane et al.	[2.307]
BE	Becker et al.	[2.299]
BM	Berko and Mader	[2.306]
FU	Fujiwara et al.	[2.305]
HA	Hasegawa et al.	[2.300]
MM	Murray and McGervey	[2.303]
MW	McLarnon and Williams	[2.301,302]
SU	Suzuki et al.	[2.308]
TH	Thompson et al.	[2.304]
TS	Triftshäuser and Stewart	[2.297]

Since 1974, additional measurements of the neck and belly radii have been performed by BERKO and MADER [2.306] on α-CuZn, by AKAHANE et al. [2.307] on α-CuAl, by SUZUKI et al. [2.308] on α-CuSi and α-CuGe, and by McLARNON and WILLIAMS [2.302] on α-CuGe. The results of all measurements on Cu alloys available at present have been collected in Figs.2.13 and 2.14 for the belly radii k_{110} and k_{100} and in Figure 2.15 for the neck radius. The abbreviations used are explained in Table 2.4. The experimental data are compared with simple theory. The curves marked 1 and 1' represent the results of the rigid-band calculation, with and without lattice expansion respectively, by BANSIL et al. [2.292]. This lattice expansion has been calculated for the case of α-CuZn where it is relatively large. Similar curves for the other alloys will therefore lie between 1 and 1'. The curves marked 2 represent the results of the sinking-conduction-band model proposed by LETTINGTON [2.309] for α-CuZn, and by REA and DE REGGI [2.310] for α-CuAl alloys to explain their optical data. In this model the conduction band sinks with respect to the Cu d bands and the Fermi level at a rate of 0.05 eV/(% change in e/a) as the solute concentration is increased. Of the three, the k_{110} radii have been determined with the highest accuracy and the results for the different alloys follow the simple theoretical models reasonably well (Fig.2.13). However, the CuAl data seem to lie consistently below the α-CuGe points. The curves marked 1 and 1' in Fig.2.14 have been corrected to bring them into agreement with the value of k_{100} for pure Cu. It seems improbable that for any of these alloys the Fermi surface will contact the (002) zone face

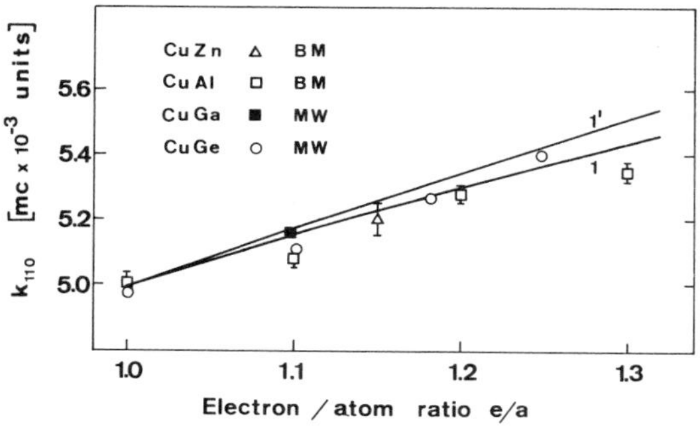

Fig.2.13. Measurements of the belly radius k_{110} of the Fermi surface in various copper-based alloys. The theoretical curves are discussed in the text, abbreviations are explained in Table 2.4

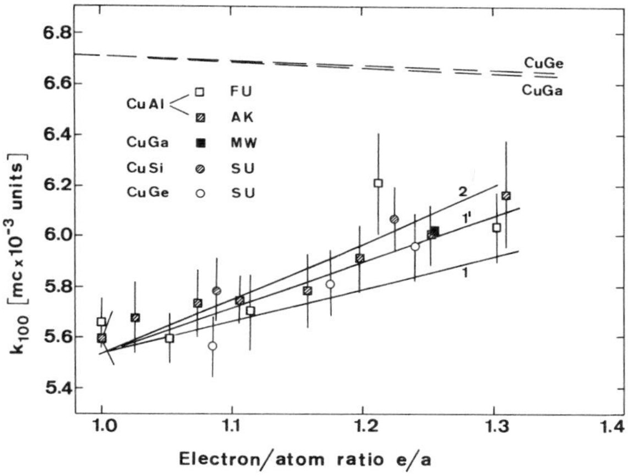

Fig.2.14. Measurements of the belly radius k_{100} of the Fermi surface in various copper-based alloys. The dashed lines represent the Brillouin zone boundary. The theoretical curves are discussed in the text, abbreviations are explained in Table 2.4

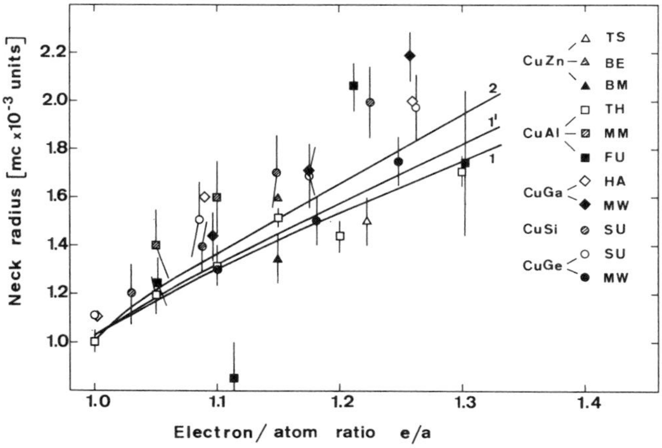

Fig.2.15. Measured neck radii of the Fermi surface in various copper-based alloys. The theoretical curves are discussed in the text, abbreviations are explained in Table 2.4

at the solubility limit, but measurements in the range of e/a = 1.3-1.4 would be of interest. Fig.2.15 contains the experimental results for the neck radius. In order not to confuse the graph, and in view of the reasons mentioned earlier, the rotating specimen data have been omitted. It is clear from this figure that the neck radii in different alloys do not depend solely on the e/a ratio, as has also become apparent from dHvA measurements in dilute alloys [2.311]. This is not surprising as the neck radius is rather sensitive to details of the band structure. The outlying datum point at e/a = 1.114 is believed to be in error [2.307].

The question as to which of two neighboring phases is the stable one at a certain alloy composition depends on the difference in energy between the two crystal structures over a range of concentrations. Knowledge of the Fermi surface of the second phase is therefore of interest. The α phase of CuGe is stable up to 9.5 at.% Ge, i.e., e/a = 1.28, while at 11.4 at.% the transformation into the hcp ζ phase is complete. SUZUKI et al. [2.312] have investigated the Fermi surface of two ζ-CuGe alloys of 12.6 and 15.0 at.% Ge (e/a = 1.38 and 1.45 respectively) with the long-slit and crossed-slit geometries. Their data suggest a severely distorted Fermi surface similar to the one proposed by MASSALSKI and COCKAYNE [2.313]. It has narrow necks along the $\langle 0002 \rangle$ axis, larger necks in the $\langle 10\bar{1}0 \rangle$ directions and large bulges towards the zone faces perpendicular to the $\langle 10\bar{1}1 \rangle$ directions. For the 15.0 at.% alloy the electrons overlap across the face perpendicular to $\langle 10\bar{1}0 \rangle$. A Fermi surface of this shape gives rise to three peaks in the density-of-states curve. This allows the electrons to have a lower total energy than in the α phase, thus explaining the relative stability of the two phases.

Besides the alloys of Cu with various B-group metals, CuNi alloys have received some attention [2.300,303,314-317]. However, the results of different studies are contradictory, probably as a result of the tendency of CuNi to segregate. The preparation of homogeneous CuNi single crystals is difficult and the in situ source technique therefore seems less suited for this alloy. The isoelectronic alloy CuPd has been investigated by HASEGAWA et al. [2.318], and HARTHOORN [2.250]. Both studies suggest a Fermi surface that shrinks roughly according to the rigid-band predictions as the Pd concentration increases from 0 to about 30-40 at.%. At higher Pd concentrations Harthoorn found no further decrease. AgPd behaves in a similar way, except that the rate of decrease with increasing Pd concentration seems to level off already at $\sim$ 30 at.%. A reconstruction of $\Gamma(\underline{p})$ from data on $Ag_{0.68}Pd_{0.32}$ single crystals shows that the Fermi surface has detached itself from the hexagonal zone face [2.250]. Other alloys of d metals studied with positrons are ζ-AgAl [2.319], α-NiZn [2.320], Fe-Ni (Invar) [2.321], and glassy and crystalline $Pd_{0.775}Cu_{0.06}Si_{0.165}$ [2.322]. Both phases of the last alloy have a much larger low-momentum fraction than pure Pd.

Another class of interesting alloys is that of the transition-metal and rare-earth hydrides. The electronic structure of these materials is of interest because of possible applications in energy storage systems. In the past, two models of the electronic structure of nonstoichiometric metal hydrides have been in use: the anionic and the protonic model. In the first model the interstitial hydrogen succeeds in binding two electrons, thereby depleting the conduction band. The second model assumes the presence of a screened proton; the hydrogen electron is donated to the rigid conduction and incompletely filled d bands, which implies that E_F and k_F increase with the hydrogen concentration. This picture is borne out by the results of angular correlation measurements on a large number of hydrides [2.323-328]. In every case the hydride curve is wider than that of the corresponding pure metal. However, ATA band-structure calculations [2.329,330], and dHvA measurements [2.331] in PdH_x present a somewhat different picture. A Pd-H molecular bonding level is formed below the Pd d band. With increasing x this level broadens into a band at the expense of the lowest Pd band. At low x the change in size of the hole pockets can be explained entirely by the lattice expansion induced by the interstitial hydrogen, so the hydrogen electron does not fill up the Pd d bands at low x. This is in agreement with CPA calculations by FAULKNER [2.332] who found a roughly constant density of states at the Fermi energy up to $x \sim 0.7$. High-resolution angular correlation studies employing the two-dimensional geometry might be able to demonstrate the filling-up of the d-band holes at higher values of x. The ATA method has also been successful in predicting Dingle temperatures for various orbits in dilute CuH [2.333]. The predicted temperatures are in excellent agreement with recent dHvA measurements [2.334].

Relatively little work has been done on alloys of nearly free-electron metals. The best studied system is LiMg, in which STEWART [2.335], and later KUBICA et al. [2.84] investigated the smearing of the Fermi break. The latter authors also measured the annihilation rate for different concentrations. The behavior of the core annihilation rate combined with an excessive smearing at low Li concentrations indicates that the positrons are localized in Li-rich regions (see also Sect.4.2.4).

2.5.2 Metallic Compounds

Momentum density studies in ordered metallic compounds have mainly been concerned with V_3Si, a high-temperature superconductor with the A-15 structure. BERKO and WEGER [2.336,337] have performed long-slit angular correlation measurements on single crystals of this material. An attempt to correlate their results with a planar structure of the Fermi surface expected from the linear-chain model of WEGER [2.338] was only partially successful. Later measurements with the point-slit geometry by BERKO and MADER [2.10] show a wealth of detail, especially in the higher zones, which indicates a strong d character of the bands at the Fermi surface. However, a detailed interpretation in terms of Fermi surface geometry is still lacking. Although a number of band-structure calculations have been performed for this and related compounds [2.339-344], a band-theoretical calculation of the momentum density is highly necessary.

Among the very few other ordered alloys studied with positron annihilation, the work on β'-CuZn by TRIFTSHÄUSER et al. [2.345] deserves mention. The structure in the derivative of the angular correlations measured along the three high-symmetry directions correlates well with the geometrical cross-sectional areas of the Fermi surface resulting from a nonlocal pseudopotential calculation by TAYLOR [2.346]. Another ordered Cu alloy, Cu_3Au, has been studied by MORINAGA [2.347] using the rotating specimen method, while ROZENFELD et al. have investigated polycrystalline Th_3As_4 and U_3As_4 [2.348] and a number of other uranium compounds [2.349].

2.6 Conclusion

Considerable progress has been made during the last decade in our understanding of the electronic structure of solids and of their momentum distribution in particular. Band-structure calculations are now able to provide an accurate description of the observed anisotropies after correction of the data for the enhancement due to $e^+ - e^-$ correlation. However, the lack of agreement between the isotropic parts of the calculated and measured curves for metals with d bands suggests that the enhancement corrections currently in use may have to be improved upon. The use of two-dimensional multidetector assemblies will allow the investigation of enhancement effects in greater detail. The superior resolution of these machines will also enable one to perform high-precision Fermi surface investigations. It will be possible to observe not only small details of the geometry of the Fermi surface, but also the amount of smearing of the Fermi breaks in disordered alloys, which can be related to the damping of the electron states at the Fermi energy. This would extend the measurements of the Dingle temperature into the range of high solute concentrations. It should be kept in mind, however, that besides the damping that one wishes to measure, and the usual optical resolution and positron thermal motion effects, also the disorder scattering of the positron and positron localization effects will contribute to the smearing. Much will also depend on the quality of the specimens, particularly on the concentration of structural lattice defects and the absence of clustering. In spite of these complications it is clear that in the field of disordered alloys positron annihilation can contribute significantly since competing experimental techniques are virtually absent. A shift of effort from pure metals to disordered alloys will therefore undoubtedly take place in the near future.

Acknowledgements. I am grateful to Professor S. Berko, who kindly read the manuscript and provided many helpful comments, and to Dr. R.N. West for an interesting discussion. I have appreciated the preprints that many colleagues have sent me. Thanks are also due to Dr. B. van Laar for a helpful discussion on resolution functions, Finally, and in particular, I thank my wife Kamlesh for her active support during the preparation of the manuscript.

References

2.1 R. Beringer, C.G. Montgomery: Phys. Rev. *61*, 222-224 (1942)
2.2 S.C. Moss, R.H. Walker: J. Appl. Cryst. *8*, 96-107 (1974)
2.3 B.M. Powell, P. Martel, A.D.B. Woods: Phys. Rev. *171*, 727-736 (1968)
2.4 A.J. McAlister, E.A. Stern, J.C. McGroddy: Phys. Rev. *140*, A2105-2109 (1965)
2.5 J.M. Tracy, E.A. Stern: Phys. Rev. B *8*, 582-593 (1973)
2.6 B. Williams (ed.): *Compton Scattering* (McGraw-Hill, London 1977)
2.7 A.T. Stewart, L.O. Roellig (eds.): *Positron Annihilation* (Academic Press, New York 1967)
2.8 R.N. West: Adv. Phys. *22*, 263-383 (1973)
2.9 I.Ya. Dekhtyar: Phys. Rep. C *9*, 243-353 (1974)
2.10 S. Berko, J. Mader: Appl. Phys. *5*, 287-306 (1975)
2.11 S. Berko: "Positron Annihilation", in [Ref.2.6, Chap.9, pp.273-322]
2.12 S. DeBenedetti, C.E. Cowan, W.R. Konneker, H. Primakoff: Phys. Rev. *77*, 205-212 (1950)
2.13 W. Brandt, L. Eder, S. Lundqvist: Phys. Rev. *142*, 165-173 (1966)
2.14 S. Berko, J.S. Plaskett: Phys. Rev. *112*, 1877-1887 (1958)
2.15 K.M. Hong, J.P. Carbotte: Can. J. Phys. *55*, 1335-1341 (1977)
2.16 L. Hedin, S. Lundqvist: In *Solid State Physics*, Vol.23, ed. by F. Seitz, D. Turnbull, H. Ehrenreich (Academic Press, New York 1969) pp.1-181
2.17 S. Kahana: Phys. Rev. *117*, 123-128 (1960)
2.18 S. Kahana: Phys. Rev. *129*, 1622-1628 (1963)
2.19 H. Weisberg, S. Berko: Phys. Rev. *154*, 249-257 (1967)
2.20 J. Crowell, V.E. Anderson, R.H. Ritchie: Phys. Rev. *150*, 243-248 (1966)
2.21 R. Bhattacharyya, K.S. Singwi: Phys. Rev. Lett. *29*, 22-25 (1972)
2.22 J.P. Carbotte, S. Kahana: Phys. Rev. *139*, A213-A222 (1965)
2.23 C.K. Majumdar: Phys. Rev. *140*, A227-A236 (1965)
2.24 J.J. Donaghy, A.T. Stewart: Phys. Rev. *164*, 396-398 (1967)
2.25 B. Bergersen: Ph. D. Thesis, Brandeis Univ. (1964, unpublished)
2.26 J.P. Carbotte: Phys. Rev. *155*, 197-207 (1967)
2.27 A.J. Zuchelli, T.G. Hickman: Phys. Rev. *136*, A1728-A1730 (1964)
2.28 J. Arponen, P. Jauho: Phys. Rev. *167*, 239-244 (1968)
2.29 H. Stachowiak: Phys. Status Solidi (b) *52*, 313-322 (1972)
2.30 H. Kanazawa, Y. Ohtsuki, S. Yanagawa: Prog. Theor. Phys. *33*, 1010-1021 (1965)
2.31 B. Bergersen: Phys. Rev. *181*, 499-505 (1969)
2.32 P. Vashishta, K.S. Singwi: Phys. Rev. B *6*, 875-887 (1972)
2.33 J. Arponen: J. Phys. C *3*, 107-125 (1970)
2.34 C.K. Majumdar, T.V. Ramakrishnan: Phys. Rev. B *7*, 1850-1854 (1973)
2.35 A.K. Rajagopal, C.K. Majumdar: Phys. Rev. B *8*, 2362-2364 (1973)
2.36 B.B.J. Hede, J.P. Carbotte: J. Phys. Chem. Sol. *33*, 727-735 (1972)
2.37 K. Fujiwara: J. Phys. Soc. Jpn. *29*, 1479-1490 (1970)
2.38 K. Fujiwara, T. Hyodo, J. Ohyama: J. Phys. Soc. Jpn. *33*, 1047-1059 (1972)
2.39 K. Fujiwara, T. Hyodo: J. Phys. Soc. Jpn. *35*, 1664-1667 (1973)
2.40 K. Fujiwara, T. Hyodo: J. Phys. Soc. Jpn. *35*, 1133-1135 (1973)
2.41 J.P. Carbotte: Phys. Rev. *144*, 309-318 (1966)
2.42 J.P. Carbotte, A. Salvadori: Phys. Rev. *162*, 290-300 (1967)
2.43 A. Salvadori, J.P. Carbotte: Phys. Rev. *188*, 550-556 (1969)
2.44 B.K. Sharma: Appl. Phys. *5*, 265-267 (1974)
2.45 T. Chiba, G.B. Dürr, W. Brandt: Phys. Status Solidi (b) *81*, 609-614 (1977)
2.46 W. Brandt, J. Reinheimer: Phys. Rev. B *2*, 3104-3112 (1970)
2.47 G.E. Lee-Whiting: Phys. Rev. *97*, 1557-1558 (1955)
2.48 J.P. Carbotte, H.L. Arora: Can. J. Phys. *45*, 387-402 (1967)
2.49 E.J. Woll, Jr., J.P. Carbotte: Phys. Rev. *164*, 985-993 (1967)
2.50 H.J. Mikeska: Phys. Lett. A *24*, 402-404 (1967)
2.51 H.J. Mikeska: Z. Phys. *232*, 159-173 (1970)
2.52 A. Perkins, J.P. Carbotte: Phys. Rev. B *1*, 101-107 (1970)
2.53 B. Bergersen, E. Pajanne: Appl. Phys. *4*, 25-35 (1974)
2.54 C.K. Majumdar: Phys. Rev. *140*, A237-A244 (1965)
2.55 A.T. Stewart, J.B. Shand: Phys. Rev. Lett. *16*, 261-262 (1966)
2.56 A.T. Stewart, J.B. Shand, S.M. Kim: Proc. Phys. Soc. *88*, 1001-1010 (1966)

2.57 S.M. Kim, A.T. Stewart, J.P. Carbotte: Phys. Rev. Lett. *18*, 385-387 (1967)
2.58 S.M. Kim, A.T. Stewart: Phys. Rev. B *11*, 2490-2499 (1975)
2.59 P. Kubica, A.T. Stewart: Phys. Rev. Lett. *34*, 852-855 (1975)
2.60 A.P. Mills, L. Pfeiffer: Phys. Rev. Lett. *36*, 1389-1393 (1976); Phys. Lett. A *63*, 118-120 (1977)
2.61 C.K. Majumdar: Phys. Rev. *149*, 406-408 (1966)
2.62 D.R. Hamann: Phys. Rev. *146*, 277-281 (1966)
2.63 B. Bergersen, E. Pajanne: Phys. Rev. *186*, 375-380 (1969)
2.64 B. Bergersen, E. Pajanne: Phys. Rev. B *3*, 1588-1598 (1971)
2.65 G. Mori: J. Phys. F *3*, 548-560 (1973); J. Phys. F *4*, 821-829 (1974)
2.66 M. Hasegawa: J. Phys. F *6*, 1433-1439 (1976)
2.67 M.A. Shulman, G.M. Beardsley, S. Berko: Appl. Phys. *5*, 367-374 (1975)
2.68 R. Hosemann, S.N. Bagchi: *Direct Analysis of Diffraction by Matter* (North-Holland, Amsterdam 1962) p.275
2.69 M.J. Stott, P. Kubica: Phys. Rev. B *11*, 1-10 (1975)
2.70 C.H. Hodges, M.J. Stott: Phys. Rev. B *7*, 73-79 (1973)
2.71 B. Donovan, N.H. March: Phys. Rev. *110*, 582-583 (1958)
2.72 D. Stroud, H. Ehrenreich: Phys. Rev. *171*, 399-407 (1968)
2.73 A.G. Gould, R.N. West, B.G. Hogg: Can. J. Phys. *50*, 2294-2301 (1972)
2.74 P. Kubica, M.J. Stott: J. Phys. F *4*, 1969-1981 (1974)
2.75 G. Kögel: Paper B2, 4th Int. Conf. on Positron Annihilation, Helsingør 1976
2.76 B. Alder, S. Fernbach, M. Rotenberg (eds.): *Methods in Computational Physics, Vol.6, Energy Bands in Solids* (Academic Press, New York 1968)
2.77 J. Hubbard: J. Phys. C *2*, 1222-1229 (1969)
2.78 J. Hubbard, P.E. Mijnarends: J. Phys. C *5*, 2323-2332 (1972)
2.79 P.E. Mijnarends: Physica *63*, 235-247 (1973)
2.80 S. Wakoh, Y. Kubo, J. Yamashita: J. Phys. Soc. Jpn. *38*, 416-422 (1975)
2.81 H. Bross, H. Stöhr: Appl. Phys. *3*, 307-311 (1974)
2.82 A. Seeger: J. Phys. F *3*, 248-294 (1973)
2.83 D.G. Lock, R.N. West: J. Phys. F *4*, 2179-2188 (1974)
2.84 P. Kubica, B.T.A. McKee, A.T. Stewart, M.J. Stott: Phys. Rev. B *11*, 11-22 (1975)
2.85 C. Koenig: Paper D6, 4th Int. Conf. on Positron Annihilation, Helsingør 1976; Phys. Status Solidi (b) *88*, 569-579 (1978)
2.86 E. Wigner, F. Seitz: Phys. Rev. *43*, 804-810 (1933); *46*, 509-524 (1934)
2.87 J. Callaway: *Energy Band Theory* (Academic Press, New York, London 1964)
2.88 J.M. Ziman: In *Solid State Physics*, Vol.26, ed. by H. Ehrenreich, F. Seitz, D. Turnbull (Academic Press, New York 1971) p.1-101
2.88a T.L. Loucks: *Augmented Plane Wave Method* (Benjamin, New York 1967)
2.89 J.O. Dimmock: In *Solid State Physics*, Vol.26, ed. by H. Ehrenreich, F. Seitz, D. Turnbull (Academic Press, New York 1971) pp.103-274
2.90 R.A. Deegan, W.D. Twose: Phys. Rev. *164*, 993-1005 (1967)
2.91 F.A. Butler, F.K. Bloom, Jr., E. Brown: Phys. Rev. *180*, 744-746 (1969)
2.92 N. Elyashar, D.D. Koelling: Phys. Rev. B *13*, 5362-5372 (1976)
2.93 T.L. Loucks: Phys. Rev. *144*, 504-511 (1966)
2.94 J. Korringa: Physica *13*, 392-400 (1947)
2.95 W. Kohn, N. Rostoker: Phys. Rev. *94*, 1111-1120 (1954)
2.96 A.R. Williams, S.M. Hu, D.W. Jepsen: "Recent Developments in KKR Theory", in *Computational Methods in Band Theory*, ed. by P.M. Marcus, J.F. Janak, A.R. Williams (Plenum Press, New York 1971) pp.157-177
2.97 F.S. Ham, B. Segall: Phys. Rev. *124*, 1786-1796 (1961)
2.98 S. Wakoh, J. Yamashita: J. Phys. Soc. Jpn. *35*, 1406-1411 (1973)
2.99 P.E. Mijnarends, A. Bansil: Phys. Rev. B *13*, 2381-2390 (1976)
2.100 J.M. Ziman: Proc. Phys. Soc. *86*, 337-353 (1965)
2.101 T. Chiba, N. Tsuda: Appl. Phys. *5*, 37-40 (1974)
2.102 T. Chiba: J. Chem. Phys. *64*, 1182-1188 (1976)
2.103 L. Hodges, R.E. Watson, H. Ehrenreich: Phys. Rev. B *5*, 3953-3971 (1972)
2.104 M. Chodorow: Phys. Rev. *55*, 675 (1939)
2.105 B. Segall: Phys. Rev. *125*, 109-122 (1962)
2.106 K.-F. Berggren, S. Manninen, T. Paakkari, O. Aikala, K. Mansikka: "Solids", in [Ref.2.6, Chap.6, pp.139-208)]

2.107 D.G. Kanhere, R.M. Singru: J. Phys. F *5*, 1146-1154 (1975)
2.108 R. Harthoorn, P.E. Mijnarends: J. Phys. F *8*, 1147-1158 (1978)
2.109 K. Fujiwara, O. Sueoka: J. Phys. Soc. Jpn. *21*, 1947-1955 (1966)
2.110 I. Epstein, B. Williams: Philos. Mag. *27*, 311-328 (1973)
2.111 P. Colombino, B. Fiscella, L. Trossi: Nuovo Cimento *27*, 589-600 (1963)
2.112 M. Hasegawa, T. Suzuki, M. Hirabayashi: J. Phys. Soc. Jpn. *43*, 89-96 (1977)
2.113 J.D. McGervey: Phys. Rev. B *9*, 2402-2405 (1974)
2.114 A.T. Stewart: Nucl. Instrum. Methods *117*, 309 (1974)
2.115 S. Berko, G.M. Beardsley, M. Haghgooie, I. Tal, J. Mader: Paper H13, 4th Int. Conf. on Positron Annihilation, Helsingør 1976
2.116 S. Berko, M. Haghgooie, J.J. Mader: Phys. Lett. A *63*, 335-338 (1977)
2.117 W. Brandt, G. Coussot, R. Paulin: Phys. Rev. Lett. *23*, 522-524 (1969)
2.118 A. Greenberger, A.P. Mills, A. Thompson, S. Berko: Phys. Lett. A *32*, 72-73 (1970)
2.119 R.N. West: Private communication
2.120 W. Triftshäuser: Proc. 2nd Int. Conf. on Positron Annihilation, Kingston, Ontario 1971, pp.4.77-86
2.121 R. Kurz, D. Protič, R. Reinartz, G. Riepe: IEEE Trans. NS-*24*, 255-259 (1977)
2.122 M.R. Howells, P.E. Osmon: J. Phys. E *4*, 929-935 (1971)
2.123 M.R. Howells, P.E. Osmon: J. Phys. F *2*, 277-288 (1972)
2.124 T.K. Chatterjee, L. Draper, M.R. Howells, P.E. Osmon: J. Phys. E *6*, 135-137 (1973)
2.125 A.A. Manuel, G.H. Bongi, Ø. Fischer, M. Peter: Helv. Phys. Acta *50*, 166 (1977)
2.126 A.P. Jeavons, G. Charpak, R.J. Stubbs: Nucl. Instrum. Methods *124*, 491-503 (1975)
2.127 A.P. Jeavons, C. Cate: IEEE Trans. NS-*23*, 640-644 (1976)
2.128 R.J. Douglas, A.T. Stewart: Paper H14, 4th Int. Conf. on Positron Annihilation, Helsingør 1976
2.129 O. Sueoka: J. Phys. Soc. Jpn. *23*, 1246-1250 (1967)
2.130 D.Ll. Williams, E.H. Becker, P. Petijevich, G. Jones: Phys. Rev. Lett. *20*, 448-450 (1968)
2.131 T. Akahane: J. Phys. Soc. Jpn. *38*, 1648-1652 (1975)
2.132 H.P. Hotz, J.M. Mathiesen, J.P. Hurley: Phys. Rev. *170*, 351-355 (1968)
2.133 K. Rama Reddy, R.A. Carrigan: Nuovo Cimento B *66*, 105-119 (1970)
2.134 J.L. Campbell, T.E. Jackman, I.K. MacKenzie, C.W. Schulte, C.G. White: Nucl. Instrum. Methods *116*, 369-380 (1974)
2.135 K.G. Lynn, A.N. Goland: Solid State Commun. *18*, 1549-1552 (1976)
2.136 K.G. Lynn, J.R. MacDonald, R.A. Boie, L.C. Feldman, J.D. Gabbe, M.F. Robbins, E. Bonderup, J. Golovchenko: Phys. Rev. Lett. *38*, 241-244 (1977)
2.137 T.E. Jackman, P.C. Lichtenberger, C.W. Schulte: Appl. Phys. *5*, 259-264 (1974)
2.138 R.M. Singru: Phys. Lett. A *46*, 61-62 (1973)
2.139 S. Berko, R.E. Kelley, J.S. Plaskett: Phys. Rev. *106*, 824-825 (1957)
2.140 T. Hyodo, O. Sueoka, K. Fujiwara: J. Phys. Soc. Jpn. *31*, 563-573 (1971)
2.141 T. Hyodo: J. Phys. Soc. Jpn. *34*, 476-478 (1973)
2.142 L.A. Page, M. Heinberg: Phys. Rev. *102*, 1545-1553 (1956)
2.143 W. Brandt, R. Paulin: Phys. Rev. B *15*, 2511-2518 (1977)
2.144 S. Dannefaer, D.P. Kerr: Nucl. Instrum. Methods *131*, 119-124 (1975)
2.145 A. Rotondi: Nucl. Instrum. Methods *142*, 499-506 (1977)
2.146 P. Paatero, S. Manninen, T. Paakkari: Philos. Mag. *30*, 1281-1294 (1974)
2.147 R.J. Weiss, W.A. Reed, P. Pattison: "Experimentation", in [Ref.2.6, Chap.3, pp.43-78]
2.148 P.U. Arifov, V.I. Goldanskii, Yu.S. Sayasov: Fiz. Tverd. Tela *6*, 3118-3123 (1964) [English transl.: Sov. Phys.-Solid State *6*, 2484-2487 (1965)]
2.149 P.E. Mijnarends: J. Appl. Phys. *40*, 3027-3033 (1969)
2.150 O.E. Mogensen: Nucl. Instrum. Methods *84*, 293-296 (1970)
2.151 S. Berko: Private communication
2.152 R.W. Hamming: *Numerical Methods for Scientists and Engineers* (McGraw-Hill, New York 1962) Chap.21
2.153 P.E. Mijnarends: Phys. Rev. *160*, 512-519 (1967)

2.154 P.E. Mijnarends: "Reconstruction of Three-Dimensional Distributions", in [Ref.2.6, Chap.10, pp.323-345]
2.155 F.M. Mueller: Phys. Rev. B *15*, 3039-3044 (1977)
2.156 H. Lipson, W. Cochran: *The Determination of Crystal Structures* (Bell and Sons, London 1953) p.291
2.157 P.E. Mijnarends: Unpublished
2.158 W.A. Reed, P. Eisenberger: Phys. Rev. B *6*, 4596-4604 (1972)
2.159 A. Bansil: Solid State Commun. *16*, 885-889 (1975)
2.160 W.R. Fehlner, S.B. Nickerson, S.H. Vosko: Solid State Commun. *19*, 83-86 (1976)
2.161 W.R. Fehlner, S.H. Vosko: Can. J. Phys. *54*, 2159-2169 (1976)
2.162 P.E. Mijnarends: Phys. Rev. *178*, 622-629 (1969)
2.163 P.E. Mijnarends: Physica *63*, 248-262 (1973)
2.164 H. Stachowiak: Phys. Status Solidi *41*, 599-604 (1970)
2.165 G. Kontrym-Sznajd, H. Stachowiak, W. Wierzchowski, K. Petersen, N. Thrane, G. Trumpy: Appl. Phys. *8*, 151-162 (1975)
2.166 P.E. Mijnarends, R.M. Singru: Appl. Phys. *4*, 303-306 (1974)
2.167 C.K. Majumdar: Phys. Rev. B *4*, 2111-2115 (1971)
2.168 A.M. Cormack: J. Appl. Phys. *34*, 2722-2727 (1963); *35*, 2908-2913 (1964)
2.169 D.G. Lock, V.H.C. Crisp, R.N. West: J. Phys. F *3*, 561-570 (1973)
2.170 G.M. Beardsley, S. Berko, J.J. Mader, M.A. Shulman: Appl. Phys. *5*, 375-378 (1975)
2.171 D.G. Lock, R.N. West: Appl. Phys. *6*, 249-256 (1975)
2.172 P. Pattison, B. Williams: Solid State Commun. *20*, 585-588 (1976)
2.173 P. Pattison, W. Weyrich, B. Williams: Solid State Commun. *21*, 967-970 (1977)
2.174 W. Schülke: Phys. Status Solidi (b) *82*, 229-235 (1977)
2.175 A.T. Stewart: Can. J. Phys. *35*, 168-183 (1957)
2.176 M.J.G. Lee: CRC Crit. Rev. in Solid State Sci. *2*, 85-120 (1971)
2.177 C.S. Barrett: Acta Cryst. *9*, 671-677 (1956)
2.178 M.J.G. Lee: Proc. Roy. Soc. (London) A *295*, 440-457 (1966)
2.179 J.M. Perz, D. Shoenberg: J. Low. Temp. Phys. *25*, 275-297 (1976)
2.180 D.L. Randles, M. Springford: J. Phys. F *6*, 1827-1844 (1976)
2.181 M.H. Cohen, V. Heine: Adv. Phys. *7*, 395-434 (1958)
2.182 J.S. Dugdale: Science *134*, 77-86 (1961)
2.183 J.J. Donaghy, A.T. Stewart: Phys. Rev. *164*, 391-395 (1967)
2.184 J. Melngailis, S. DeBenedetti: Phys. Rev. *145*, 400-405 (1966)
2.185 J.J. Paciga, D.L. Williams: Can. J. Phys. *49*, 3227-3233 (1971)
2.186 W.E. Rudge: Phys. Rev. *181*, 1024-1035 (1969)
2.187 M. Rasolt, S.H. Vosko: Phys. Rev. Lett. *32*, 297-301 (1974)
2.188 M. Rasolt, S.B. Nickerson, S.H. Vosko: Solid State Commun. *16*, 827-830 (1975)
2.189 S.B. Nickerson, S.H. Vosko: Phys. Rev. B *14*, 4399-4406 (1976)
2.190 J.A. Arias-Limonta, P.G. Varlashkin: Phys. Rev. B *1*, 142-146 (1970)
2.191 D.R. Gustafson, G.T. Barnes: Phys. Rev. Lett. *18*, 3-5 (1967)
2.192 I.K. MacKenzie, R. LeBlanc, B.T.A. McKee: Phys. Rev. Lett. *27*, 580-582 (1971)
2.193 P. Kubica, M.J. Stott: Can. J. Phys. *53*, 450-454 (1975)
2.194 D.R. Gustafson, J.D. Willenberg: Phys. Rev. B *13*, 5193-5198 (1976)
2.195 A.T. Stewart, J.B. Shand, J.J. Donaghy, J.H. Kusmiss: Phys. Rev. *128*, 118-119 (1962)
2.196 S. Berko: Phys. Rev. *128*, 2166-2168 (1962)
2.197 J.B. Shand: Phys. Lett. A *30*, 478-479 (1969)
2.198 E.H. Becker, E.M.D. Senicki, A.G. Gould, B.G. Hogg: Can. J. Phys. *50*, 2520-2522 (1972)
2.199 E.H. Becker, R.N. West, A.G. Gould, E.M.D. Senicki, B.G. Hogg: Solid State Commun. *16*, 1175-1178 (1975)
2.200 J.C. Kimball, R.W. Stark, F.M. Mueller: Phys. Rev. *162*, 600-608 (1967)
2.201 J.H. Kusmiss, J.W. Swanson: Phys. Lett. A *27*, 517-519 (1968)
2.202 R.J. Douglas, A.T. Stewart: Paper B14, 4th Int. Conf. on Positron Annihilation, Helsingør 1976
2.203 O. Mogensen, K. Petersen: Phys. Lett A *30*, 542-543 (1969)
2.204 G. Kontrym-Sznajd, H. Stachowiak: Appl. Phys. *5*, 361-365 (1975)

2.205 J. Pajak, S. Chabik, B. Rozenfeld, G. Kontrym-Sznajd: Acta Phys. Pol. A *50*, 623-632 (1976)
2.206 B. Rozenfeld, S. Chabik: Appl. Phys. *13*, 81-85 (1977)
2.207 J.J. Burton, G. Jura: Phys. Rev. *171*, 699-701 (1968)
2.208 M. Ross, K.W. Johnson: Phys. Rev. B *2*, 4709-4714 (1970)
2.209 T. Okada, H. Sekizawa, N. Shiotani: J. Phys. Soc. Jpn. *41*, 836-840 (1976)
2.210 Y. Kubo, S. Wakoh, J. Yamashita: J. Phys. Soc. Jpn. *41*, 830-835 (1976)
2.211 J. Mader, S. Berko, H. Krakauer, A. Bansil: Phys. Rev. Lett. *37*, 1232-1236 (1976)
2.212 J.H. Terrell, H.L. Weisberg, S. Berko: "On positron lifetimes vs. two gamma correlations in the alkali metals", in [Ref.2.7, pp.269-275]
2.213 P. Hautojärvi: Solid State Commun. *11*, 1049-1052 (1972)
2.214 O.E. Mogensen, G. Trumpy: Phys. Rev. *188*, 639-644 (1969)
2.215 G. Kontrym-Sznajd: Paper B6, 4th Int. Conf. on Positron Annihilation, Helsingør 1976
2.216 C.V. Briscoe, G.M. Beardsley, A.T. Stewart: Phys. Rev. *141*, 379-380 (1966)
2.217 A.P. Cracknell: *The Fermi Surfaces of Metals* (Taylor & Francis, London 1971)
2.218 I.Ya. Dekhtyar, V.S. Mikhalenkov: Doklady Akad. Nauk SSSR *133*, 60-63 (1960) [English transl.: Sov. Phys.-Doklady *5*, 739-742 (1960)]
2.219 M. Szuszkiewicz: Acta Phys. Pol. A *45*, 873-883 (1974)
2.220 W.A. Reed, P. Eisenberger, K.C. Pandey, L.C. Snyder: Phys. Rev. B *10*, 1507-1515 (1974)
2.221 D.M. Schrader, J.K. Kim: Appl. Phys. *4*, 249-256 (1974)
2.222 D.M. Schrader, C.M. Wang: J. Phys. Chem. *80*, 2507-2518 (1976)
2.223 P. Colombino, B. Fiscella, L. Trossi: Nuovo Cimento *31*, 950-960 (1964)
2.224 J.C. Erskine, J.D. McGervey: Phys. Rev. *151*, 615-620 (1966)
2.225 N.F. Mott, H. Jones: *The Theory of the Properties of Metals and Alloys* (Clarendon, Oxford 1936) p.154
2.226 P.E. Mijnarends: Phys. Rev. B*4*, 2820-2822 (1971)
2.227 G. Dlubek, O. Brümmer: Phys. Status Solidi (b) *73*, K107-K110 (1976)
2.228 W. Schülke: Phys. Status Solidi (b) *62*, 453-460 (1974)
2.229 A. Seth, D.E. Ellis: J. Phys. C *10*, 181-194 (1977)
2.230 R.M. Singru: Phys. Status Solidi (a) *30*, 11-38 (1975)
2.231 S. Berko, S. Cushner, J.C. Erskine: Phys. Lett. A *27*, 668-669 (1968)
2.232 S. Cushner, J.C. Erskine, S. Berko: Phys. Rev. B *1*, 2852-2854 (1970)
2.233 M.R. Halse: Phil. Trans. Roy. Soc. (London) A *265*, 507-532 (1969)
2.234 E.M.D. Senicki, E.H. Becker, A.G. Gould, B.G. Hogg: Phys. Lett. A *41*, 293-294 (1972)
2.235 E.M.D. Senicki, E.H. Becker, A.G. Gould, R.N. West, B.G. Hogg: J. Phys. Chem. Sol. *34*, 673-677 (1973)
2.236 F. Herman, S. Skillman: *Atomic Structure Calculations*(Prentice Hall, Englewood Cliffs 1963)
2.237 J.H. Wood: Phys. Rev. *117*, 714-718 (1960)
2.238 R.M. Singru: Pramana *2*, 299-303 (1974)
2.239 R.M. Singru, P.E. Mijnarends: Phys. Rev. B *9*, 2372-2380 (1974)
2.240 E.H. Becker, A.G. Gould, E.M.D. Senicki, B.G. Hogg: Can. J. Phys. *52*, 336-339 (1974)
2.241 B. Rozenfeld, S. Chabik, J. Pajak, K. Jerie, W. Wierzchowski: Acta Phys. Pol. A *44*, 21-35 (1973)
2.242 H. Morinaga: Phys. Lett. A *38*, 513-514 (1972)
2.243 O. Sueoka: J. Phys. Soc. Jpn. *26*, 864 (1969)
2.244 H. Morinaga: J. Phys. Soc. Jpn. *33*, 996-1002 (1972)
2.245 T. Akahane, K. Fujiwara: J. Phys. Soc. Jpn. *35*, 1660-1663 (1973)
2.246 J. Melngailis: Phys. Rev. B *2*, 563-565 (1970)
2.247 R.M. Singru: J. Phys. Chem. Sol. *35*, 33-35 (1974)
2.248 J. Pajak, B. Rozenfeld: Acta Phys. Pol. A *50*, 611-622 (1976)
2.249 B. Rozenfeld, M. Szuszkiewicz, W. Wierzchowski: Acta Phys. Pol. A *40*, 3-15 (1971)
2.250 R. Harthoorn: Ph.D. Thesis, Univ. of Amsterdam (1977, unpublished); Report ECN-29, Netherlands Energy Research Foundation ECN, Petten, The Netherlands
2.251 D.J. Roaf: Philos. Trans. A*255*, 135-152 (1962)
2.252 S. Berko, P.E. Mijnarends: To be published

2.253 S. Berko, J. Zuckerman: Phys. Rev. Lett. *13*, 339a-341a (1964); Phys. Rev. Lett. *14*, 89 (1965)
2.254 S. Berko, A.P. Mills: J. Physique *32*, C1-287-289 (1971)
2.255 C.G. Shull, Y. Yamada: J. Phys. Soc. Jpn. *17*, Supp. B-III,1-6 (1962)
2.256 K.J. Duff, T.P. Das: Phys. Rev. B *3*, 192-208 (1971); Phys. Rev. B *3*, 2294-2306 (1971)
2.257 O. Gunnarson, B.I. Lundqvist, S. Lundqvist: Solid State Commun. *11*, 149-153 (1972)
2.258 Y. Nakao: Appl. Phys. *7*, 81-82 (1975)
2.259 N. Sakai, K. Ôno: J. Phys. Soc. Jpn. *42*, 770-778 (1977)
2.260 S. Wakoh, Y. Kubo: J. Magnetism Magn. Mat. *5*, 202-211 (1977)
2.261 B. Rozenfeld, S. Szuszkiewicz: Bull. Acad. Pol. Sci. *22*, 1289-1294 (1974)
2.262 S. Szuszkiewicz, B. Rozenfeld, G. Kontrym-Sznajd: Acta Phys. Pol. A *50*, 719-729 (1976)
2.263 T.W. Mihalisin, R.D. Parks: Phys. Lett. *21*, 610-611 (1966); Phys. Rev. Lett. *18*, 210-211 (1967); Solid State Commun. *7*, 33-35 (1969)
2.264 N. Shiotani, T. Okada, H. Sekizawa, T. Mizoguchi, T. Karasawa: J. Phys. Soc. Jpn. *35*, 456-460 (1973)
2.265 P. Eisenberger, W.A. Reed: Phys. Rev. B *9*, 3242-3247 (1974)
2.266 J. Rath, C.S. Wang, R.A. Tawil, J. Callaway: Phys. Rev. B *8*, 5139-5142 (1973)
2.267 C.S. Wang, J. Callaway: Phys. Rev. B *11*, 2417-2420 (1975)
2.268 N. Shiotani, T. Okada, H. Sekizawa, S. Wakoh, Y. Kubo: J. Phys. Soc. Jpn. *43*, 1229-1236 (1977)
2.269 G.S. Goodbody, B.R. Watts, R.N. West: Paper B10, 4th Int. Conf. on Positron Annihilation, Helsingør 1976
2.270 N. Shiotani, T. Okada, T. Mizoguchi, H. Sekizawa: J. Phys. Soc. Jpn. *38*, 423-430 (1975)
2.271 S. Wakoh, Y. Kubo, J. Yamashita: J. Phys. Soc. Jpn. *40*, 1043-1047 (1976)
2.272 S. Wakoh, T. Fukamachi, S. Hosoya, J. Yamashita: J. Phys. Soc. Jpn. *38*, 1601-1606 (1975)
2.273 D.R. Gustafson, A.R. Mackintosh: J. Phys. Chem. Sol. *25*, 389-394 (1964)
2.274 R.W. Williams, T.L. Loucks, A.R. Mackintosh: Phys. Rev. Lett. *16*, 168-170 (1966)
2.275 R.W. Williams, A.R. Mackintosh: Phys. Rev. *168*, 679-686 (1968)
2.276 G. Coussot: Phys. Rev. B *3*, 1048-1049 (1971)
2.277 P.G. Mattocks, R.C. Young: J. Phys. F *7*, 1219-1228 (1977)
2.278 R.P. Gupta, T.L. Loucks: Phys. Rev. *176*, 848-850 (1968)
2.279 K. Iyakutti, C.K. Majumdar, R.S. Rao, V. Devanathan: J. Phys. F *6*, 1639-1645 (1976)
2.280 C. Hohenemser, J.M. Weingart, S. Berko: Phys. Lett. A *28*, 41-42 (1968)
2.281 D.R. Gustafson, J.D. McNutt, L.O. Roellig: Phys. Rev. *183*, 435-440 (1969)
2.282 R.F. Gempel, D.R. Gustafson, J.D. Willenberg: Phys. Rev. B *5*, 2082-2085 (1972)
2.283 S.M. Kim, W.J.L. Buyers: Can. J. Phys. *50*, 1777-1781 (1972)
2.284 D.G. Kanhere, R.M. Singru: Phys. Lett. A *53*, 67-68 (1975)
2.285 S.S. Barshay, R.E. Leber, R.M. Lambrecht: Appl. Phys. *5*, 67-70 (1974)
2.286 I.K. MacKenzie, T.E. Jackman, P.C. Lichtenberger: Appl. Phys. *9*, 259-260 (1976)
2.287 W. Hume-Rothery: *The Metallic State* (Oxford University Press, London 1931) pp.328-336
2.288 H. Jones: Proc. Phys. Soc. *49*, 250-257 (1937)
2.289 W. Hume-Rothery, D.J. Roaf: Philos. Mag. *6*, 55-59 (1961)
2.290 D. Stroud, N.W. Ashcroft: J. Phys. F *1*, 113-124 (1971)
2.291 V. Heine, D. Weaire: In *Solid State Physics*, Vol.24, ed. by H. Ehrenreich, F. Seitz, D. Turnbull (Academic Press, New York 1970) pp.249-463
2.292 A. Bansil, H. Ehrenreich, L. Schwartz, R.E. Watson: Phys. Rev. B *9*, 445-465 (1974)
2.293 R.B. Dingle: Proc. Roy. Soc. (London) A *211*, 517-525 (1952)
2.294 B.L. Gyorffy, G.M. Stocks: J. Physique *35*, C4 - 75-80 (1974)
2.295 G.M. Stocks, B.L. Gyorffy, E.S. Giuliano, R. Ruggeri: J. Phys. F *7*, 1859-1866 (1977)
2.296 P.E. Mijnarends, A. Bansil: Phys. Rev. (to be published)

2.297 W. Triftshäuser, A.T. Stewart: J. Phys. Chem. Sol. *32*, 2717-2722 (1971)
2.298 D.L. Williams: In *Solid State Physics, Vol. I, Electrons in Metals* (Gordon and Breach, New York 1968) pp.343-346
2.299 E.H. Becker, P. Petijevich, D.L. Williams: J. Phys. F *1*, 806-814 (1971)
2.300 M. Hasegawa, T. Suzuki, M. Hirabayashi, S. Yajima: Acta Crystallogr. A *28*, S102 (1972)
2.301 J.G. McLarnon, D.L. Williams: Solid State Commun. *13*, 1469-1471 (1973)
2.302 J.G. McLarnon, D.L. Williams: J. Phys. Soc. Jpn. *43*, 1244-1246 (1977)
2.303 B.W. Murray, J.D. McGervey: Phys. Rev. Lett. *24*, 9-13 (1970)
2.304 A. Thompson, B.W. Murray, S. Berko: Phys. Lett. A *37*, 461-462 (1971)
2.305 K. Fujiwara, O. Sueoka, T. Imura: J. Phys. Soc. Jpn. *24*, 467-476 (1968)
2.306 S. Berko, J. Mader: Phys. Condensed Matter *19*, 405-416 (1975)
2.307 T. Akahane, O. Sueoka, H. Morinaga, K. Fujiwara: J. Phys. Soc. Jpn. *36*, 135-141 (1974)
2.308 T. Suzuki, M. Hasegawa, M. Hirabayashi: Appl. Phys. *5*, 269-274 (1974)
2.309 A.H. Lettington: Philos. Mag. *11*, 863-869 (1965)
2.310 R.S. Rea, A.S. De Reggi: Phys. Lett. A *40*, 205-206 (1972)
2.311 P.T. Coleridge, I.M. Templeton: Can. J. Phys. *49*, 2449-2461 (1971)
2.312 T. Suzuki, M. Hasegawa, M. Hirabayashi: J. Phys. F *6*, 779-788 (1976)
2.313 T.B. Massalski, B. Cockayne: Acta Metall. *7*, 762-768 (1959)
2.314 L.J. Rouse, P.G. Varlashkin: Phys. Rev. B *4*, 2377-2380 (1971)
2.315 S. Tanigawa, S. Nanao, K. Kuribayashi, M. Doyama: J. Phys. Soc. Jpn. *31*, 1689-1694 (1971)
2.316 S. Nanao, K. Kuribayashi, S. Tanigawa, M. Doyama: Phys. Lett. A *38*, 489-490 (1972)
2.317 M. Hasegawa, T. Suzuki, M. Hirabayashi: J. Phys. Soc. Jpn. *37*, 85-91 (1974)
2.318 M. Hasegawa, T. Suzuki, M. Hirabayashi: J. Phys. Soc. Jpn. *43*, 89-96 (1977)
2.319 T. Suzuki, M. Hasegawa, S. Koike, M. Hirabayashi: Paper D4, 4th Int. Conf. on Positron Annihilation, Helsingør 1976
2.320 B. Rozenfeld, J. Rudzińska-Girulska, W. Światkowski: Bull. Acad. Pol. Sci. *22*, 733-737 (1974)
2.321 Y. Tanji, M. Matsui, F. Itoh, H. Moriya, Y. Nakagawa, S. Cikazumi: Phys. Status Solidi (a) *25*, K85-K87 (1974)
2.322 S.Y. Chuang, S.J. Tao, H.S. Chen: J. Phys. F *5*, 1681-1686 (1975)
2.323 J. Wesolowski, B. Rozenfeld, M. Szuszkiewicz: Acta Phys. Pol. *24*, 729-734 (1963)
2.324 M.P. Chouinard, D.R. Gustafson, R.C. Heckman: J. Chem. Phys. *51*, 3554-3558 (1969)
2.325 M.P. Chouinard, D.R. Gustafson: J. Chem. Phys. *54*, 5082-5084 (1971)
2.326 V.I. Sabin, R.A. Andrievskii, V.V. Gorbachev, A.D. Tsyganov: Fiz. Tverd. Tela *14*, 3320-3323 (1972) [English transl.: Sov. Phys.-Solid State *14*, 2815-2817 (1973)]
2.327 B. Rozenfeld, E. Debowska: Acta Phys. Pol. A *47*, 37-43 (1975)
2.328 I.Ya. Dekhtyar, V.I. Shevchenko: Phys. Status Solidi (b) *49*, K11-K13 (1972); Phys. Status Solidi (b) *83*, 323-330 (1977)
2.329 C.D. Gelatt, J.A. Weiss, H. Ehrenreich: Solid State Commun. *17*, 663-666 (1975)
2.330 A. Bansil, S. Bessendorf, L. Schwartz: Inst. Phys. Conf. Ser. *39*, 493-497 (1978)
2.331 R. Griessen, W.J. Venema, J.K. Jacobs, F.D. Manchester, Y. de Ribaupierre: J. Phys. F *7*, L133-L138 (1977)
2.332 J.S. Faulkner: Phys. Rev. B *13*, 2391-2397 (1976)
2.333 L. Huisman, J.A. Weiss: Solid State Commun. *16*, 983-985 (1975)
2.334 W.R. Wampler, B. Lengeler: Phys. Rev. B *15*, 4614-4622 (1977)
2.335 A.T. Stewart: Phys. Rev. *133*, A1651-A1653 (1964)
2.336 S. Berko, M. Weger: Phys. Rev. Lett. *24*, 55-58 (1970)
2.337 S. Berko, M. Weger: Positron Annihilation Experiments and the Band Structure of V_3Si", in *Computational Solid State Physics*, ed. by F. Herman, N. Dalton, T. Koehler (Plenum Press, New York 1972) pp.59-77
2.338 M. Weger: Rev. Mod. Phys. *36*, 175-177 (1964); J. Phys. Chem. Sol. *31*, 1621-1639 (1970)
2.339 L.F. Mattheiss: Phys. Rev. *138*, A112-A128 (1965); Phys. Rev. B *12*, 2161-2180 (1975)

2.340 M. Weger, I.B. Goldberg: In *Solid State Physics*, Vol.28, ed. by H. Ehrenreich, F. Seitz, D. Turnbull (Academic Press, New York 1973) pp.1-177
2.341 G. Barak, I.B. Goldberg, M. Weger: J. Phys. Chem. Sol. *36*, 847-857 (1975)
2.342 I.B. Goldberg: J. Phys. C *8*, 1159-1180 (1975)
2.343 B. Klein, D. Papaconstatopoulos, L.L. Boyer: Ferroelectr. *16*, 299-302 (1977)
2.344 T. Jarlborg, G. Arbman: J. Phys. F *6*, 189-197 (1976); J. Phys. F *7*, 1635-1649 (1977)
2.345 W. Triftshäuser, A.T. Stewart, R. Taylor: J. Phys. Chem. Sol. *32*, 2711-2716 (1971)
2.346 R. Taylor: Proc. Roy. Soc. (London) A *312*, 495-517 (1969)
2.347 H. Morinaga: Phys. Lett. A *32*, 75-76 (1970); Phys. Lett. A *34*, 384-385 (1971)
2.348 B. Rozenfeld, E. Debowska, Z. Henkie: J. Solid State Chem. *17*, 101-105 (1976)
2.349 B. Rozenfeld, E. Debowska, Z. Henkie, A. Wojakowski, A. Zygmunt: Acta Phys. Pol. A *51*, 275-281 (1977)

3. Positron Studies of Lattice Defects in Metals

R. N. West

With 12 Figures

The *conscious* study of lattice defects with positrons is now some ten years old. Although sharing the techniques and guided by the results of the longer established studies described in Chap.2, defect studies are based on a somewhat different philosophy. Electronic structure studies rely for their credence on a belief that, in sufficiently well-annealed single-crystal samples, the positrons annihilate from a common, well-defined, and delocalised state in which they can freely sample non-local properties of the medium. The perturbing effect of the positron charge on the electronic system is in general an unwelcome complication. In defect studies the situation is radically different. Here the charged positron's interactions with the surrounding medium play the vital role in providing for the possibility of a variety of additional localised positron states at the structural defects. The investigation of defect-induced changes in the commonly measured characteristics of annihilation (Chap.1) and the interpretation of such changes in terms of such things as the concentration, nature, and origins of the defects responsible are the essential ingredients of our study.

The realization of the positron's strong affinity for structural defects in metals and the initial feverish burst of activity it provoked prompted several early reviews [3.1-4]. Today, notwithstanding significant advances, we are in the midst of a predictable period of agonising reappraisal. In this chapter we shall try to reflect both these aspects of the field. Because of the content of Chap.4, the emphasis will be on the experimental rather than the theoretical side of the subject.

In Sect.3.1 we describe the effects that can occur when positrons are trapped at defects, the various methods of data analysis that are employed in such situations and, with hindsight, their relative merits. In Sect.2.3 we consider the many investigations and problems concerned with the measurement of vacancy formation parameters. The length of this section bears witness to both the considerable activity in this area in recent years and its relevance to our understanding of the positron defect trapping problem. Nonequilibrium defect problems are considered in Sect.3.3. Here and in the subsequent section on defects in alloys we see how the synthesis of positron and other more traditional defect-probing techniques and ideas extends the understanding. Finally in Sect.3.5 we turn to the subject of liquid and amorphous metals. Here positron activity has been more modest but the strong analogy between positron trapping in solids and positron behavior in liquids suggests the possibility of much more work in the future.

3.1 Annihilation Parameters for Defect Studies

3.1.1 The Defect Trapping Phenomenon and Its Effect

There exists today an overwhelming body of evidence [3.5-8] to suggest that positrons entering metals are reduced to essentially thermal energies ("thermalized") in a time ($\sim 10^{-12}$ s) much less than the typical lifetimes against annihilation ($\sim 10^{-10}$ s). During the latter stages of thermalization and following it (no real distinction really exists) various stable or metastable localized states may be formed as a result of positron trapping at lattice defects.

The tendency for such trapping is easily understood. A thermalized positron in a metal is strongly repelled by the positive ions because of its positive charge. The positron density distribution has a "swiss cheese" character [3.9] with holes around each ion. This squeezing of the positron into interstitial regions provides a positive contribution to the ground state of energy of as much as 5 eV in some metals [3.10]. Thus should additional space become available in the form of a lattice vacancy, dislocation core, or other hole or void, the positron can relax therein with considerable energy advantage [3.11].

The effect on observable characteristics of the annihilation process of appreciable trapping of this type is equally easily argued in qualitative terms. The positron annihilation rate in any state reflects the positron density "seen" by the positron in that state (Chap.1). Contributions arise both from the itinerant electron gas and more tightly bound ion core electrons. Around a lattice defect (here and throughout we and the positron exclude the interstitial) the electron density and particularly the core electron density will be reduced with a consequent decrease in the annihilation rate. The electron-positron momentum density changes as measured by either the angular correlation or the Doppler broadening technique can also be usefully, albeit approximately, described in terms of conduction and core electron terms.

The Doppler broadened energy and common long-slit angular correlation curves (Chap.1) for the simpler metals can be well described (Fig.3.1) as made up of a central parabolic conduction electron contribution sitting upon a broader Gaussian type of core electron curve. The curves for the less simple metals (i.e., transition and rare earths) are not so simply defined. Nevertheless, in either case, we may assert that large momentum events arise predominantly from the deeper-lying electron states and vice versa. A positron in a defect sees a reduced electron

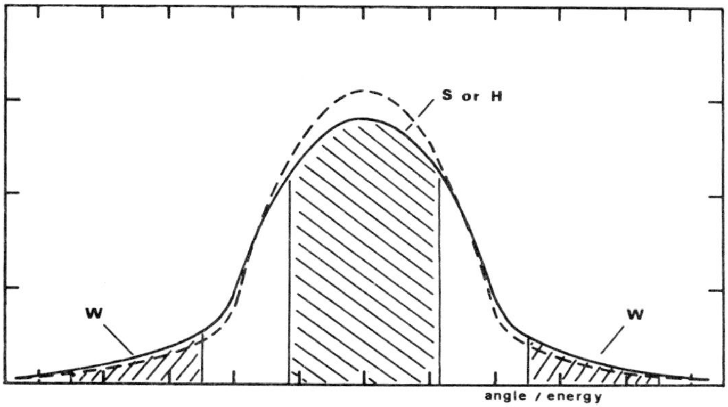

Fig.3.1. A schematic indication of the type of momentum density narrowing that results from positron-defect trapping in metals. Solid line - well annealed specimen; broken line - defected sample. The cross-hatched areas illustrate the typical regions of integration involved in the peak height S or H parameters and in the tail or wing W parameter

density and, as previously noted, the greater deficit is normally that for the core electrons. Thus inevitably, the major effect of appreciable positron trapping is a narrowing (Fig.3.1) of the momentum curves.

The crude descriptions of observables given above do scant service to some 25 years of detailed theoretical analyses and predictions of positron results for well-characterised metal specimens. Nevertheless, in situations where positrons can interact with several varieties of ill-defined local structure, or in investigations involving radically or rapidly changing defect situations, they provide a valuable basis for restrained and rational interpretation and have stimulated particular parametric approaches to data analysis to which we now turn our attention.

3.1.2 Positron States and Lifetime Spectra

Positron lifetime spectra (Chap.1) can provide an invaluable guide to the pre-annihilation behavior of positrons in condensed matter. An adequate understanding of the various ways in which single-exponential component, multi-exponent or still more complex forms of spectra can arise is an important factor in the successful interpretation of many defect investigations. A fairly detailed discussion has been given by this author elsewhere [3.4] but as an aid to completeness in subsequent sections the essential ideas are briefly recounted here.

The annihilation rate for a positron in a stationary state is independent of time, is determined by the electron density "seen" by the positron in that state, and the decay is simple exponential. More generally, the positrons may exist in a spectrum of quasi-stationary states s, each characterized by a particular annihilation rate $\lambda(s)$, with relative probabilities given by a suitable normalized distribution function $P(s)$. The form of the consequent lifetime spectrum then depends on the persistence, $\tau(s)$, of these states relative to the corresponding annihilation lifetimes $\lambda(s)^{-1}$.

If $\tau(s)\lambda(s) << 1$ for all s, each positron samples many states in an average lifetime and the measured spectrum approximates to a single component of "rate"

$$\bar{\lambda} = \int \lambda(s)P(s)ds \quad . \tag{3.1}$$

If, in the other extreme $\tau(s)\lambda(s) >> 1$ the spectrum has the form

$$I(t) = \int P(s) \exp[-\lambda(s)t]ds \quad . \tag{3.2}$$

If P(s) is a smoothly varying function embracing a significant range of $\lambda(s)$ the commonly employed multi-exponent fitting procedures [3.12] may or may not converge but in neither case is useful physics likely to emerge. More useful are those cases in which contributions to the integral are confined to restricted ranges of $\lambda(s)$ and the spectrum approximates to a finite number of discrete components:

$$I(t) = \sum_{i}^{N} I_i \exp(-\lambda_i t) \quad . \tag{3.3}$$

In this case each component intensity I_i provides a measure of the relative positron population, and each rate λ_i the effective electron density in the corresponding positron state. Such components will arise whenever essentially stationary states are formed in times very much less than any $\tau_i = \lambda_i^{-1}$, and comparable to that of the thermalization process.

Of particular interest to us here are those situations in which some transitions between the various possible states occur in times comparable with that for annihilation. A complete analysis requires in principle, an understanding or a microscopic model of the mechanisms by which such transitions occur. Complex and comprehensive analyses have been presented [3.13,14] but a simple rate equation approach involving, with obvious notation, time-independent transition rates $K_{ij}(K_{ij} \neq K_{ji}; K_{ii} = 0)$ has been found more than sufficient for most analyses thus far. Then any initial positron population decays as

$$n(t) = \sum_{i}^{N} n_i(t) \tag{3.4}$$

where the $n_i(t)$ are determined by a set of coupled differential equations

$$\frac{dn_i(t)}{dt} + \left(\lambda_i + \sum_{j\neq i}^{N} K_{ij}\right)n_i(t) = \sum_{j\neq i}^{N} K_{ji}n_j(t) \quad . \tag{3.5}$$

The lifetime spectrum

$$I(t) = \sum_\nu I_\nu \exp(-\Gamma_\nu t) \tag{3.6}$$

where the I_ν and Γ_ν are complicated functions of the λ_i and K_{ij}. Two approximations provide useful insight into the totality of possibilities.

If $K_{ij} >> \lambda_i$ for all i,j approximate solutions analogous to (3.1) obtain and for example in the case N = 2 [3.15,16]

$$n(t) = n_0 \exp\left[-\frac{(\lambda_1 K_{21} + \lambda_2 K_{12})}{K_{21} + K_{12}}\right]t \quad . \tag{3.7}$$

The simple trapping models [3.17-19] assume that at some time, $t = 0(<< \lambda_i^{-1})$ all the positrons exist in a common delocalized state in which they can either annihilate or make transitions to other states in or around structural or other defects. If the probability of further transitions (i.e., escape from the traps) is negligibly small we may rewrite (3.5) as

$$\frac{dn_1(t)}{dt} + (\lambda_1 + \sum_{j \neq 1} K_{1j})n_1(t) = 0 \quad ,$$

and

$$\frac{dn_j(t)}{dt} + \lambda_j n_j(t) = K_{1j} n_1(t) \tag{3.8}$$

for all $j \neq 1$. Application of the appropriate boundary conditions, $n_j(0) = n_0 \delta_{1,j}$, implied above yields as in (3.3) a discrete spectrum

$$n(t) = n_0 \sum_j I_j \exp(-\Gamma_j t)$$

or more specifically,

$$n(t) = n_0\left[1 - \sum_{j \neq 1} \frac{K_{1j}}{(\lambda_1 - \lambda_j + \varepsilon)}\right]\exp -(\lambda_1 + \varepsilon)t$$

$$+ \sum_{j \neq 1} \frac{n_0 K_{1j}}{(\lambda_1 - \lambda_j + \varepsilon)} \exp(-\lambda_j t) \tag{3.9}$$

where $\varepsilon = \sum_j K_{1j}$.

Again as in (3.3) the decay rate of any resolvable component $j \neq 1$, is simply the annihilation rate in the corresponding trap or centre. For the first state [3.20],

$$\lambda_1 = \sum_j I_j \Gamma_j \quad . \tag{3.10}$$

Observable effects of positron trapping will occur in lifetime spectra whenever the corresponding K_{1j} is of the same order as λ_1 and the λ_j is sufficiently different from λ_1. Should the first of these conditions obtain at a sufficiently low concentration of traps, c_j, the relation,

$$K_{1j} = \nu_{1j} c_j \tag{3.11}$$

may well be applicable. A change in the concentration of traps can then be followed through the consequent changes in Γ_1 or the appropriate component intensity.

The potential value of positron lifetime spectrum analyses in defect trapping studies is perfectly clear. However, the realization of this potential is somewhat less straightforward. Three major problems exist.

I) Any spectrum, deriving from the random nuclear events always involves a finite statistical precision which makes, other complications apart, the purely mathematical problem of the analysis of sums of exponentials extremely difficult [3.21].

II) The measured spectra amount to a convolution of an "ideal" spectrum with an instrumental resolution function which can never be exactly defined.

III) There is almost always present in measured spectra a small intensity of component or components arising from annihilation in the ^{22}Na positron source and associated foils and or interfaces that are unavoidable with the normally adopted sample-source-sample sandwich arrangements (Chap.1)

Significant advances in the struggle against these problems have been made in recent years largely as a result of the stimulus provided by the defects application.

Sophisticated computer fitting procedures, of which the POSITRONFIT programs [3.12] have received the closest to universal acceptance, incorporate both resolution function and source components in the fit. Several developments in the basic measurement technique provide data to match the analyses. The introduction of temperature and digital stabilization techniques [3.22] or computer based fast sampling systems [3.23] has allowed orders of magnitude increase in data accumulation times without the previous inevitable degradation of spectra by electronic drift. Progressive increases in the specific activity of commercially available $^{22}NaCl$ solutions have resulted in corresponding decreases in source component intensities. A radically new method of source fabrication recently described by HERLACH and MAIER [3.24] provides "sealed" sources of so small self-absorption as to almost rule out the source component as a significant problem in spectrum analysis.

The result of these developments in both measurement and analyses has been dramatic. A decade ago the measurement of what then appeared to be single component lifetime spectra for metals stretched the contemporary technique to its limits. Today with lifetime spectra that can and should ideally embrace 10^6-10^7 events, good resolution and centroid stability, constraint free analyses [3.25-27] involving 2, 3, or exceptionally 4 components can be attempted with some success. Nevertheless, the basic problems I) and II) remain and the usual significance tests and common sense and restraint in interpretation must remain an essential part of the practice. A few illustrations of the problems that can arise may be valuable at this point.

Whenever appreciable positron trapping at defects occurs short lived ($\leq$ 100 ps) components will be present in the spectra [note the first term in (3.9)]. The *apparent* intensity and rate of such a component depends critically on the assumptions or deductions made about the prevailing resolution function. Long-lived components of small intensity can be easily distorted by or confused with similar small intensity source components. Analyses involving rigid assumptions about the number of components present are fraught with even greater dangers in that complete erroneous patterns of change in a series of samples or sample condition may easily emerge [3.28]. Constrained analyses involving particular versions of trapping model must always be viewed with caution.

Detailed lifetime spectrum analyses involve both considerable data accumulation and computing time and are not always feasible when the problem at hand requires the monitoring of a large number of different but related defect situations. In such cases the measurement of a single parameter expressing some average property of the spectrum can be valuable. The spectrum routing technique [3.22,29] enables a rapid and precise measurement of positron *mean* lifetime.

For the trapping model result (3.9), the mean lifetime,

$$t = \left(1 + \sum_j K_{1j}/\lambda_j\right)\Big/\left(\lambda_1 + \sum_j K_{1j}\right) \quad , \tag{3.12}$$

which superficially does not seem a particularly valuable result. However, in any situation in which only one of the K_{1j} is subject to change — let us denote it by K_{12} — and thus can influence t, a significant simplification results.

That is,

$$t = t_1(\chi + K_{12}t_2)/(\chi + K_{12}t_1) \quad , \tag{3.13}$$

where $\chi = 1 + \sum_{j\neq 2} K_{1j}/\lambda_j$, and t_1 and t_2 are the asymptotic values of t for $K_{12} = 0$ and $K_{12} = \infty$, respectively. Thus

$$K_{12} = \chi t_1^{-1}(t - t_1)/(t_2 - t) \tag{3.14}$$

which simply defines the variable trapping rate in terms of the observable changes in mean lifetime. Equation (3.14) is well known for the particular case of the simple two-state trapping model [3.18,19] but can be seen to be common to all situations in which the general N-state model (3.8) applies. In this respect its utilization in the determination of *relative* changes (χ cannot be determined in the general case) in K_{12} has advantages over analyses based on (3.9) and observed changes in component intensities since there, additional assumptions about the number of components present must always feature.

3.1.3 Momentum Density Parameters

The shape of angular correlation or Doppler broadened energy curves depends, instrumental factors apart, only upon the final relative number of annihilations in each positron state, i.e.,

$$N_j = \int_0^\infty \lambda_j n_j(t)dt \quad . \tag{3.15}$$

In the case of the trapping model result (3.9)

$$N_1 = \frac{\lambda_1 n_0}{(\lambda_1+\varepsilon)} \quad , \quad N_j = \frac{K_{1j} n_0}{(\lambda_1+\varepsilon)} \quad . \tag{3.16}$$

Now if any characteristic of annihilation, F, is a linear function of positron state in the sense that

$$F = \sum_j F_j N_j / \sum_j N_j = \sum_j F_j P_j \quad , \tag{3.17}$$

then its measurement can also provide quantitative information about trapping rates, concentrations and sites. The positron lifetime or reciprocal rate is one such example already considered in the preceding section. Measured momentum density curves provide a variety of additional parameters.

The choice of any particular parameter depends on several easily argued criteria:

a) *Convenience* in the light of the apparatus available and the problem at hand.

b) The combination of its *sensitivity* to the trapping phenomenon and the *statistical precision* with which it can be measured.

c) Its *relevence* to interpretation in terms of the underlying positron states.

In those cases where essentially all of the positrons annihilate from a single state, angular distributions give information about the types and distributions of electrons seen by the positron in that state. Such distributions are themselves state parameters in the sense of (3.17) and can provide invaluable information (Chap.4) complementary to that obtained from lifetime studies. In a defected solid where the positron population is distributed between several such states the angular distribution is an appropriately weighted combination of the individual "pure state" curves. Then in favourable cases curve fitting procedures [3.30] can be used to determine the various weights. The procedure is analogous to lifetime spectrum analysis but is generally less satisfactory since the functional form of the individual components is here less clearly defined. Furthermore the measurements and analyses take considerable time and thus where trapping rates and related quantities are the major interest simpler parameters can have greater value.

In conventional long-slit angular correlation measurements (Chap.1) the angular distribution is built up by repeated scans of a moving detector through the necessary angular range. The H-parameter [3.16],

$$H = \int_{-\ell}^{+\ell} N(p_z)dp_z \Big/ \int_{-\infty}^{+\infty} N(p_z)dp_z \tag{3.18}$$

where the restricted integral is taken over a small angular range around the peak of the distribution, is a particularly convenient choice in that its measurement requires little or no modification to the basic apparatus.

An explicit evaluation of the denominator in (3.18) can often be omitted as long as systemtatic variations in equipment performance factors are minimized and monitored via side-channel counting rates [3.31].

Defect trapping inevitably produces an increase in the H parameter. The magnitude of the effect depends on the region of integration embraced and the difference between the curves for the delocalised and trapped states. An appropriate compromise between sensitivity and statistical precision is usually a matter of trial and error.

Other parameters having the property (3.17) can be utilized. The wing parameter W [3.32] involves an integral (Fig.3.1) over large angle (momentum) parts of a curve. Here, however, a sometimes greater sensitivity to trapping effects can be offset by poorer statistical precision and care is needed in the evaluation and proper treatment of accidental background effects.

Doppler broadening line-shape measurements [3.33,34] give information equivalent to that obtained from angular distributions although with somewhat poorer momentum resolution. However, the simpler single detector system, the possibility of continuous monitoring of the instrumental performance via coincidence routing techniques [3.35] or via a neighbouring energy gamma ray line [3.36,37], and the orders of magnitude greater rate of data accumulation, more than compensate for this single limitation. Developments in deconvolution of Doppler broadened curves [3.38,39] result in momentum distributions with an information content approaching that of precision long-slit angular distributions. Such procedures when applied to distributions obtained by the low-energy background dual-detector method developed by LYNN et al. [3.40] suggest great promise for the Ge(Li) technqiue in detailed studies of momentum spectra.

The simultaneous accumulation of data throughout the entire momentum range allows considerable flexibility in the choice of "single number" line-shape parameters for Doppler broadening studies. CAMPBELL [3.41] has considered the statistical properties and sensitivity to trapping of various parameters. The Doppler line-shape equivalent of the H parameter is MACKENZIE's [3.34] S parameter. CAMPBELL considers the relative merits of S, the wing parameter W (Fig.3.1), and functions of S and W combined. The ratio, S/W [3.42], though a sensitive measure of defect trapping is not linear in the sense of (3.17). An alternative combination, D = S - W, does satisfy (3.17) and of those considered is suggested to make best use of available data in the purely statistical sense. Other parameters currently under consideration [3.43] involve reweighted versions of S, W, and D and can have even greater statistical merit. However, their merit in physical interpretation may be open to greater question.

In situations involving positron trapping by a single type of center a linear relation between the resultant changes in H or S or W should ideally obtain. The actual numerical relationship will depend on the center in question and as we shall see in a later section (Sect.3.3.3) provides a useful additional guide to the type of center involved.

All of the parameters discussed above have the virtue that they can be quickly "on line" recorded by any multi-channel analyser having simple region of interest integration facilities. Thus extended runs embracing a very large number of individual short duration spectrum accumulations can be performed without recourse to expensive on-line computing facilities. Digital stabilization and simple corrections for residual instrumental drift can be readily incorporated [3.44] by simultaneous measurements on a suitable reference peak. The result is a cheap (in equipment and computing time) and efficient method for rapid monitoring of continuously changing situations which, to this author's mind, compares favorably with any defect probing technique available today.

3.2 Monovacancies in Equilibrium

3.2.1 The Naive Approach to Temperature Effects

The strong and nonlinear temperature dependence of equilibrium lattice vacancy concentrations provides, through the consequent strong temperature dependence of positron parameters in some metals, one of the most direct and striking demonstrations of the positron trapping phenomenon. Early experiments [3.45-47] established the essential facts. As the temperature of many solid metals is raised the temperature dependence of the parameters t, H, or S, which is initially relatively weak, increases dramatically at or around $T = 0.6\ T_m$ and so continues until T approaching T_m when it eases to suggest an approach to some saturation effect (Fig.3.2). Careful examination of the angular distribution effects underlying the change in H show that the dominant effect is a relative decrease in the proportion of the broad "Gaussian" core electron part of the distribution. A simple two-state version of the trapping model (3.8) provides a transparent interpretation of these effects [3.17-19].

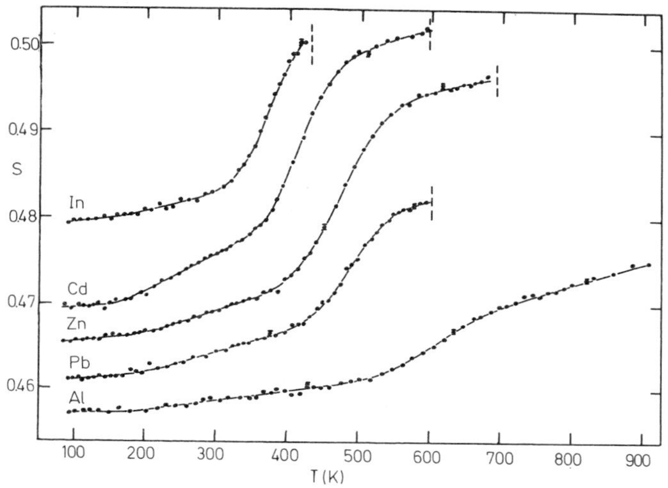

Fig.3.2. S parameter vs temperature curves for several metals [3.110]. The solid curves have no theoretical significance but perhaps reflect the author's prejudice in respect to prevacancy temperature effects (Sect.3.2.2)

At low temperature all of the thermalized positrons in the metal annihilate from a delocalized state with a characteristic rate $\lambda_1 = \lambda_f$ (f for "free") which for the present we assume is effectively temperature independent. As the temperature is raised the monovacancy concentration increases as $C_{1v} = \exp(S_{1v}^F/k)$ $\exp(-H_{1v}^F/kT)$ where k is the Boltzmann's constant, and S_{1v}^F and H_{1v}^F are the monovacancy formation entropy and enthalpy, respectively. For reasons already noted in Sect.3.1.1 a monovacancy is an attractive site for a positron. Thus, should the two interact (the precise mechanism need not concern us here — see Chap.4), the positron may well become trapped, experience therein a reduced electron density and, unless it subsequently escapes, give rise to annihilation parameters F_{1v}, different from that, F_f, that would have arisen from the delocalized state. As the vacancy concentration rises the probability of such trapping also increases and (3.16,17)

$$F = (\lambda_f F_f + K_{1v} F_{1v})/(\lambda_f + K_{1v}) \tag{3.19}$$

where (3.11)

$$K_{1v} = \nu_{1v} \exp(S^F_{1v}/k) \exp(-H^F_{1v}/kT) \quad . \tag{3.20}$$

Eventually when $K_{1v} \gg \lambda_f$, all the positrons are trapped (the actual instantaneous positron density in any experimental situation << 1), and a saturation, $F \to F_{1v}$ obtains (Fig.3.2). The s-shaped F vs T dependence predicted is essentially that observed. Further simple manipulation provides the equation

$$\ln[(F - F_f)/(F_{1v} - F)] = -(H^F_{1v}/kT) + S^F_{1v}/k + \ln(\nu_{1v}/\lambda_F) \tag{3.21}$$

on which Arrhenius plots may be based.

The essential correctness of (3.19) was apparent in the earliest studies. H parameter measurements in several low melting point metals [3.16,48] gave straight line Arrhenius plots and plausible values for H^F_{1v}. Lifetime spectrum analyses [3.25,49,50] demonstrated the superiority of the no-escape model over the alternative (3.7) large-escape or "enhancement" model [3.16,49] which predicts a similar variation of t or H with temperature. The original measurements on In, Cd, Zn, Pb, and Al were soon reinforced by additional studies on the noble metals [3.51,52] and Doppler broadening data [3.33,38,53]. Recent times have seen the introduction of imaginative and sophisticated variations on the simple theme of sandwich-source-sample assemblies enabling measurements on high melting point metals [3.54-56]. For the present we defer a catalogue of the many excellent H^F_{1v} investigations and devote our discussion to the various additional more subtle effects and considerations that have emerged as the techniques have developed.

3.2.2 Prevacancy Effects

A temperature dependence of t or H below the onset of the vacancy trapping effects was recognized in some of the earliest vacancy studies and was incorporated in the H^F_{1v} analyses [3.16,51,52]. In some cases, for example copper and silver, both the change in H [3.57] and in t [3.58] was found to closely match the volume expansion ($\Delta F/F \sim \Delta V/V$ for the same T). In other metals [3.16,58,59] the temperature dependence, although still approximately linear, increased at a quite different rate. In the case of mean lifetime where the temperature coefficient often may be no more than 2 or 3 ps per 100 K there is always room for dispute but in cadmium at least a temperature dependence of t or H or S very much larger than volume expansion is absolutely clear (Fig.3.2).

LICHTENBERGER et al. [3.60] were the first to draw attention to the particularly large prevacancy effect in the S parameter for cadmium which they attributed to positron trapping in transient dilatations. Their results are not presented here since statistical fluctuations apart they are indistinguishable from those shown in Fig.3.2. SEEGER [3.61] has proposed a model of metastable self-trapping to explain these results. The theoretical details and plausibility of this model are adequately discussed in Chap.4. Here we content ourselves with an examination of the experimental evidence.

There is in the curve of Fig.3.2 and in that of [3.57] indisputable evidence of an "intermediate" temperature region with an onset at 100 → 200 K. Above 350 K the usual vacancy trapping effects are apparent. SEEGER argued an onset at ~ 230 K which would clearly rule out a purely thermal expansion effect. The position and sharpness of this onset and perhaps we may venture to add, a suggestion of convex upward curvature in the LICHTENBERGER et al. results do much for the fit to the self-trapping theory [Ref.3.61, Fig.3]. However, these details are not so clear in other cadmium results. Angular correlation measurements by KIM and BUYERS [3.62] do show smearing of the angular distribution consistent with positron localization in a self-trapped state but, apart from two datum points, provide inadequate evidence as to where the onset actually is. Lifetime measurements [3.23,37,39,63] are equally vague on this point and an onset as low as 90 K cannot be ruled out. Indium S parameter measurements by SEGERS et al. [3.64] provided a remarkable picture of the type of intermediate temperature behavior that the self-trapping theory predicts, but a subsequent study by RICE-EVANS et al. [3.59] showed a linear rise in good agreement with the results presented in Fig.3.2.

None of these observations rule out the self-trapping phenomenon but other possibilities can and should be considered. There is no particular reason why such prevacancy effects should scale with thermal expansion. All the measured parameters depend on both conduction and core electron annihilations which in turn involve both many-electron and single-particle wave-function effects [3.4]. The very existence of positron trapping at lattice defects is evidence of strong positron-lattice coupling. At the very least we can expect a thermalized positron to react to the already present phonon field and thus be sensitive to not only thermal expansion but also local density fluctuations. Indeed the positron may not merely correlate strongly with appropriate excitations but may, indeed must to some extent, perturb the phonon field. In the extreme a fortuitous combination of intrinsic and stimulated excitations may conspire to trap the positron in a self-created hole.

Calculations by JAMIESON et al. [3.58] show that the volume dependence of positron-core electron wave-function overlap can often result in an enhanced thermal expansion effect. More recent calculations [3.65] of the effects of both thermal expansion and lattice vibration on the annihilation rate predict prevacancy changes in simple and noble metals which correlate strongly with those observed. The effect of lattice vibration is found to be at least as great and frequently greater than that of thermal expansion, a result supported by recent pressure studies [3.66] which reflect the volume changes alone. The use of a simple model of lattice vibration in [3.65] precludes the assessment of onset temperatures for the prevacancy effects but such onsets are argued in qualitative terms as also is the angular distribution smearing discussed in [3.62]. Prevacancy changes in momentum density parameters are not assessed in [3.65] since they represent a much more difficult problem. Attempts which emphasise thermal expansion have been [3.67] and are being made [3.68].

The combined effects of thermal expansion and lattice vibration provide a plausible explanation of the observed intermediate temperature dependence of the commonly measured F_f at least in the close-packed metals. The observed reversal in slope at still lower temperatures [3.37] is unlikely to have significant implications in the study of vacancy trapping phenomena but warrants more investigation. The very low temperature ($T < 50$ K) involved superficially suggest positron trapping at very weak (binding energy: $\sim 10^{-2} - 10^{-1}$ eV) traps rendered inactive at the higher positron temperatures. The complex low ($T \ll T_m$) temperature dependence of the W parameter for tantalum [3.56] may indicate still more subtle effects but here the pioneering nature of the study and techniques involved in this single study advises against premature theoretical speculation.

To conclude this section we would make but one more observation. In the first two decades of positron studies in metals almost all the observed phenomena were attributed to electronic structure effects. The last decade has seen an almost exclusive preoccupation with the positron's role in the problem. Perhaps it is time to restore the balance in a time when experimental studies are exposing increasingly more subtle effects.

3.2.3 Other Complications

The implications of the results and theories discussed above for the analyses of the higher temperature vacancy trapping curves is not entirely clear. A linear extrapolation of the observed prevacancy temperature rise of any parameter for the free state into the vacancy trapping region cannot be easily justified on purely theoretical grounds. Most of the existing information about lattice vibrations and their effect on particle states is confined to relatively modest ($T \ll T_m$) temperatures. The temperature dependence of S in gallium [3.67] is essentially linear throughout the solid phase but here the absence of vacancy trapping effects may indicate a relatively weaker positron-lattice coupling. Direct evidence of the lifetime temperature dependence of the free state in principle can be obtained by lifetime spectrum analysis but here again other complications arise.

High-resolution lifetime spectrum studies in lead [3.26] suggest, through application of (3.10) to deduced rates and intensities, a strong and nonlinear decrease in λ_f as $T \to T_m$. An essentially similar treatment of aluminum data [3.27] provides the same result. In neither case do variations in the details of spectrum analysis, such as choice of resolution function [3.69], starting channel for the analysis [3.25,26,69], background, etc., appear to significantly affect the result in question. In a subsequent paper [3.70] the lead results are reinterpreted in terms of a trapping model which allows for enhanced trapping of incompletely thermalized positrons. An adequate explanation of the [3.26] results is obtained. A similar "direct trapping" hypothesis is also mentioned briefly in [3.27].

Notwithstanding the success of the model of [3.70] we would suggest that the apparent behavior of λ_f could easily be an artefact of spectrum analysis. The deviation from linearity (Fig.3.3) correlates strongly with an inevitable decrease in precision caused by the combination of the rapidly decreasing intensity and increasing rate of the shortest lived component. In this author's experience of spectrum analyses the deduced rate and intensity of such a component are strongly correlated quantities and the effect of both this and the linearising approximation implicit in the analyses [3.71] on functions of rate and intensity like $\lambda_1 = \sum_j I_j \Gamma_j$ need further investigation. Extensive simulation exercises would be of value here. The temperature dependence of λ_{1v}, the annihilation rate in the vacancy trap, can also be studied by spectrum analysis [3.26,27,69] but here again the uncertainties are large when the corresponding intensity is small. Where the precision is better the deduced temperature dependence of λ_{1v} is comparatively modest and that of λ_1 is consistent with that observed in the prevacancy region.

The intrinsic value of lifetime spectrum analyses in exposing the essential facts about the trapping phenomena such as the relative merits of no-escape (3.9) or large-escape (3.7) assumptions is obvious. Their further utility in monovacancy formation enthalpy measurements is also apparent in the various studies [3.26,27, 50,69] mentioned above to which the reader is referred for additional and invaluable discussions about both the details of measurement and analysis procedures and further possible refinements in the basic theoretical models. However, in this latter respect the alternative single-parameter studies and purely theoretical considerations play an equally important role.

The dominant temperature dependence in the vacancy trapping region is clearly that explicit in (3.20). A possible additional but weaker T^n dependence in ν_{1v} has been considered in various theoretical works [3.72-76]. The theoretical situation is reviewed in Chap.4 and here it is sufficient to note that most contemporary theory favours at T^0 dependence. The experimental evidence is largely in agreement with this prediction. Early nonequilibrium quenching studies of gold gave inconsistent results ranging from at T^0 [3.77] to a T^1 [3.78] dependence.

More recent investigations of Au [3.37], Al, Ni, and Cu specimens [3.79] containing a nonequilibrium vacancy concentration achieved by low-temperature electron irradiation come down firmly in favor of the T^0 dependence at least for $T < 250$ K. The subject is of considerable intrinsic interest in respect of the underlying physics but is largely academic from the point of view of monovacancy enthalpy measurements since all the various theories offered suggest a temperature dependence for ν_{1v} which is very much weaker than the exponential dependence of c_{1v}. The effect of an inappropriate choice in the theoretically offered range $T^{-\frac{1}{2}} \rightarrow T^{+\frac{1}{2}}$ is but a few percent in the deduced value of H^F_{1v} [3.16,27].

All of the temperature effects considered thus far lie within the scope of the simple no-escape two-state trapping model. Other possible effects which demand a generalization of this simple theory should now be considered.

At the threshold of the vacancy trapping phenomenon, $T \sim 0.6\ T_m$ (Fig.3.2) the concentration of monovacancies can be plausibly argued as orders of magnitude greater than that of any multivacancy center. At the higher temperatures where $T \rightarrow T_m$, positron trapping at divacancies may start to play a role [3.2,80-85]. Again at the lower temperatures both theoretical positron-vacancy binding energies [3.86] and observations of discrete components support the no-escape assumptions but as $T \rightarrow T_m$ in the higher melting point metals the assumption becomes more fragile. Following SEEGER [3.80] we consider these higher-temperature effects together as they may well in practice occur.

Analytic solutions of (3.5) are extremely complicated when detrapping terms are present, even for small N, but the evaluation of the weighting factors $N_i/\sum_i N_i = P_i$ that determine the simple F parameters (3.15,17) is relatively straightforward. Simple integration, $t = 0 \rightarrow \infty$, of (3.5) yields the equations

$$\Big(1 + \sum_{j \neq i} K_{ij}/\lambda_i\Big)P_i = \sum_{j \neq i} (K_{ji}/\lambda_j)P_j + \delta_{i,1} \tag{3.22}$$

where as usual $\delta_{ij} = 1$ if $i = j$ and zero otherwise.

In the present discussion we are concerned with a three-state problem with $K_{12} = K_{1v}$, $K_{13} = K_{2v}$ as trapping rates and $K_{21} = \varepsilon_{1v}$, $K_{31} = \varepsilon_{2v}$ as "detrapping" rates for monovacancies and divacancies respectively. We may expect the detrapping rates to reflect the positron binding energies in the traps. Since the binding energy in a divacancy should be substantially greater than that in a monovacancy we shall for simplicity set $\varepsilon_{2v} = 0$. For further simplicity we take the positron mean lifetime as the vehicle for our discussion.

With the assumptions outlined above we may readily deduce from (3.22), (3.17) with $F = t$, (3.11) and an obvious generalization of our earlier used notation, a previously reported [3.80] result

$$t = t_f \frac{(1+t_{2v}\nu_{2v}c_{2v})(1+\varepsilon_{1v}t_{1v}) + t_{1v}\nu_{1v}c_{1v}}{(1+t_f\nu_{2v}c_{2v})(1+\varepsilon_{1v}t_{1v}) + t_f\nu_{1v}c_{1v}} \quad . \tag{3.23}$$

The equilibrium monovacancy concentration (3.20)

$$c_{1v} = \exp(S^F_{1v}/k)\exp(-H^F_{1v}/kT) \tag{3.24}$$

and with an obvious extension of the notation

$$c_{2v} = (Z/2)\ \exp(S^F_{2v}/k)\ \exp(-H^F_{2v}/kT) \tag{3.25}$$

where Z is the coordination number for the lattice in question.

$$S^F_{2v} - 2S^F_{1v} = \Delta S_{2v} \quad , \quad \text{and} \quad 2H^F_{1v} - H^F_{2v} = H_{2v}$$

where ΔS_{2v} and H_{2v} are, respectively, the association entropy and binding enthalpy of divacancies [3.72,74].

A proper quantum mechanical theory of detrapping, in the spirit of those (Chap.4) applied to the trapping process, has not to this author's knowledge as yet been attempted. Existing analyses make use of the simple and intuitively appealing expression

$$\varepsilon_{1v} = \Gamma_0\ \exp(-E^B_{1v}/kT) \tag{3.26}$$

where E^B_{1v} is the positron-vacancy binding energy.

From (3.23) we see that detrapping will be an important consideration if ε_{1v} is of order or greater than $\lambda_{1v}(t^{-1}_{1v})$. An estimation of ε_{1v} is made difficult by uncertainties in both E^B_{1v} and Γ_0. The present uncertainty in positron-vacancy binding energies, even in the most studied case of aluminum where we might optimistically suggest $E^B_{1v} = 2 \pm 0.5$ eV (Chap.4) gives values for the exponential factor in (3.26) ranging from 10^{-14} to 10^{-9} at T = 900 K. Various prescriptions and values for the pre-exponential factor Γ_0 have also been offered. SEEGER et al. [3.14,80] adopted a molecular reaction rate result, $kT/H \sim 2 \times 10^{10}$ T s^{-1}. In the earliest comprehensive discussion GOLAND and HALL [3.87] also assumed a T^1 dependence but anticipated the necessity for an appropriate light-particle bound state model in a quantitative estimate of Γ_0. In its crudest terms the problem involves the frequency of oscillation of the positron in the potential well of the vacancy, DOYAMA and HASIGUTI [3.2] combined a positron thermal velocity $\sim(kT/m)^{\frac{1}{2}}$ with a localization length $a \sim 10^{-10}$ m giving $\Gamma_0 \sim 2 \times 10^{13} T^{\frac{1}{2}} s^{-1}$. TAM et al. [3.88, 89] refined the prescription by recognising the importance of localization in raising the positron kinetic energy. Adopting a Gaussian model for the positron-density profile they estimate values of $\Gamma_0 \sim 10^{16}\mathrm{s}^{-1}$ essentially temperature independent but implicitly a function of binding energy.

Let us adopt the TAM value of 10^{16} s^{-1} as a plausible upper bound on Γ_0. Then for aluminum at 900 K we estimate $10^2 < \varepsilon_{1v} < 10^7$ comfortably less than a $\lambda_{1v} \sim 4 \times 10^9$ s^{-1}. Thus thermal detrapping effects in aluminum and the other close-packed low melting point metals where the predicted binding energies are similar (Chap.4) seem unlikely. In the noble metals the situation is less clear. If, as is suggested in some studies $E_{1v}^B \sim 1$ eV or less (Chap.4) detrapping may be important.

The bcc metals should be considered apart. Positron-vacancy binding energies in the alkali metals are believed to be small (Chap.4). Thus the monovacancy trapping rate should be small [3.11], the detrapping rate should be large [3.90], and the effect of vacancies in equilibrium experiments unobservable [3.91]. An almost as severe situation may also obtain in the high melting point bcc metals. Theoretical positron-vacancy binding energies correlated strongly with the Wigner-Seitz or zero-point positron energy E_0 in the perfect lattice (Chap.4). E_0 for the transition metals is larger than that for the simple metals but only slightly so (Tables 4.6, 7). Thus E_0/T_m is, and E_B^{1v}/T_m may well be, much smaller than that for the close-packed metals and consequently ε_{1v} much larger. We may easily deduce from (3.23) that if $\varepsilon_{1v}t_{1v} = \chi >> 1$ we can expect an *apparent* trapping rate $\nu_{1v}' = \nu_{1v}/\chi$ and an *apparent* formation enthalpy $H_{1v}^F - E_{1v}^B$ [3.80]. That such may be the case in tantalum is strongly asserted in [3.56] and other studies of vanadium [3.55], zirconium and titanium [3.92] add credence of this view.

The possibility that additional detrapping might be caused by monovacancy migration has been raised by KURIBAYASHI and DOYAMA [3.90]. The jumping rate of monovacancies

$$\Gamma_v = Z\nu \exp(S_{1v}^M/k) \exp(-H_{1v}^M/kT)$$

where Z is the coordination number, $\nu \sim$ the Debye frequency, and S_{1v}^M and H_{1v}^M are the vacancy migration entropy and enthalpy, respectively. In many metals close to melting, Γ_v may become comparable to or greater than λ_{1v}. Thus there arise two questions.

I) At high temperatures where $\Gamma_v > \lambda_{1v}$, can a vacancy with a positron move as freely as an empty vacancy?

II) In a vacancy jump does the positron get detrapped?

TAM and SIEGEL [3.89] argue plausibly that if the latter is the case, the effective vacancy migration energy barrier is raised by an amount E_{1v}^B. Thus the superficial criterion for an observable effect, $\Gamma_v > \lambda_{1v}$, is modified to $\Gamma_v' = \Gamma_v \exp(-E_{1v}^B/kT) > \lambda_{1v}$, which is much less easily satisfied. They conclude that such detrapping is much less likely than the ordinary thermal detrapping earlier considered.

Returning to (3.23) we see that divacancies become important when $t_{2v}\nu_{2v}c_{2v}$ is of order or greater than $t_{1v}\nu_{1v}c_{1v}/(1 + \varepsilon_{1v}t_{1v})$. Following SEEGER [3.80] we note that divacancy effects come in at lower temperatures if positron detrapping from monovacancies is appreciable. If $\varepsilon_{1v}t_{1v} << 1$ and if divacancy trapping is appreciable, i.e., $t_{2v}\nu_{2v}c_{2v} >> 1$ (3.23) may be rewritten

$$t \sim t_{1v} \frac{(1+t_{2v}\nu_{2v}c_{2v}/t_{1v}\nu_{1v}c_{1v})}{(1+\nu_{2v}c_{2v}/\nu_{1v}c_{1v})} \quad . \tag{3.27}$$

Like its simpler counterparts, (3.27) predicts an S shaped transition between asymptotes t_{1v} and t_{2v}. The "midpoint" of the transition $t = (t_{1v} + t_{2v})/2$ is when (3.24-26)

$$\frac{\nu_{2v}c_{2v}}{\nu_{1v}c_{1v}} = \frac{\nu_{2v}}{\nu_{1v}} \cdot \frac{Z}{2} \exp \frac{(S^F_{2v} - S^F_{1v})}{k} \exp \frac{(H^F_{1v} - H^F_{2v})}{kT} = 1 \quad . \tag{3.28}$$

Now $t_{2v} - t_{1v}$ is presumably somewhat smaller than $t_{1v} - t_f$. Further, $H^F_{2v} \sim 2H^F_{1v}$ and $\exp(-H^F_{1v}/kT) \sim 10^{-4}$ at the melting point. Thus for a significant divacancy effect we require that

$$(\nu_{2v}/\nu_{1v})(Z/2)\exp[(S^F_{2v} - S^F_{1v})/k] \sim 10^3 \quad .$$

Theoretical estimates suggest that $S^F_{2v} - S^F_{1v} = S^F_{1v} + \Delta S_{2v}$ can sometimes be a few times k [3.93], thus as long as ν_{2v} is somewhat greater than ν_{1v} measurable divacancy effects are quite possible.

Much of the experimental data seem to support this contention. S and H parameter curves for aluminum [3.94] and Fig.3.2, and for the noble metals [3.81,83,84] show a behavior close to melting strongly indicative of divacancy effects. It must be admitted that lifetime studies [3.95] seem not to show the phenomenon to the same extent. However, it is quite possible that t_{2v}/t_{1v} is generally somewhat smaller than H_{2v}/H_{1v} etc.

In this and in the preceding section we have dealt at length with a large variety of possible temperature effects. The list is by no means exhausted. For example, temperature dependence in the vacancy formation enthalpies can also be considered, but as KIM and BUYERS [3.96] have argued it will, in first order, be largely nullified by a compensating change in the entropy. We have paid but little attention to the temperature dependence of the annihilation parameters for the positron trapped in a vacancy. But aside from a single calculation [3.97] there is little theory to guide us and experimental investigations are hard to almost impossible.

It is probably fair to say that theoretical speculation is currently running somewhat ahead of experimental verification. In order not to exacerbate the situation further we shall now return directly and exclusively to the experimental situation.

3.2.4 Vacancy Formation Enthalpy Measurements

A positron measurement of monovacancy formation enthalpy involves three distinct but interrelated steps.

I) The preparation of a pure and well-annealed specimen and, in the case of Doppler broadening and lifetime studies, the incorporation of a suitable positron source.

II) A precision measurement of the temperature dependence of some selected parameter (Sect.3.1) over the widest possible temperature range in the solid phase.

III) Numerical analysis of the resulting data in terms of one or more of the simpler versions of the positron trapping model.

The majority of studies thus far have employed pure (~ 5N) polycrystalline samples. Prior annealing and chemical polishing is usual and is advisable to decrease the probability of surface and other nonequilibrium defect trapping effects. The choice of positron source and source-sample geometry depends on both the parameter to be measured and the temperature range involved.

In lifetime studies simple evaporated ^{22}NaCl sandwich-source-sample assemblies, with or without metallic or other interfacing foils [3.4], can be used at temperatures up to ~ 900 K as long as the possibility of undesirable chemical reaction or diffusion effects is recognised and guarded against. At higher temperatures sealed assemblies involving other geometries [3.84,85] or electron beam welding techniques [3.24,65] can be employed. The wider selection of sources available for Doppler broadening studies allows greater discrimination. Financial considerations apart, sources such as ^{68}Ge and ^{44}Sc which emit higher energy positrons than the cheaper and more common isotopes may be preferable in minimizing surface and self-absorption effects [3.103,104]. Metallic foil sources such as ^{68}Ge-Ni (1100 K) [3.105] and ^{58}Co-Rh (1300 K) have been utilized and others could no doubt be devised. For work at significantly higher temperatures two main lines of approach can be discovered. MAIER et al. [3.24,37,56] have employed both electron beam welding for sealed source preparation and electron beam heating of samples to achieve temperatures in excess of 3000 K. CAMPBELL et al. [3.54,55] employ a ^{19}Ne gas source continuously generated by an electron linac and have achieved temperatures ~ 2000 K [3.55,106].

In angular correlation studies employing an external positron source the necessarily much higher source activities generally rule out the use of the more expensive isotopes. Here, however, the availability of relatively cheap and attack-resistant sources such as long-lived ceramic ^{22}Na or "throwaway" sources such as ^{64}Cu and ^{58}Co, makes the achievement of relatively high temperatures somewhat easier. When sample evaporation is no problem a combination of high vacuum and simple radiation shielding allows the easy attainment of temperatures ~ 1500 K [3.107]. The achievement of significantly higher temperatures than this is made difficult by the opposing demands of an unimpeded positron flux and thermal isolation of the source and sample. However, the problem is not insuperable and some advances can be expected.

The precision measurement of the often quite modest temperature-induced changes in positron annihilation parameters under consideration is not easy. Adequate precautions against both short-term fluctuations and particularly long-term drifts in equipment performance are always necessary [3.22,23,31,35-37,44]. Complete cycling through the necessary temperature range is also desirable in providing a check on both the equilibrium or otherwise nature of the sample condition and the long-term equipment stability.

A comparison of the ultimate precision and sensitivity of the three common measurement techniques is not easily made in raw spectrum terms. However, single number parameters such as t, H, S, etc., provide some scope for comparison. $t_{1v} - t_f$ [see, for example, (3.19)] varies from ~ 50 ps in low melting point metals to ~ 100 ps in the noble metals. Thus a precision of $\Delta t \sim \pm 1$ ps in individual measurements which is common today amounts to 1-2% of the total lifetime change observed in the solid phase. The major limiting factor here is spectrum centroid stability and a significant improvement in precision would seem unlikely without a major development or change in the basic measurement technique. Vacancy trapping induced changes in Doppler broadening parameters are generally modest but the already noted facility for simultaneous equipment stability checks and corrections (Sect.3.1.3) can provide a resultant precision close to that predicted by the counting statistics alone [3.44]. The situation in angular correlation studies where the electronics techniques are undemanding and systematic errors caused by sample thermal expansion, etc., can be readily avoided by side-channel normalization procedures is similar. Thus in either case the ultimate precision is largely determined by statistical considerations and a $\Delta F/(F_{iv} - F_f)$ better than 1% is not hard to achieve.

F parameters versus temperature curves from different groups of workers, which in the past have shown quite alarming divergences, are today for the metals most studied, displaying much closer agreement, and give confidence in the basic data. Only in the final stage of analysis and application of trapping theory do clear differences in both approach and conclusions emerge.

The majority of trapping model analyses make use of (3.17) with P_i's, (3.22) derived for a particular but necessarily arbitrary choice of trapping model, and F_i's that are constant or have a weak dependence on temperature. A small minority of studies involve detailed spectrum analyses. Such an approach does have the advantage that it gives some guidance towards the most appropriate choice of trapping model but whether or not it contains any particular *additional* virtue over the single parameter studies is not entirely clear, at least to this author. Any analysis that makes more use of all the available spectrum data than simple integrals (H, S, or W) or first moment (t) in principle should be superior. However, when such analyses are attempted they are often constrained by some model assumptions which if inadequate may outweigh their superficial advantage. The discussion following (3.14) is of relevance here. Whatever the inherent limitations of lifetime spectra analyses, they do allow for a variety of subsequent routes to the specific trapping rate of interest. Some may be better than others. For example in the two-state case the use of $K_{12} = I_2(\Gamma_2 - \Gamma_1)$ is evidently superior to $K_{12} \propto I_2/I_1$ [3.26,27] which is only valid for temperature-independent annihilation rates.

The "fit" of any set of data to a particular trapping model result is usually expressed in "Chi squared per degree of freedom" or equivalent statistical terms. For the present we defer consideration of the relevance of this measure and merely note that most authors appear content when this is reduced to an appropriate minimum ($\chi^2/\nu \sim 1$ when normal statistics apply). We have already noted that published F versus temperature data today seem reliable. The least we can ask of the resulting deduced formation enthalpies is that they display a similar trend at least when deduced from roughly similar theoretical models. In Tables 3.1,2 we show results from a variety of analyses embracing the simpler and most easily justified of the refinements discussed in Sects.3.2.2,3.

The results for aluminum (Table 3.1) are numerous and in good agreement with respect to both H^F_{1v} and ν_{1v} $\exp(S^F_{1v}/k)$. The mean $H^F_{1v} \sim 0.66$ eV is evidently also consistent with that obtained through quenching experiments and with total vacancy concentration measurements at higher temperatures [3.27]. For most of the other low melting point metals there is either a paucity of results or quite marked disagreement. All the indium studies involve the H or S parameters and very similar model analyses. Thus the pronounced variation in deduced H^F_{1v} values is somewhat hard to explain. Here and in zinc and cadmium, where comparable uncertainties may well exist and where information from other techniques is scarce, additional studies would be desirable.

In the noble metals and in nickel (Table 3.2) the positron results for H^F_{1v} are reasonably consistent ($\sim \pm 0.05$ eV) although generally higher than some "preferred" values arrived at by syntheses of a variety of other equilibrium, quenching, annealing, and diffusion data [3.57,84,85,101,102,108 and references therein]. Consideration of detrapping effects can reduce the H^F_{1v} value by 0.1 eV or so but discrepancies still remain. Some authors incorporate combinations of "complications" (Sect.3.2.3) in their analyses to arrive at "preferred" values of H^F_{1v} and other defect parameters [3.36,83,84,108] but sometimes without too obvious justification. This author feels much less competent than some others to judge the plausibility of any particular result in the light of all available information on point defect phenomena but does feel compelled to re-emphasise the purely practical situation.

Table 3.1. Monovacancy formation enthalpies and trapping rates as deduced by conventional s curve trapping model analyses: simple metals

Metal	$\nu_{1v} \exp(S^F_{1v}/K)$ $[s^{-1}]$	H^F_{1v} [eV]	$\nu_{2v} \exp(S^F_{2v}/k)$ $[s^{-1}]$	H^F_{2v} [eV]	Comments	Method	Ref.
Al	1.2×10^{15}	0.66	-	-		H	[3.48]
Al	0.8×10^{15}	0.64	-	-		Lifetime spectrum analysis	[3.69]
Al	0.6×10^{15}	0.66	-	-		H, W	[3.31]
Al	1.3×10^{15}	0.67	-	-		H	[3.98]
Al	1.9×10^{15}	0.68	-	-		H	[3.85]
Al	0.8×10^{15}	0.66	-	-		Lifetime spectrum analysis	[3.27]
Zn	3.6×10^{15}	0.54	-	-		H	[3.48]
Cd	1.9×10^{14}	0.39	-	-		H	[3.48]
In	1.5×10^{17}	0.55	-	-		H	[3.48]
In	1.5×10^{16}	0.48	-	-		H	[3.31]
In	-	0.39	-	-		H, t	[3.99]
In	-	0.59	-	-		S	[3.59]
Sn	1.5×10^{14}	0.50	-	-	Limited temp. range	t	[3.100]
Sn	1×10^{15}	0.51	-	-		S	[3.100]
Pb	6.2×10^{14}	0.50	-	-		H	[3.48]
Pb	1.6×10^{16}	0.54	-	-		H	[3.31]
Pb	$0.2\text{-}1.0 \times 10^{16}$	0.58-0.65	-	-	Constraint free analysis	Lifetime spectrum analysis	[3.26]

Table 3.2. Monovacancy formation enthalpies and trapping rates as deduced by conventional s curve trapping model analyses: noble metals and nickel

Metal	$\nu_{1v}\exp(S^F_{1v}/k)$ $[s^{-1}]$	H^F_{1v} [eV]	$\nu_{2v}\exp(S^F_{2v}/k)$ $[s^{-1}]$	H^F_{2v} [eV]	Comments	Method	Ref.
Cu	-	1.17	-	-		H	[3.52]
Cu	-	1.28	-	2.26		H	[3.83]
Cu	$\sim 4 \times 10^{15}$	1.20	-	-	Divacancy analysis not entirely successful	H	[3.81]
Cu	1.3×10^{16}	1.29	-	-		H	[3.57]
Cu	2×10^{16}	1.26	-	-			[3.101]
Cu	1.5×10^{15}	1.16	3.3×10^{16}	2.0	Divacancy and detrapping effects assumed	H	[3.84]
Ag	1.74×10^{16}	1.16	-	-		H	[3.57]
Ag	-	1.20	-	-		S, W, D	[3.102]
Au	1.5×10^{15}	0.98	-	-		Lifetime spectrum analysis	[3.69]
Au	1.2×10^{15}	0.97	-	-		H	[3.57]
Au	-	0.92	-	-	Self-trapping included	W	[3.37]
Au	0.5×10^{15}	0.96	1.2×10^{16}	1.69	Divacancy and detrapping effects assumed	H	[3.84]
Ni	-	1.65	-	2.90		H	[3.83]
Ni	-	1.72	-	-		H	[3.85]
Ni	-	1.73	-	-		S, W, D	[3.102]

We have already noted in a previous paragraph the use of χ^2/ν and similar statistical tests to justify complex multi-parameter fits. A simple statement of the numerical value of χ^2/ν, etc., is common in such cases but is seldom really sufficient. For example, when normal statistics apply, $\chi^2/\nu \approx 1$ is a tempting target for cessation of effort but one that can often be achieved well away from the real solution. The χ^2 or other optimizing parameters can be viewed as a function in the N-dimensional space of the fitting parameters [3.109]. A complete search through this N-dimensional space is not always practicable but can be performed when N is modest [3.109]. In other cases the approach to the fitting minimum should at least be studied in the projected space of some of the fitting parameters. This is particularly true when the uncertainties in the individual datum values cannot be determined exactly. The following example is of some relevance to both this and the more general problem of the analysis of vacancy trapping curves.

In the aluminum data of Fig.3.2 there is a clear indication as $T \rightarrow T_m$ of a saturation not to a constant value of H but to an approximately linear region. A parallel study by HOOD and SCHULTZ [3.94] provides the same conclusion. In zinc and cadmium (Fig.3.2) a similar but weaker effect is indicated. The author and colleagues [3.110] have analysed these results with two comparatively simple versions of trapping model, henceforth referred to as (a) and (b). We consider the picture emerging from the cadmium data analysis.

In (a) the simple two state result (3.19,20)

$$S = \frac{S_f\lambda_f + S_{1v}\nu_{1v}\exp(S^F_{1v}/k)\exp(-H^F_{1v}/kT)}{\lambda_f + \nu_{1v}\exp(S^F_{1v}/k)\exp(-H^F_{1v}/kT)} \tag{3.29}$$

is applied with $S_f = S^0_f(1+\alpha_1 T)$, as suggested by the data for the prevacancy region, to the data from 300 K to the melting temperatur. A plot of the sum of the squared residuals reveals a clear minimum $\sum r_i^2 \sim 2.5\times10^{-6}$ at $\alpha_1 \sim 3.5\times10^{-5}$ deg^{-1} similar to but slightly less than that ($\sim 4.4\times10^{-5}$ deg^{-1}) observed in the prevacancy region. The corresponding deduced value of H^F_{1v} is 0.52 ± 0.03 eV.

In (b) an additional temperature dependence $S_{1v} = S^0_{1v}(1+\alpha_2 T)$ is included. Values of $\sum r_i^2$ and H^F_{1v} are then obtained for a large selection of both α_1 and α_2. The resulting fitting surface is shown in Fig.3.3. The new minimum $\sum r_i^2 \sim 1.4\times10^{-6}$ is somewhat lower than the original 2.5×10^{-6} noted above but cannot be argued a priori because of uncertainty in the absolute precision of the original datum points. From Fig.3.3 we could conclude that H^F_{1v} = 0.57 ± 0.08 eV, $\alpha_1 = 3.6\times10^{-5}$ deg^{-1}, and $\alpha_2 = 1.6\times10^{-5}$ deg^{-1}. The dramatic increase in both the deduced H^F_{1v} and the associated possible error as compared with the corresponding results from (a) is clear. Similar trends are found in the analyses of the remaining curves of Fig.3.2 and in the work of [3.94]. Of course, we need not confine ourselves to simple linear temperature coefficients for S_F and S_{1v}. However, such a dependence for S_F seems the most plausible choice at present (Sect.3.2.2) and quenching studies on aluminum [3.111] show a low temperature dependence of line shape in quenched samples very similar to that observed in the equilibrium measurements [3.94]. We could also perhaps invoke divacancy trapping and/or monovacancy detrapping effects in an attempt to explain the anomalous high temperature part of the aluminum curve. However, the inclusion of the necessary additional fitting parameters would undoubtedly have even more serious effects on the deduced values of H^F_{1v} and their associated precision. Inclusion of divacancy trapping effects usually results in a decrease of the apparent formation entropy whereas inclusion of a positive temperature coefficient in F_t appears to increase H^F_{1v}. The implication of an inappropriate choice is clear!

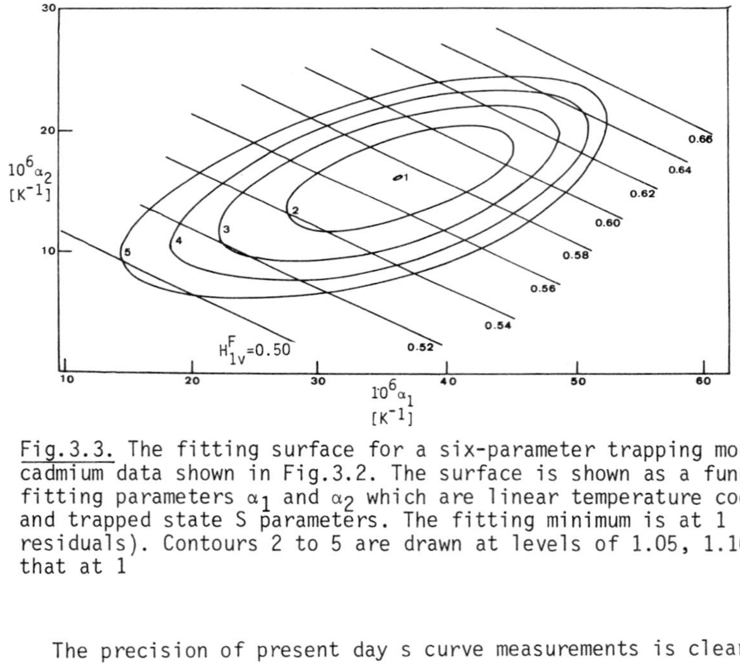

Fig.3.3. The fitting surface for a six-parameter trapping model analysis of the cadmium data shown in Fig.3.2. The surface is shown as a function of two of the fitting parameters α_1 and α_2 which are linear temperature coefficients for the free and trapped state S parameters. The fitting minimum is at 1 (sum of the squared residuals). Contours 2 to 5 are drawn at levels of 1.05, 1.10, 1.15 and 1.20 of that at 1

The precision of present day s curve measurements is clearly not sufficient to resolve the remaining uncertainties concerning the various possible high temperature effects considered in this and preceding sections. Thus analyses which focus attention on the lower temperature F_f and monovacancy trapping dominated parts of the characteristic s curves can be both useful and potentially less likely to embrace serious misinterpretation.

3.2.5 Characteristic or Threshold Temperatures

Many workers have employed simple plausibility tests to their deduced values for H^F_{1v}. A requirement for the fcc metals that the monovacancy concentration at the melting point, $C_{1v}(T_m)$, should be $\sim 10^{-4} \rightarrow 10^{-3}$ or $H^F_{1v} \sim 9\ kT_m$ is quite common. Many empirical relations between various physical parameters which depend on the binding energy of lattice atoms such as T_m, H^F_{1v}, the monovacancy migration enthalpy H^M_{1v}, and the activation energy for self-diffusion $Q(= H^F_{1v} + H^M_{1v}$ for a simple vacancy diffusion mechanism [3.112]) have been proposed and experimentally verified in recent years (see, for example [3.113] and references therein).

KURIBAYASHI et al. [3.114] were the first to point out explicitly a further correlation between the H^F_{1v} obtained from conventional analyses of vacancy trapping curves and the characteristic or threshold temperatures, T_c, at which the onset of the vacancy trapping effects is first apparent. Subsequently MACKENZIE and LICHTENBERGER [3.115] demonstrated a linear relationship between T_c and Q, and with the aid of the further assumption, $H^F_{1v} \sim Q/2$ [3.112], deduced formation enthalpy values from a catalogue of T_c's.

The relevance of a suitably defined T_c to H^F_{1v} is easily established. Definitions of T_c vary in detail from worker to worker [3.107,113-115] but do not significantly alter the arguments presented below. Following the definition implied in [3.115] we define T_c by the intersection of straight line approximations to the prevacancy and vacancy trapping parts of the usual temperature curve. We follow the excellent analysis given in [3.107].

We consider the simple two-state result (3.29) for a general parameter

$$F(T) = \frac{F_f + F_{1v} t_f \nu_{1v} \exp(S^F_{1v}/k) \exp(-H^F_{1v}/kT)}{1 + t_f \nu_{1v} \exp(S^F_{1v}/k) \exp(-H^F_{1v}/kT)} \quad . \tag{3.30}$$

The tangent at the "midpoint" of the vacancy trapping region, $F(T) = (F_f + F_{1v})/2$, is

$$F_L(T) = (F_f + F_{1v})/2 - [(F_{1v} - F_f)\ln(A)]/4 + [(F_{1v} - F_f)(\ln A)^2 kT]/4H^F_{1v} \quad , \tag{3.31}$$

where

$$A = t_f \nu_{1v} \exp(S^F_{1v}/k) \quad . \tag{3.32}$$

T_c is defined by the intersection of (3.31) with $F = F_f$. Thus

$$H^F_{1v} = \{(\ln A)^2/[\ln(A)-2]\} kT_c \quad . \tag{3.33}$$

Now $A \sim 10^5$ in most metals and thus the coefficient of T_c in (3.33) is a weakly varying function of A. Furthermore, as HOOD and McKEE [3.113] have shown, when quantum mechanical transition limited positron trapping obtains (Chap.4), variations in t_f tend to be cancelled by opposing variations in ν_{1v}. It follows that $H^F_{1v} = (13\text{-}15)kT_c$ if $A = 4 \times 10^4 \rightarrow 4 \times 10^5$ as is suggested by most data.

Other implicit temperature effects in (3.30) to little to alter this result. Temperature variations in A have a minimal effect for the reasons already noted above. Linear temperature dependence in F_f and F_{1v} are considered in [3.107] and again are found to little affect the picture.

In Fig.3.4 a compilation of threshold temperatures reported by various workers or taken by the writer directly from published curves are plotted against the melting temperatures. The sources for most of the metals are those given in Tables 3.1,2. Additional data includes that for fcc iron [3.106,107], vanadium [3.106] and tantalum [3.56].

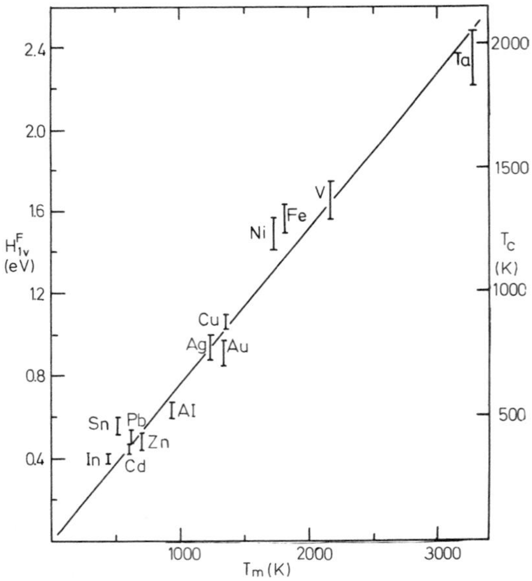

Fig.3.4. Threshold temperatures T_c, and mor vacancy formation enthalpies, H^F_{1v}, deduced from the MACKENZIE relation $H^F_{1v} = 14\ kT_c$ (Sect.3.2.5) vs melting temperatures for a variety of metals. Sources are indicated in the text. Suggested uncertainties are based both on the intrinsic uncertainty in defini T_c from any particular set of results and t number of independent measurements availabl

The simultaneous presentation of both H^F_{1v} and T_c vs T_m in Fig.3.4 is based on the commonly adopted MACKENZIE and LICHTENBERGER result $H^F_{1v} = 14\ kT_c$ [3.56,102,115]. The "best" straight line gives $T_c = 0.63\ T_m$ or $H^F_{1v} \sim 8.8\ kT_m$, as deduced by HOOD and McKEE [3.113] albeit via a somewhat different route. The data scatter about the line most probably reflects the limited correlation between H^F_{1v} and T_m rather than a serious breakdown in the earlier derived relationship between H^F_{1v} and T_c. The numerical data of Table 3.3 suggest some interesting trends. As compared with the more conventionally deduced H^F_{1v}'s (Tables 3.1,2) both the measured T_c's

Table 3.3. Threshold temperatures (Sect.3.2.5) and related monovacancy formation enthalpies as deduced from Fig.3.4. H_{1v}^F values in the right hand column are as suggested by SEEGER

Metal	T_c [K]	H_{1v}^F [eV] T_c analysis	H_{1v}^F [eV] SEEGER [3.114]
Al	520 ± 10	0.63 ± 0.04	0.66
Zn	395 ± 15	0.48 ± 0.04	
Cd	370 ± 10	0.45 ± 0.02	0.41
In	330 ± 10	0.40 ± 0.02	
Sn	465 ± 20	0.56 ± 0.04	
Pb	415 ± 15	0.50 ± 0.03	0.50
Cu	876 ± 10	1.06 ± 0.04	1.06
Ag	775 ± 25	0.93 ± 0.06	0.99
Au	755 ± 20	0.91 ± 0.06	0.92
Ni	1235 ± 25	1.49 ± 0.07	
Fc (fcc)	1290 ± 20	1.56 ± 0.07	
V	1370 ± 20	1.65 ± 0.10	
Ta	1940 ± 40	2.34 ± 0.13	

(2-3% uncertainty) and the related H_{1v}^F's are better defined. For the low melting point metals there is general agreement between the H_{1v}^F derived from either type of analysis. In the case of the noble metals and nickel the $H_{1v}^F(T_c)$ are significantly smaller than those listed in Tables 3.1,2 but are, at least for copper and gold, closer to preferred values based on other criteria [3.114]. In vanadium and iron the limited temperature range embraced inhibits [3.106] or seriously limits the viability of a conventional unconstrained trapping model analysis. A constrained but otherwise conventional two-state analysis [3.107] of data for α and γ iron yields H_{1v}^F values of 1.4 eV(α - Fe) and 1.7 eV(α - Fe) a little larger than those implied ($\sim$ 1.3 eV, $\sim$ 1.6 eV) by T_c of 1070 K and 1300 K, respectively.

The general tendency of the T_c derived H_{1v}^F to fall below the more conventionally obtained results for the higher melting point metals is perhaps significant and suggestive of detrapping effects (Sect.3.2.3) [3.80,84]. A remarkable agreement between the characteristic temperature result for tantalum and a conventional curve fitting analysis embracing detrapping gives further support for this detrapping hypothesis but unfortunately little "insight" into the early noted complex prevacancy effects in the data [3.56]. Temperature studies in zirconium and titanium and a zirconium-titanium alloy [3.92] reveal effects associated with the $\alpha \rightarrow \beta$ (hcp $\rightarrow$ bcc) phase transitions, at least in titanium and the alloy, but are disappointing in regard to vacancy trapping phenomena. There is a suggestion of the familiar s shape vacancy trapping form in the curve for zirconium but the effect is weak and ill-defined. The comparative weakness of the trapping effect both here and in bcc vanadium [3.106] and tantalum [3.56] as already noted in Sect.3.2.3 indicate perhaps generally weaker (relative to kT_m) positron-vacancy binding energies in the bcc metals.

In 1969 experimental values for H_{1v}^F were confirmed to some half dozen metals. Today the positron results extend over some dozen metals and a comparable number of binary alloy systems. To this extent at least much has been achieved. However, problems remain particularly in respect to the absolute values of derived formation enthalpies. The correlation between T_c and T_m is encouraging (Fig.3.4) but the derived values of H_{1v}^F (Table 3.3) may be seriously in error through an inappropriate choice of H_{1v}^F vs T_c relation. Published values of H_{1v}^F obtained by full trapping model analyses are frequently quoted with very modest associated errors (~ ±0.02 -0.03 eV). It is likely the spread of values through Tables 3.1-3 provides a better measure of the current state of the art.

3.2.6 Pressure Experiments

In temperature studies at ordinary pressures the monovacancy formation enthalpy H_{1v}^F and energy E_{1v}^F are little different and terminology can be and is often loose. At elevated pressure the difference as expressed in the definition of H_{1v}^F as

$$H_{1v}^F = E_{1v}^F + pV_{1v}^F$$

involves the formation volume V_{1v}^F.

Thus the vacancy concentration as a function of both temperature and pressure is

$$c_v = \exp(S_{1v}^F/k)\ \exp-|(E_{1v}^F + pV_{1v}^F)/kT|$$

and pressure studies at constant temperature can provide for the measurement of V_{1v}^F.

The extension of positron measurements to high pressures is not simple. True hydrostatic pressure studies embracing pressure vessels imply appreciable gamma ray scattering and loss of resolution and efficiency in Doppler line-shape or lifetime measurements. The Bridgemann anvil technique provides a good approximation to the ideal hydrostatic pressure condition which is reasonably convenient and tenable in in situ source angular correlation experiments. DICKMAN et al. [3.116, 117] have employed this combination of techniques in pioneering studies of the vacancy formation volume in several metals. Linear Arrhenius plots of $(F - F_f)/(F_{1v} - F_f)$ vs p [cf. (3.21)] again establish the essential correctness of the simple trapping model and suggest a new and important application of the positron technique.

3.3 Nonequilibrium Studies

3.3.1 The "Many Defects" Problem

The effect of deformation on the angular distribution for a metallic system was first reported by DEKHTYAR et al. [3.118] in 1964. A narrowing of the angular distributions for Fe-Ni alloys was interpreted in terms of electronic structure changes brought about by the deformation. Contemporary [3.45] and later [3.46] reports of the essentially similar narrowing of angular distribution that accompanies heating in many metals (Sect.3.2) led BERKO and ERSKINE [3.119] to correctly interpret similar observations in plastically deformed aluminum in terms of positron trapping at dislocations and vacancies. Lifetime studies soon followed. Various modes of strong deformation in aluminum disclosed increases in mean positron lifetime [3.120-122] of as much as 40% and superficially somewhat more striking than the corresponding changes in angular distribution [3.119]. Lifetime spectrum analyses in aluminum [3.121] and in copper [3.123,124] suggested spectrum changes roughly consistent with the predictions of the simple two-state no-escape trapping model (Sect.3.1.2) and thus strong positron trapping by some or other of the various defects present.

Several of the earliest studies attempted to distinguish between the effects of vacancies and more complex positron traps. In [3.125] the narrowing of the angular distributions for platinum samples occasioned by vacancy inducing prior electron irradiation was compared with the similar but slightly more pronounced effect of sample deformation. Similar relative effects were also found in iron [3.126] wherein was also discerned the neutralization of vacancy traps by migrating carbon atoms. Annealing procedures in both deformed [3.126,127] and quenched [3.122,128] metals showed that a significant fraction of the initial angular distribution or lifetime changes annealed out at $T \sim T_m/2$ and thus could be attributed to single vacancies or small vacancy clusters. The remaining more "permanent" contribution can be plausibly attributed to dislocation trapping.

The extent to which positron studies can in general distinguish between various types of defect or trap is of course of considerable significance. Vacancies versus dislocations provide a useful and important basis for an initial discussion. The distinction between a point defect and a line defect is clear to most of us. Regrettably it has become clear that the positron may quite often be less discriminating.

The characteristic annihilation parameters corresponding to any particular defect state are most easily measured when saturated trapping in this state obtains. Lifetime measurements suggest that the characteristic lifetime or reciprocal rate for the positron dislocation state in most metals is not too different from that vor vacancies. Thus in aluminum the measured dislocation lifetime, τ_D, lies in the range 230-250 ps [3.120,122,123] as compared with the vacancy lifetime $\tau_V \sim 240$ ps (Chap.4). In copper, $\tau_D \sim 180$-190 ps [3.124,129], slightly smaller than the vacancy lifetime [3.130] but probably insufficiently different for real discrimination in other than the simplest situations. Angular correlation measurements on various metals [3.125-127,131] suggest the momentum distribution for positrons in vacancies is somewhat narrower than for dislocations. Here, however, the quantification of the effect must necessarily be more complex than in the lifetime case.

Fairly successful attempts at theoretical predictions of the momentum density curves for positrons in a defect have already been made for the particular cases of voids and for vacancies in aluminum (Chap.4). For less simple metals and more complex defect types such as dislocations adequate theories are still some way off and the analysis of experimental data is usually based on fairly empirical methods. Momentum density models involving admixtures of Gaussian [3.30] or admixtures of Gaussians and parabolas are common and today with the advent of fairly successful deconvolution procedures [3.38-40] can be applied to Doppler broadened energy curves [3.132,133] as well as the more traditional angular distributions [3.134]. The Doppler broadening technique has proved an invaluable asset in the rapid investigations of momentum distribution systematics in both annealed and deformed metals [3.36,38,53,129,135]. Momentum density parameters studied include the relative weights of the parabolic conduction electron and Gaussian core electron contributions and the position and relative sharpness of the Fermi cut-off (Chap.1) all of which can provide useful information about the nature of the underlying positron state. Extensive surveys of the effects of temperature and deformation on both momentum distributions and positron lifetimes in metals have been compiled by MACKENZIE and his colleagues. The momentum parameters obtained by line-shape deconvolution are in the main in good agreement with extensive but well dispersed angular correlation data. Further experimental work and refinements in data analysis are currently being undertaken [3.136] and a comprehensive library relevant to specific positron-defect complexes now seems not too distant.

When mere identification of a specific defect type, rather than a detailed picture of the trapped positron state is the essential requirement, the defect specific R-parameter [3.31,137] can be particularly useful.

In situations where one type of trapping site predominates we deduce from (3.17)

$$F = F_f(1 - P_t) + F_tP_t \tag{3.34}$$

where the subscripts f and t refer to the free and trapped positron states respectively. Now (3.34) applies equally to both the H (or S) and W parameters. On the other hand, the H and W parameters deriving as they do from quite different regions of the momentum density curve reflect rather different aspects of the positron states involved. Thus the ratio

$$R = \left|\frac{H - H_f}{W - W_f}\right| = \left|\frac{H_t - H_f}{W_t - W_f}\right| \qquad (3.35)$$

which is independent (in principle at least) of P_t and thus the actual trapping rate, nevertheless provides a parameter characterizing the particular trapping site in question. We shall return to R again in our discussion of annealing studies.

Thus far we have been largely concerned with the quantification of results corresponding to trapping dominated by a single type of defect center. The extension of quantitative analyses to situations involving simultaneous positron trapping by a variety of defect species is straightforward in principle but usually difficult in practice. Lifetime spectrum analyses involving several components of very similar rate seldom provide consistent results and as noted previously (Sect.3.1.3.) the separation of momentum density curves into specific defect components is made even more difficult by the generally far from simple functional form of these individual contributions. Nevertheless the momentum distributions form an indispensible part of the available information and quantitative analyses of both these and lifetime spectra can be valuable provided only that they are supported by intelligent sample treatment procedures based on well established metallurgical practices and pictures.

3.3.2 Deformation, Quenching, and Irradiation Experiments

Various methods of deformation have been employed for positron studies including simple compression [3.121,123,127], cold rolling [3.125,126,131-133,138], tensile stress [3.130,139,140], impact methods [3.129,141,142], and cyclic deformation [3.120,143]. The consequent increases in parameters such as t, H, or S show the following common pattern. At low deformations there is a strong but smooth dependence on strain. At larger deformations (5 → 10%) the dependence weakens to approach a saturation which although dependent on sample and treatment generally sets in at ~ 10 → 20% deformation [3.123,125,139]. Prior to saturation the observed changes are consistent with the predictions of the simple two-state trapping model [3.121, 123,124,132,140,141] which thus allows a measure of the relative positron population in trapped states albeit with some uncertainty as to the nature and variety of the effective traps.

Irrespective of the microscopic details, the remarkable ability of the positron technique to monitor the more gross features of the deformed metal is clear. A close correlation between the degree of positron trapping and sample hardness [3.138,142] suggests a potential application. LYNN et al. [3.143] have combined positron measurements on nickel and a nickel-cobalt alloy subjected to cyclic deformation with X-ray line broadening measurements of particle size. A saturation in positron lifetime is found to occur well before the saturation in X-ray particle size and the authors of [3.143] suggested the technique as "a useful and conservative indicator of fatigue damage". The potential of positron studies in the non-destructive testing field is clear [3.144] and considerable activity is now developing [3.145].

Analyses of data for metals in the as-deformed state in terms of dislocation trapping alone should always be viewed with caution because of the likelihood of other potential positron traps. LYNN et al. [3.140] have found that both the pre-deformation and subsequent deformation induced changes in positron lifetimes in copper depend on the initial grain size. LEIGHLY [3.146] has used their data to argue a linear relationship between grain boundary surface area per unit volume and the positron lifetime in annealed specimens. PENDRYS et al. [3.147] reported effects attributable to grain boundary trapping in evaporated copper and lead films. The very low temperature effects discussed in Sect.3.2.2 may be a further example of trapping in shallow grain boundary traps.

Many authors suggest appreciable trapping by vacancies or vacancy clusters in the as-deformed state [3.127,131,139]. In studies of aluminum at room temperature [3.119-121,133,134] such effects are understandably small [3.120,121]. In the case of higher melting point metals various studies have shown that an appreciable part of the initial effects of deformation can be removed by modest temperature anneals [3.126,127,130,131]. A comparison of positron and resistivity measurements by SAIMOTO et al. [3.139] suggests this soft component is probably due to small vacancy clusters rather than single vacancies.

Confirmation that the more permanent contribution to trapping in deformed metals indeed arises from dislocations emerges from both systematic annealing studies (Sect.3.3.3) and observed correlation between positron parameters and the degree of deformation. In fcc metals a quadratic dependence of the dislocation density on applied stress [3.139,141] and sometimes strain can be invoked. In a transition limited trapping regime (Chap.4) the trapping rate is proportional to the density of traps. Then a quadratic dependence of trapping rate on applied stress or resultant strain is implied. Such a dependence is found to be approximately the case for low deformations in the noble metals and nickel [3.129,132,139]. The utilization of various suggested quantitative relationships between dislocation density, hardness, stress and strain then allows for the quantitative estimation of positron-dislocation trapping rates and related quantities [3.129,131,132] via the simple two-state model.

A reasonable amount of comparative data exists in the case of copper. Comparison with the trapping rate data for vacancy trapping (Table 3.2) is facilitated by regarding a dislocation as a row of vacancylike centers [3.122,129,131,132]. Estimates of ν_D (cf. ν_{1v} Table 3.2) for copper obtained from positron lifetime [3.129,130] or Doppler broadening studies [3.129,131,132] are remarkably consistent at $\sim 2.5 \times 10^{15}\ s^{-1}$. Utilization of the simple collision cross-section idea through the equation

$\nu = \sigma\bar{v}\rho$

[3.19,122], where ρ is the defected site density [cm^{-3}] and $\bar{v}$ the positron thermal velocity can provide values for cross section σ plausibly $\sim 10^{-15}$ cm^2. The remarkable similarity of these dislocation parameters to the corresponding vacancy values is at first sight surprising but is generally believed to arise from the compensating effects of the long-range strain field around the dislocation and a final state in the dislocation core.

The picture of the positron trapped in the dislocation core gets more support from analyses of saturated dislocation trapping data. In an early analysis HAUTOJÄRVI [3.134] showed how quantities such as the positron kinetic, potential and binding energies could be estimated from lifetime and angular correlation data. He and others have shown [3.133,134] that the binding energies implied are $\sim$ 0.5-1.0 eV and incompatible, as are the lifetimes, with continuum models of dislocations (Chap.4).

Notwithstanding considerable effort, this author has never been able to conquer the theories of positron trapping at extended defects. As already noted above, a trapping probability proportional to the density of traps is a common assumption. BRANDT et al. [3.140] suggested a more complex dependence appropriate to diffusion limited trapping for dislocations in copper. Their analysis has some appeal but as yet their conclusions depend on a comparatively limited amount of data. The temperature dependence of the trapping rate is evidently a significant factor and has prompted several studies.

MACKENZIE et al. [3.148] have studied the temperature dependence of positron trapping in nickel from 100 k-300 K following various levels of deformation. A very weak positive temperature coefficient is found in material of 99.8% purity by none in 99.99% purity. A similar study in copper with essentially similar results has been reported by RICE-EVANS and HLAING [3.149]. A more recent study by this same group [3.150] over a larger low-temperature range shows an eventual broadening in line shape or apparent decrease in trapping as the sample temperature approaches 4.2 K from above. However, as MACKENZIE has recently pointed out [3.151] this, like the opposite effect observed in annealed specimens [3.37] may merely indicate thermal detrapping from shallow grain boundary traps.

Quenching and irradiation experiments can provide additional information about defect trapping rates. We have already noted in earlier sections investigations of vacancy trapping parameters by both quenching [3.78] and electron irradiation [3.79] treatment. The range of temperatures embraced in such studies is always restricted because of annealing effects. Nevertheless the latter effects can sometimes be useful. COTTERILL et al. [3.122] achieved controlled production of both vacancies and dislocations in single crystals of aluminum by careful quenching and annealing studies. Both MANTL et al. [3.79], and GAUSTER and DOLCE [3.152] combined neutron irradiation and annealing to isolate dislocation loop trapping. Evidence for both a significant dependence of line shape [3.79] and a trapping probability somewhat smaller than that found in deformed specimens [3.153] was found. A similar temperature dependence of line shape for small vacancy clusters is also reported in [3.79]. Annealing studies over the complete temperature range from the initially deformed, quenched, or irradiated state to full recovery, produce even more striking effects to which we now turn our attention.

3.3.3 Annealing Studies

Isochronal and isothermal annealing studies have now been performed in several metals. Recovery of deformed aluminum [3.154], copper [3.129,135], nickel [3.154,156], and iron [3.157] is dominated by a prominent recrystallization stage or possibly stages [3.158] whose position and sharpness depend on both the degree of deformation and the sample purity. Quantitative analyses emphasising both the degree of recrystallization at any temperature [3.160] or the rate of recrystallization [3.161] are possible.

Most experiments thus far reported have been confined to room temperatures and above and thus there is at present little information concerning the early stages of recovery in any deformed metal. However, the already noted ambiguity concerning vacancy and dislocation trapping effects suggests that extension of measurements to lower temperatures is unlikely to bring much reward.

If large deformation can be achieved without appreciable sample heating, vacancy effects may be more apparent at the higher temperatures [3.158]. In annealing stage III effects attributable to vacancy migration and clustering are sometimes observed. In pure ($\sim$ 5N) samples the annealing curves are relatively smooth [3.135, 148] but in less pure samples subjected to large deformation, effects attributable to the formation of vacancy clusters probably at impurity sites, can sometimes be discerned [3.159]. The critical role played by impurities in various aspects of the annealing process is considered in several works [3.155,158,159,162]. In [3.158, 159] long and detailed analyses of positron and other studies of nickel of various purities are used to argue the origin of recovery stages III and IV.

Annealing studies of neutron and electron irradiated metals are accompanied by equally detailed discussions. PETERSEN et al. [3.163-165], in a series of studies of neutron irradiated molybdenum, utilize quantitative analyses of both positron lifetime spectra and momentum density curves in investigations of the mechanisms underlying the various annealing stages. A progressive growth of vacancy clusters is shown to result eventually in the formation of large voids. Some of the interpretation of the results of such neutron irradiation studies is made difficult by the complexity of the initial damage produced. ELDRUP et al. [3.166] employed the "cleaner" method of electron irradiation in their molybdenum studies. Studies of stage III annealing again gave strong evidence of vacancy migration and clustering leading eventually to the formation of voids. More recent work in electron irradiated copper [3.135,167] and in quenched aluminum samples provides an essentially similar picture. The particularly transparent R-parameter analysis of copper given by MANTL and TRIFTHÄUSER [3.135] provides a useful demonstration of the current state of the art.

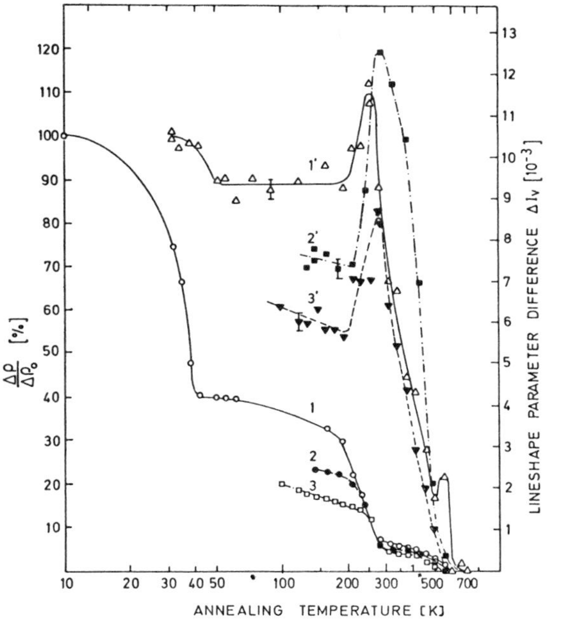

Fig.3.5. Line-shape parameter (upper curves) and resistivity (lower curves) developments following isochronal annealing of copper specimens subjected to electron irradiation of various (1,2,3 etc.) degrees of severity [3.135]

Isochronal annealing studies of copper subjected to various doses embrace both positron and complementary resistivity measurements. The positron measurements are performed at 10 K and the resistivity measurements at 4.2 K. In the case of the highest dose sample annealing commenced at 31 K in stage I. Here a decrease in both line shape I_V (roughly our H parameter) and in resistivity (Fig.3.5) is attributed [3.135] to a progressive initial loss of vacancies presumably due to recombination with migrating interstitials [3.168]. At lower doses a small decrease in I_V is also observed in stage II. More dramatic is the pronounced increase in I_V in stage III where the resistivity decreases sharply. As the authors [3.135] point out, if the positrons are here affected by the same type of traps as in stages I and II, I_V suggests a significant change in the nature of the dominant trapping sites.

This particular belief is indeed confirmed by the associated R-parameter measurements (Fig.3.6). During stages I and II the R-parameter remains constant and, as verified by supplementary studies, is appropriate to vacancy trapping. The increase in R in stage III is plausibly argued by MANTL and TRIFTHÄUSER to reflect the formation of three-dimensional vacancy clusters which eventually collapse into vacancy loops observable in electron microscope studies.

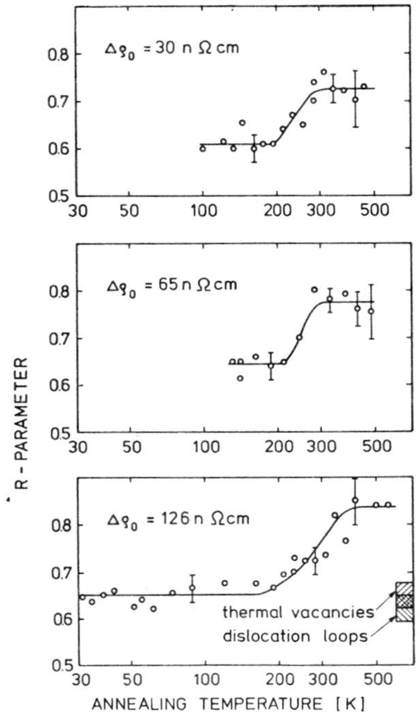

Fig. 3.6. R-parameter (3.35) changes in annealing stages II and III for electron irradiated copper specimens as in Fig. 3.5

A companion study of electron irradiated aluminum provides no evidence of the vacancy cluster effect notwithstanding its observation in quenching studies. In [3.169] WAMPLER and GAUSTER argued a strong dependence of the vacancy clustering effect on the density of vacancy sinks in the sample.

3.3.4 Positron Studies of Voids

The creation of voids in laboratory annealing studies of irradiated and otherwise defected metals is of considerable intrinsic interest but their accidental creation in other situations can have more serious technological implications. Void production in heavily irradiated metals at temperatures greater than about 0.4 T_m has serious implications in respect of the choice of reactor cladding materials in fast breeder reactors because it is invariably accompanied by swelling and eventual mechanical breakdown.

The first reports of the dramatic effect of positron trapping at voids appeared appropriately in "Nature" in 1972 in a related pair of letters [3.170,171]. Positron measurements on molybdenum samples neutron irradiated to fluences $\sim 10^{22}$ n cm^{-2} in the Dounreay high flux reactor, disclosed extreme narrowing (Fig.3.7) of the angular distributions [3.170] and lifetime components with $\tau \sim 500$ ps [3.171]. Up to that time such lifetimes and angular distributions had been exclusively associated with positronium (Chap.1) formation in insulators. The theoretical possibility of positronium formation in bulk metals was known to be remote but the possibility of positronium in large voids a priori could not be entirely ruled out. The suggestion of little or no positronium in a magnetic quenching study (see [3.4]) preserved the puzzle and other experimental and theoretical studies (Chap.4) were soon in evidence.

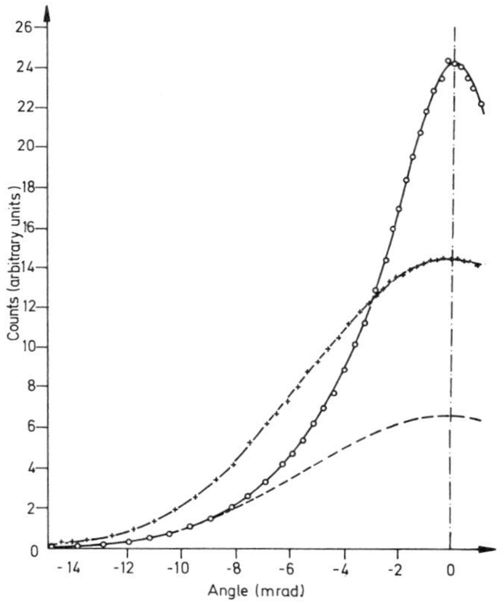

Fig. 3.7. Angular correlation curves for polycrystalling molybdenum samples containing voids (circles) and for annealed polycrystalline molybdenum (crosses). The two curves are normalized to equal areas. [3.170]

Void production by self-irradiation alpha particle damage in plutonium has been reported by GUSTAFSON and BARENS [3.172]. Voids in neutron irradiated aluminum [3.173-175] and nickel [3.176,177] show effects in lifetime [3.174,175] and in angular distribution [3.173-177] essentially the same as those in molybdenum. A search for voids in a specimen of actual reactor material has been carried out by KIM et al. [3.178]. A null result is attributed to the inhibition of vacancy clustering by vacancy sinks in the form of both the numerous silicon atoms in the Al-0.15% Si alloy and the high dislocation density in the initially cold worked metal.

The factor affecting the initial stages of void creation and growth and the extent to which the positron technique can be used to study these initial stages of growth clearly are of importance. The stabilization of void nucleation sites by gas atoms has been studied by SNEAD and GOLAND [3.179]. The annealing studies discussed in Sect.3.3.3 show a progressive increase in positron lifetime and narrowing of momentum density curves as the vacancy clusters grow. In electron irradiated molybdenum the initial value of lifetime for the relevant component is $\sim$ 200 ps [3.166] and is probably the characteristic monovacancy lifetime. In the neutron irradiated samples the value is closer to 300 ps [3.163,165]. In either case the τ values increase between 100 and 400°C to values close to 450 ps. The lifetime is then essentially constant until at $\sim$ 600°C it increases further to $\sim$ 600 ps. An interpretation of these final stages in terms of void growth and coarsening [3.166] is consistent with electron microscopy studies but hard to explain theoretically. Theoretical calculations of the dependence of positron parameters on void size [3.180,181] based on the surface trapped state concept originally introduced by HODGES and STOTT [3.182] provide realistic pictures of the momentum density curves [3.180] but predict a positron lifetime saturation $\tau \sim$ 450 ps for voids of order of 10 Å or greater size [3.180,181]. An explicit experimental investigation of various molybdenum specimens containing voids of a mean size varying from 9 to 45 Å diameter [3.183] discloses a constant value of the positron lifetime $\tau \sim$ 465 ps. The discrepancy between these and the earlier annealing experiment results [3.163,165,166] has been shown by THRANE and EVANS [3.184] to be almost certainly due to impurity effects in the annealing studies [3.163,165, 166]. For smaller void sizes (dia. $\lesssim$ 10 Å) the positron lifetime indeed reflects the void diameter and thus the positron technique appears to be a sensitive probe of small void growth and also eventual swelling [3.185].

Experimental and theoretical studies of the factors influencing positron trapping at voids have considerable significance both to positron trapping theories in general and to potential studies of void densities in heavily irradiated materials and in annealing situations. Experimental studies of the temperature dependence of Doppler broadened line shape in voided aluminum show a strong dependence [3.79] similar to that recently observed in positron lifetime [3.186]. A theoretical analysis also presented in [3.186] suggests both transition limited and diffusion limited regimes (Chap.4). As already implied in earlier paragraphs the behavior of positrons in the larger voids is essentially a surface problem and finds analogies in a large variety of other positron studies in powders, foils, etc. In some cases positronium (Chap.1) effects may appear. This is a rapidly and expanding area of positron physics we regard as lying outside our brief and accordingly refer the reader to the recent review by BRANDT [3.187].

3.4 Defect Studies in Alloys

3.4.1 Defect vs Impurity Problems

We have already noted at various points in preceding sections examples of work where arguable defect-impurity interactions have influenced and sometimes aided and deepened the interpretation of the results in question. The utility of the deliberate introduction of metallic and sometimes nonmetallic impurities into a sample is apparent in almost all areas of metal physics. In the extreme, the study of alloys as a function of composition, solute and solvent differences, etc., often brings considerable rewards in respect of our general understanding of some physical phenomenon notwithstanding an inevitable increase in the complexity of the actual microscopic processes involved.

Early studies of dilute solid solutions by DEKHTYAR et al. [3.188,189] disclosed effects in angular correlation for iron impurities in various hosts which were interpreted in terms of positron trapping by the magnetic iron impurity centers and local electronic structure. SEDOV et al. [3.190] investigated the effect of precipitations in single crystals of Al-0.36 at.% Zn alloys following quenching and ageing procedures. Effects in angular correlation were again interpreted in terms of positron trapping at zinc enriched zones and the local electronic environment. A comparison of the angular distribution for the initially quenched sample with those for samples allowed various periods of ageing showed that the large-angle parts of the angular distributions increased as the ageing progressed. SEDOV et al. interpreted the changes as due to the formation and development of precipitation or Guinier-Preston (GP) zones. However, as DOYAMA and HASIGUTI [3.2] have noted the relatively smaller core contribution in the angular distribution for the initially quenched sample can be equally well and perhaps even better explained in terms of positron trapping in the quenched defect population.

This uncertainty in interpretation can be easily understood. Positron trapping at lattice defects is, as we have already seen in earlier sections, a well-established phenomenon as also are pronounced defect-impurity interactions and affinities. Impurity atoms may also attract positrons and further confuse the issue. We find in these three and potentially even more numerous sets of implied interactions both the strength in potential applications and weakness in ultimate interpretation of positron alloy studies.

The relative importance of positron-defect versus positron-impurity effects in any system will clearly depend on both the relative concentrations of impurities and defects and the relative strengths of their positron affinities. The first ingredient lies within the control of the investigator and the second remains with the gods but can, to some extent, be studied in a systematic way.

In an alloy system, and here we include the very dilute alloy or metal with impurity system, the positron may show preference for one type of atom at the expense of the others. If such preference is for the minority species and the preference is particularly strong we have a positron trapping situation. The theory of the equilibrium positron density distribution in such systems is usually based on the positron pseudopotential approach [3.9,190]. The current theoretical situation is described in Chap.4. Here we need merely outline the general picture. The relative positron densities at A and B type cells in a binary AB alloy depends on the positron pseudopotential difference between the two types of cell. In the case of a uniform positron density throughout the alloy the positron annihilation parameters for a particular composition can be easily represented in terms of an annihilation rate weighted mixture of those for the constituent metals [3.191-193]. In the case of a positron preference for say A-type atoms the composition dependence is changed and the annihilation parameter values are biased towards those for the A-type metal. Experimental studies of the composition dependence of core electron annihilation rate [3.192] or angular distribution H parameter [3.193] show trends in general agreement with theoretical predictions based on calculated positron pseudopotentials. The theoretical calculations suggest that when solute atoms are well dispersed positron density enhancement [3.193] effects should be small [3.192, 194]. When more serious enhancement effects occur it is generally agreed that they probably arise from clustering or segregation effects [3.192-194]. Thus the ability of an isolated impurity atom to trap a positron is very much less than that of a monovacancy as might easily be argued a priori. On the other hand, where appreciable sized impurity clusters exist pronounced effects in annihilation parameters can be expected and indeed are often found.

DOYAMA et al. [3.195] have investigated the effect of a range of impurities in well-annealed dilute copper alloys (impurity content ~ 0.5 at.%). Transition metal impurities are found to have a much greater effect on angular distribution than impurity atoms such as germanium. A trapping model analysis suggests an impurity "trapping rate" some two orders of magnitude smaller than that for monovacancies consistent with the weaker trapping effect argued above. In this respect the trapping model approach is useful but its general inapplicability at higher impurity concentrations is also clear [3.187].

Most of the work discussed above has been concerned with specimens in which chemical clustering effect and lattice defect concentrations were minimized by appropriate sample pretreatment. Thus there, the positron-atom interactions could be isolated. In the section that follows we shall consider the developments that occur when vacancies are introduced by sample heating and interactions between this simple defect and the neighboring atoms play the dominant role. Finally in Sect.3.4.3 we shall turn to still more complex problems where all the various interactions between positrons, defects, and impurities and impurity clusters feature and must be considered in what have been as yet essentially qualitative interpretations.

3.4.2 Vacancy Studies

The extension of the methods and models presented in Sects.3.2.1-3.2.6 to binary alloy systems is straightforward in principle but somewhat harder in practice. In concentrated alloys there are as many types of monovacancy as there are vacancy environments. A vacancy affinity for a particular type of atom may imply a reduced positron affinity for the resulting vacancy centres [3.196]. The atomic disorder may also affect the positron trapping rate. The situation is a little easier in the case of dilute alloys where a significant vacancy-impurity binding energy may be the dominant effect.

The vacancy-impurity binding energy concept is most easily applied to very dilute situations where the probability of more than a single impurity atom next to a vacancy is vanishingly small. In such cases [3.197] the monovacancy concentration contains two parts. The normal or unassociated vacancy concentration

$$c_v = (1 - Zc_i)\exp(S^F_{1v}/k)\exp(-H^F_{1v}/kT) \tag{3.36}$$

where Z is the coordination number for the structure in question, c_i is the impurity concentration, and S^F_{1v} and H^F_{1v} are as before (Sect.3.2.1). The vacancy-impurity pair concentration

$$c_{v,i} = Zc_i \exp[(S^F_{1v} - S_B)/k]\exp[(E_B - H^F_{1v})/kT] \quad , \tag{3.37}$$

where S_B and E_B are the vacancy-impurity binding entropy and energy, respectively. Now in general the specific trapping rates ν_v and $\nu_{v,i}$ are different [3.69,198, 199] for the two vacancy types. Thus, in principle, a three-state trapping model is required. Some authors, recognising the fraily of three-state trapping model analyses, have made the approximation $\nu_v = \nu_{v,i}$ [3.89], others [3.69,199] embrace more complex possibilities.

SNEAD et al. [3.69,201] have applied analyses of varying complexitiy to their lifetime data for an Al-1.7 at.% Zn alloy. Vacancy-impurity energies varying from -0.09 to +0.04 eV are deduced. In Al-0.27 at.% Si KIM et al. [3.200] deduced $E_B = 0 \rightarrow 0.12$ eV from angular correlation data. DOYAMA et al. [3.199] get $E_B \sim 0.27$ eV for the case of germanium atoms in copper.

The real significance of these different results is somewhat hard to assess. The results from the somewhat simpler pure metal measurements suggest an overall reproducibility in formation enthalpy values $\sim \pm 10\%$. Thus in this respect only the result of [3.199] can be regarded as firm evidence of a vacancy-impurity binding effect. However, the complexity of the analysis adopted in [3.199] may even here advise against too firm a conclusion.

Even if vacancy-impurity binding effects are hard to establish in equilibrium studies they are clearly in evidence in quenching studies. DOYAMA et al. [3.196, 198,202] have studied the effect of quenching in various dilute copper based alloys. In [3.202] quenching of Cu-0.5 at.% Mn is followed by angular correlation studies employing the short-slit geometry (Chap.2). An apparent extinction of the characteristic copper <111> zone face neck structure is interpreted in terms of positron localization at vacancy-Mn impurity pairs formed during the quenching process. Further studies of other dilute copper based alloys [3.196,198] suggest that the possibility of identifying this vacancy-impurity effect depends on the position of the solute element relative to copper in the periodic table. Thus when Cu-0.6 at.% Ni is quenched the resulting angular distribution differs from that of copper in a manner consistent with positron localization at the impurity atoms (Fig.3.8). Similar results for Cu-Cr and Cu-Fe alloys are also noted [3.198]. For solute atoms to the right of copper in the periodic table the interpretation is more difficult. Thus in Cu-Zn and Cu-Ge (both of 0.5 at.% impurity) the narrowing of the angular distributions (Fig.3.8) is merely indicative of positron localization somewhere but the particular environment is not clear. In the case of Cu-Ge [3.199] and Cu-Al [3.198] dilute alloys equilibrium studies do suggest vacancy-impurity binding effects.

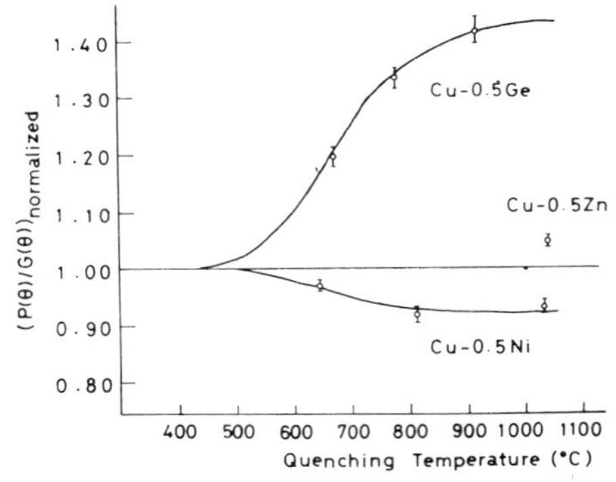

Fig.3.8. The ratio of the fractions of the parabolic part $P(\theta)$ ot the Gaussian part $G(\theta)$ of angular correlation curves for copper alloys quenched from various temperatures [3.196]. Used with kind permission of Pergamon Press Ltd and authors

Formation energies and vacancy concentrations in concentrated alloys have relevence to the metallurgical treatment of commercial alloys. Here (3.36,37) are quite inadequate because of the infinity of potential vacancy types. DOYAMA and KOEHLER [3.203] analysed the problem of the apparent formation energy in random alloys with the aid of the chemical bond concept. One may deduce from their work that the apparent formation energy [3.204]

$$E^F_{v,(AB)} = c_A E^F_{v,A} + c_B E^F_{v,B} + Zc_Ac_BW_{AB} \quad . \qquad (3.38)$$

$$W_{AB} = E_{AB} - 1/2(E_{AA} - E_{BB}) \qquad (3.39)$$

where the $E^F_{v,i}$ and c_i have their usual meaning and E_{AB}, E_{AA}, E_{BB} are the pairwise energies of interaction between A-B, A-A, and B-B atom pairs, respectively. $Zc_Ac_BW_{AB}$ is the heat of solution and is positive for clustering-type alloys and negative for alloys having a tendency to order. The analysis given in [3.203] also implies, as might be expected from the simple bond picture embraced, linear relationships between $E^F_{v,A}$ and other manifestations of the atomic binding such as cohesive energy and melting temperature.

If, in (3.38) W_{AB} is zero, the apparent vacancy formation energy changes linearly with composition. FUKUSHIMA and DOYAMA [3.205] have studied various concentrated copper alloys. Their initial report on the Cu-Ge system suggests a composition dependence of E^F_v which is essentially linear. SUEOKA [3.206] has also reported a linear E^F_v vs composition relation for α phase Cu-Al alloys. However, DOYAMA [3.204] has recently reported nonlinearities in both the Cu-Ge and the Cu-Ni systems.

The work of [3.205] discloses various interesting correlations between E^F_v and other simply measured or deduced alloy parameters. A semi-quantitative relationship between E^F_v and melting temperature is clearly in evidence in the phase diagram plots of Fig.3.9. A further relation between E^F_v and electron to atom ratio, e/a, is shown in Fig.3.10. The resulting linear relationship deduced

$$E^F_v = 1.24 - 1.22\ \Delta Z$$

where ΔZ is the alloy–solvent valence difference [3.204] indicates the importance of valency difference in such alloy problems.

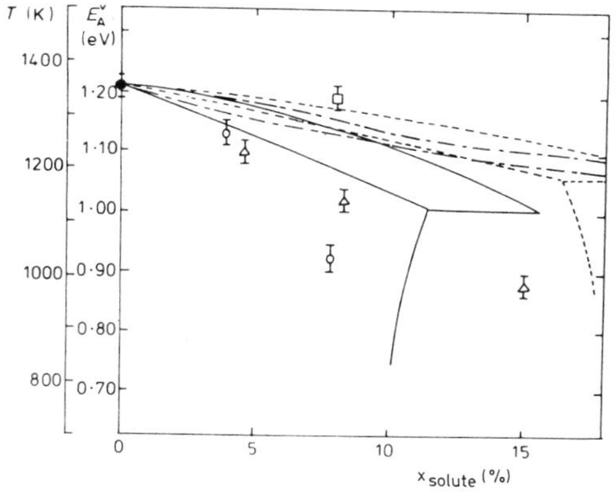

Fig.3.9. The apparent formation energies E^v_A and phase diagrams of the Cu-Si (circles), Cu-Ga (triangles) and Cu-Mn (squares) systems. [3.205]. Copyright The Institute of Physics

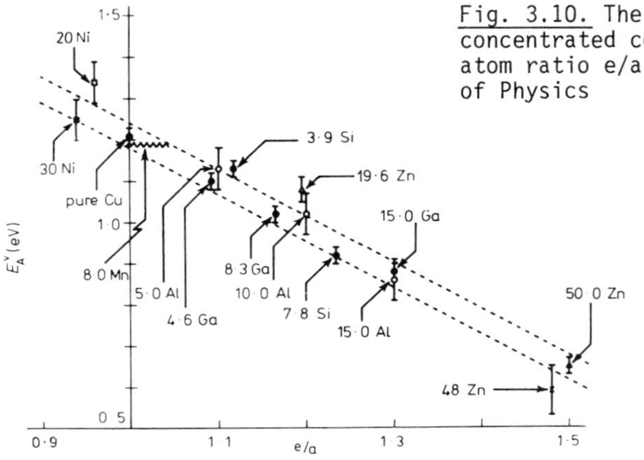

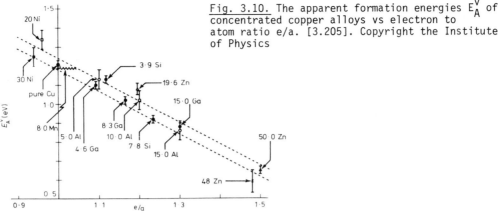

Fig. 3.10. The apparent formation energies E_A^V of concentrated copper alloys vs electron to atom ratio e/a. [3.205]. Copyright the Institute of Physics

Some of the data used in Fig.3.10 derives from work by MACKENZIE et al. [3.207] and as noted in [3.205] superficially suggests that the same relation between E_V^F and e/a persists across phase boundaries. However, more recent studies by the same group [3.208] suggest less simple behavior.

Throughout the α-phase (Fig.3.11) the form of the momentum density "weighted W parameter" vs temperature curves for the Cu-Zn system changes smoothly with increasing zinc concentration and a linear decrease for both T_c and H_{1v}^F is observed. The 40 wt.% Zn and 45 wt.% Zn alloys are mixed phase specimens with almost identical values of T_c and H_{1v}^F but with radically different vacancy saturated high-temperature line shapes. However, if this effect is provocative even more so is the curve for the 48 wt.% Zn alloy. A variety of tentative explanations of the anomalous 20-100° C part of the curve ranging from structural vacancy to antiphase boundary effects are considered in [3.208] and also feature in the interpretation of other phase boundary studies.

3.4.3 Phase Transitions and Boundary Effects

The possibility that positron measurements might be sensitive to phase transitions was recognised by DEKHTYAR et al. [3.209] as long ago as 1963 and long before the birth of the defect application. Using the angular correlation technique they investigated the ordered and disordered states of CuAu, Cu_3Au, and Ni_3Mn. In CuAu and Ni_3Mn small differences between the angular distributions for the ordered and disordered phases were found while in Cu_3Au no effect could be discerned. Subsequent studies of Cu_3Au by MORINAGA [3.210] who used the comparatively sensitive rotating specimen angular correlation technique (Chap.2) were more successful in that momentum density changes attributable to the additional Brillouin zone boundaries implied for the ordered alloy were then resolved.

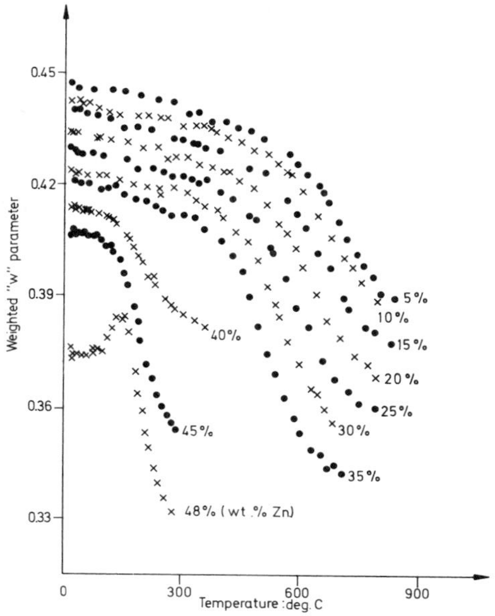

Fig.3.11. The temperature dependence of a weighted W parameter for Cu-Zn alloys. [3.208]

We have already seen in previous sections how the effects of impurities, structure, composition, etc., on positron parameters become much greater when accompanied by defect trapping phenomena occasioned by heating, quenching, or other defect inducing techniques. This pattern persists in the case of phase transitions.

DOYAMA et al. [3.211], and KURIBAYASHI et al. [3.212] have found anomalous effects in the temperature dependence of H parameter attributable to order-disorder transitions in the Cu_3Au, CuZn, and Cu-Pd systems. In an Cu_3Au specimen a small inflection or hump is found in the usual s-shaped vacancy trapping curve close to the order-disorder temperature. In CuZn and Cu-Pd alloys there is a much sharper structure suggesting a discontinuous change in H. In [3.212] the different effects could be apparently classified according to the order of the transition [3.213]. However, a more recent study of much better resolution and detail shows that this simple association is not universal and the effects observed may involve a variety of different defect trapping phenomena.

The simplest considerations suggest that the kinetics of most phase transformations are controlled by the atomic diffusion process [3.214]. Thus the vacancies present at equilibrium directly affect the transition. S-curve vacancy trapping studies, again by DOYAMA and his colleagues, reflect vacancy influences at both eutectoid decomposition [3.215,216] and precipitation temperatures in the Cu-Al and Al-Zn zinc systems. Similar s-curve anomalies are found at martensitic transformations [3.217,218] and pretransformation and hysteresis effects stimulate varied and complex discussions.

An assessment of the discussions and analyses presented in the various papers noted above is beyond the competence of this author. Nevertheless it is clear that such studies contain a potentially enormous wealth of detail about technologically important materials and metallurgical processes. Today, with the entry of a considerable number of competent metallurgists into the positron field, considerably growth in such specialized but pertinent studies can be expected. The recent study by MEURTIN and LESBATS [3.219] of anomalous vacancy concentrations in Fe-Al alloys is a case in point. Nevertheless interesting discoveries can still be made in a state of blissful ignorance. Mg-In alloys of $\sim$ 30 at.% Mg are unstable and exhibit catastrophic corrosion, indium whisker growth and eventual total mechanical breakdown in periods ranging from a few weeks to a few months following preparation. Positron measurements made before the appearance of significant macroscopic effects disclose anomalies which provide an early indicator of the subsequent material breakdown and additional insight into the microscopic situation [3.220]. Thus the next author in this field could well be required to end his account with a section on corrosion. Today we conclude with a brief discussion of the liquid and amorphous state.

3.5 Liquid and Amorphous Metals

Early studies of the effects of melting involved angular correlation measurements [3.46,221-224] and were to a large extend prompted by electronic structure considerations and in particular the effect of disorder on the conduction electron states [3.225]. Significant changes in angular distribution on melting were found in a variety of metals including tin, gallium, mercury and bismuth. The changes, which are most pronounced in mercury, (Fig.3.12) are qualitatively similar to those arising from vacancy trapping in solid metals. A logical association of these effects with positron trapping in low-density vacancy-like regions of the disordered liquids was soon supported by the complementary observation of appropriately large changes in positron lifetimes in mercury [3.226] and in gallium [3.227] and comparisons of the similar effects of irradiation and melting of copper [3.228].

The analogy with vacancy trapping is made clearer by a wider survey. Metals that demonstrate significant vacancy trapping effects in the solid usually show relatively minor additional changes on melting [3.4]. The alkali metals show little or no change on either heating or melting [3.46,229] other than smearing attributable to purely electronic effects [3.4]. Thus large melting effects only occur in metals (and particularly semimetals) where vacancy trapping might be expected but is either missing or very weak in the solid phase.

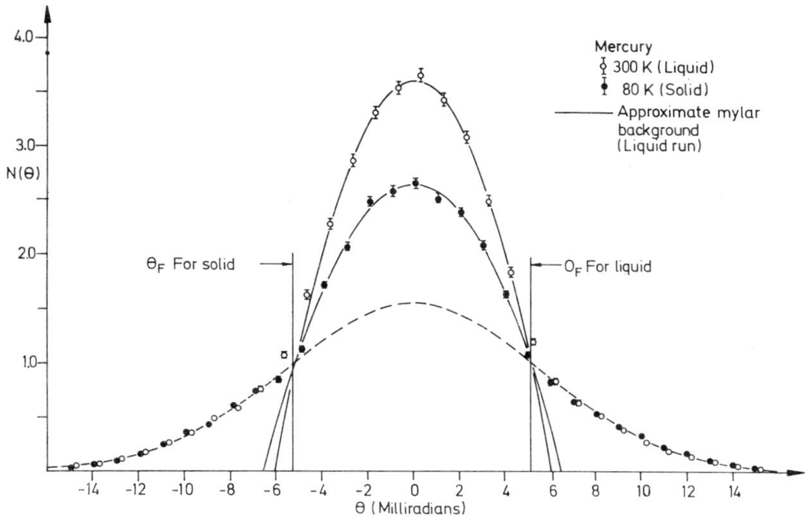

Fig.3.12. Long-slit angular distribution for solid and liquid mercury [3.222]

When large changes occur they are, as already noted above essentially similar to those produced by vacancy trapping. The smearing and narrowing of the conduction electron part of the angular distribution is as observed in vacancy trapping and may be similarly interpreted in terms of positron localization and local density effects (Chap.4). Early investigations suggested a close similarly between the core distributions for each phase [3.222,223] but subsequent studies have shown differences attributable to either positron localization in the simplest sense or to more subtle positron wave-function effects [3.65,230].

The nature of the effective traps is of considerable interest but is as yet far from clear. The investigation of positron traps in solid metals is made somewhat easier by the persistence of these traps relative to positron lifetimes. In liquid metals, a size distribution of potential positron traps [3.4,231], their transient nature, and the possibility that the positron may not merely perturb the traps but may even create them [3.4,232,233] all need consideration.

Positron lifetime spectrum analyses can in principle provide information about the spectrum of positron traps but always require great care. Early studies of positron lifetime in mercury gave very variable results [3.226,234,235]. Subsequent comparative studies [3.236] involving both conventional sandwich sources and homogenised Hg-^{22}Na amalgam resolved some of the problems and allowed fairly realistic spectrum analyses at least for the solid phase. For the solid amalgam an essentially single component spectrum was found. Multicomponent (1-3) spectrum analyses for the liquid amalgam consistently failed, suggesting a more complex type of spectrum perhaps of the form (3.2). However, more recent lifetime studies of liquid indium by the author and his colleagues, albeit marred by a more conventional source geometry, suggest a single component spectrum in the liquid phase.

Lifetime studies in liquid metals and particularly comparisons of solid and liquid spectra are made difficult by source-sample interface problems. Wetting, or meniscus effects and the high reactivity of most liquid metals demand particular care in source-sample assembly design. Should these problems be satisfactorily overcome, lifetime studies embracing a reasonable temperature range could provide significant information.

Naive considerations suggest, that at sufficiently high temperatures, where both the positron and the surrounding atoms are very hot, the lifetime spectrum should become single component as in (3.1). At lower temperatures lifetime spectrum analyses might distinguish between positron trapping in a wide variety of different sized and comparatively short-lived traps (3.2) or pronounced stabilization or creation of some ideal trap as evidenced by a single component spectrum. Again at high temperatures one might expect the annihilation rate to have decreased from a traplike value to something closer to that appropriate to the mean density of the system. Both lifetime [3.227,236] and angular correlation studies [3.31] give some suggestion of a movement towards a weaker trapping effect as the temperature is raised although one study in gallium [3.231] suggests an opposite trend. A single component spectrum at low temperatures close to T_m would be strongly suggestive of positron self-trapped states and whether present or not might well qualify similar hypothesis about the solid phase (Sect.3.2.2). An adequate theory of positron trapping in liquid metals must clearly be quite complex. ITOH [3.137] has made a brave first step.

Positron studies in liquid alloys are still today quite rare. An early study in a bismuth-mercury alloy [3.223] provided an angular distribution of a form very close to that of a composition weighted mixture of the pure metal curves. Thus no particular positron affinity for one of the two components was in evidence. Subsequent studies of liquid Cu-Bi and Cu-Sn alloys [3.237] have nevertheless shown large positron affinities for bismuth and tin attributable to the atomic arrangement in these alloys. Recent studies of Ge-noble and transition metal alloys show complex trends in angular distribution [3.240] more suggestive of electronic structure influences than positron affinity effects.

Amorphous or glassy alloys have been studied by several groups [3.233,238,239]. In all cases the angular distributions for the amorphous solid state are very similar to those for the crystalline state. This is clearly a very different situation from that obtaining in liquid metals and as DOYAMA [3.233] has pointed out is very suggestive of positron "hole-digging" effects in the liquid disordered and mobile phase.

References

3.1 A. Seeger: J. Phys. F *3*, 248 (1973)
3.2 M. Doyama, R. Hasiguti: Cryst. Lattice Defects *4*, 139 (1973)
3.3 I.Ya. Dekhtyar: Phys. Rep. C *9*, 243 (1974)
3.4 R. N. West: Adv. Phys. *22*, 263 (1973)
3.5 H.J. Mikeska: Z. Phys. *232*, 159 (1970)
3.6 A. Perkins, J.P. Carbotte: Phys. Rev. B *1*, 101 (1970)
3.7 B. Bergersen, E. Pajanne: Appl. Phys. *4*, 25 (1974)
3.8 P. Kubica, A.T. Stewart: Phys. Rev. B *11*, 2490 (1975)
3.9 M.J. Stott, P. Kubica: Phys. Rev. B *11*, 1 (1975)
3.10 C.H. Hodges, M.J. Stott: Phys. Rev. B *7*, 73 (1973)
3.11 C.H. Hodges: Phys. Rev. Lett. *25*, 284 (1970)
3.12 P. Kirkegaard, M. Eldrup: Computer Phys. Commun. *3*, 240 (1972); *7*, 401 (1974)
3.13 W. Brandt, R. Paulin: Phys. Rev. B *5*, 2430 (1972)
3.14 W. Frank, A. Seeger: Appl. Phys. *3*, 66 (1974)
3.15 M. Berlotaccini, A Dupasquier: Phys. Rev. B *1*, 2896 (1970)
3.16 D.C. Connors, V.H.C. Crisp, R.N. West: J. Phys. F *1*, 355 (1971)
3.17 W. Brandt: In *Positron Annihilation*, ed. by A.T. Stewart, L.O. Roellig (Academic Press, New York 1967) p.155
3.18 B. Bergersen, M.J. Stott: Solid State Commun. *7*, 1203 (1969)
3.19 D.C. Connors, R.N. West: Phys. Lett. A *30*, 24 (1969)
3.20 M. Bertolaccini, A. Bisi, G. Gambarani, L. Zappa: J. Phys. C *4*, 734 (1971)
3.21 C. Lanzcos: *Applied Analysis* (Pitman, London 1957) p.272
3.22 V.H.C. Crisp, I.K. MacKenzie, R.N. West: J. Phys. E *6*, 1191 (1973)
3.23 L.C. Smedskjaer, M.J. Fluss, M.K. Chason, D.G. Legnini, R.W. Siegel: J. Phys. F *7*, 1261 (1977)
3.24 D. Herlach, K. Maier: Appl. Phys. *11*, 199 (1976)
3.25 V.H.C. Crisp, D.G. Lock, R.N. West: J. Phys. F *4*, 830 (1974)
3.26 S.C. Sharma, S. Berko, W.K. Warburton: Phys. Lett. A *58*, 405 (1976)
3.27 M.J. Fluss, L.C. Smedskjaer, M.K. Chason, D.G. Legnini, R.W. Siegel: To be published
3.28 M. Eldrup, Y.M. Huang, B.T.A. McKee: Proc. 4th Intern. Conf. on Positron Annihilation (Helsingør 1976) unpublished
3.29 D. Knapton, B.T.A. McKee: Proc. 3rd Intern. Conf. on Positron Annihilation (Kingston 1971) unpublished
3.30 P. Kirkegaard, O. Mogensen: Risø-M Rpt. 1615 (1973)
3.31 W. Trifthäuser: Phys. Rev. B*12*, 4634 (1975)
3.32 W. Trifthäuser, J.D. McGervey: Appl. Phys. *5*, 177 (1974)
3.33 I.K. MacKenzie: Phys. Lett. A *30*, 115 (1969)
3.34 I.K. MacKenzie, J.A. Eady, R.R. Gingerich: Phys. Lett. A *33*, 279 (1970)
3.35 I.K. MacKenzie, P.C. Lichtenberger, J.L. Campbell: Can. J. Phys. *52*, 1389 (1974)
3.36 Stef. Charalambous, M. Chardalas, Sp. Dedoussis: Phys. Lett. A *59*, 235 (1976)
3.37 D. Herlach, H. Stoll, W. Trost, H. Metz, T.E. Jackman, K. Maier, H.E. Schaefer, A. Seeger: Appl. Phys. *12*, 59 (1977)
3.38 T.E. Jackman, P.C. Lichtenberger, C.W. Schulte: Appl. Phys. *5*, 259 (1974)
3.39 S. Dannefaer, D.P. Kerr: Nucl. Instrum. Methods *131*, 119 (1975)
3.40 K.G. Lynn, J.R. MacDonald, R.A. Boie, L.C. Feldman, J.D. Gabbe, M.F. Robbins, E. Bonderup, J. Golovchenko: Phys. Rev. Lett *38*, 241 (1977)
3.41 J.L. Campbell: Appl. Phys. *13*, 365 (1977)
3.42 K. Petersen, N. Thrane, R.M.J. Cotterill: Philos. Mag. *29*, 1 (1974)
3.43 I.K. MacKenzie: Private Communication
3.44 D.P. Kerr, P.D. Fellows, D.J. Sullivan, R.N. West: Phys. Lett. A *61*, 418 (1977)
3.45 I.K. MacKenzie, G.F.O. Langstroth, B.T.A. McKee, C.G. White: Can. J. Phys. *42*, 1837 (1964)
3.46 J.H. Kusmiss, A.T. Stewart: Adv. Phys. *16*, 63 (1967)

3.47 I.K. MacKenzie, T.L. McKhoo, A.B. McDonald, B.T.A. McKee: Phys. Rev. Lett. *19*, 946 (1967)
3.48 B.T.A. McKee, W. Trifthäuser, A.T. Stewart: Phys. Rev. Lett. *28*, 358 (1972)
3.49 B.T.A. McKee, A.G.D. Jost, I.K. MacKenzie: Can. J. Phys. *50*, 415 (1972)
3.50 C.L. Snead, T.M. Hall, A.N. Goland: Phys. Rev. Lett. *29*, 62 (1972)
3.51 J.D. McGervey, W. Trifthäuser: Phys. Lett. A *44*, 53 (1973)
3.52 S. Nanao, K. Kuribashi, S. Tanigawa, M. Mori, M. Doyama: J. Phys. F *3*, L5 (1973)
3.53 T.E. Jackman, C.W. Schulte, J.L. Campbell, P.C. Lichtenberger, I.K. MacKenzie, M.R. Wormald: J. Phys. F *4*, L1 (1974)
3.54 J.L. Campbell, C.W. Schulte: Phys. Lett A *58*, 335 (1976)
3.55 C.W. Schulte, J.L. Campbell, R.R. Gingerich: Nucl. Instrum. Methods *138*, 647 (1978)
3.56 K. Maier, H. Metz, D. Herlach, H.E. Schaefer, A. Seeger: Phys. Rev. Lett. *39*, 484 (1977)
3.57 W. Trifthäuser, J.D. McGervey: Appl. Phys. *6*, 177 (1975)
3.58 H.C. Jamieson, B.T.A. McKee, A.T. Stewart: Appl. Phys. *4*, 79 (1974)
3.59 P. Rice-Evans, Tin Hlaing, I. Chaglar: Phys. Lett. A *60*, 368 (1977)
3.60 P.C. Lichtenberger, C.W. Schulte, I.K. MacKenzie: Appl. Phys. *6*, 305 (1975)
3.61 A. Seeger: Appl. Phys. *7*, 85 (1975)
3.62 S.M. Kim, W.J.L. Buyers: J. Phys. F *6*, L67 (1976)
3.63 K.P. Singh, R.N. West: J. Phys. F *6*, L267 (1976)
3.64 D. Segers, L. Dorikens-Vanpraet, M. Dorikens: Appl. Phys. *13*, 51 (1977)
3.65 M.J. Stott, R.N. West: J. Phys. F *8*, 635 (1978)
3.66 J.D. McGervey, P. Sen, I.K. MacKenzie, T. McMullen: J. Phys. F *7*, L255 (1977)
3.67 D. Segers, C. Dauwc, M. Dorikens, L. Dorikens-Vanpraet: Appl. Phys. *10*, 121 (1976); Phys. Status Solidi (a) *40*, 153 (1977)
3.68 T. McMullen, B.T.A. McKee: Private Communication
3.69 T.M. Hall, A.N. Goland, C.L. Snead: Phys. Rev. B *10*, 3062 (1974)
3.70 W.K. Warburton, M.A. Schulman: Phys. Lett. A *60*, 448 (1977)
3.71 K.G. Lynn: Private Communication
3.72 A. Seeger: Phys. Lett. A *40*, 135; A *41*, 267 (1972)
3.73 C.H. Hodges: J. Phys. F *4*, L230 (1974)
3.74 B. Bergersen, D.W. Taylor: Can. J. Phys. *52*, 1594 (1974)
3.75 T. McMullen, B. Hede: J. Phys. F *5*, 669 (1975)
3.76 T. McMullen: J. Phys. F *7*, 3041 (1977)
3.77 B.T.A. McKee, H.C. Jamieson, A.T. Stewart: Phys. Rev. Lett. *31*, 634 (1973)
3.78 T.M. Hall, A.N. Goland, K.C. Jain, R.W. Siegel: Phys. Rev. B *12*, 1613 (1975)
3.79 S. Mantl, W. Kesternich, W. Trifthäuser: J. Nucl. Mater. *69-70*, 593 (1978)
3.80 A. Seeger: Appl. Phys. *4*, 183 (1974)
3.81 O. Sueoka: J. Phys. Soc. Jpn. *36*, 464 (1974); *37*, 875 (1974)
3.82 S. Nanao, K. Kuribayashi, S. Tanigawa, M. Doyama: J. Phys. F *3*, L225 (1973)
3.83 S. Nanao, K. Kuribayashi, S. Tanigawa, M. Doyama: J. Phys. F *7*, 1403 (1977)
3.84 G. Dlubek, O. Brümmer, N. Meyendorf: Appl. Phys. *13*, 67 (1977)
3.85 G. Dlubek, O. Brümmer, N. Meyendorf: Phys. Status Solidi (a) *29*, K95 (1977)
3.86 M. Manninien, R. Nieminen, P. Hautojärvi, J. Arponen: Phys. Rev. B *12*, 4012 (1975)
3.87 A.N. Goland, T.M. Hall: Phys. Lett. A *45*, 397 (1973)
3.88 S.W. Tam: J. Met. *27*, A23 (1975)
3.89 S.W. Tam, R.W. Siegel: J. Phys. F *7*, 877 (1977)
3.90 K. Kuribayashi, M. Doyama: J. Phys. F *5*, L92 (1975)
3.91 I.K. MacKenzie, T.W. Craig, B.T.A. McKee: Phys. Lett. A *36*, 227 (1971)
3.92 G.M. Hood, R.J. Schultz, G.J.C. Carpenter: Phys. Rev. B *14*, 1503 (1976)
3.93 A. Seeger, H. Mehrer: In *Vacancies and Interstitials in Metals*, ed. by A. Seeger, D. Schumacher, W. Schilling, J. Diel (North-Holland, Amsterdam 1970) pp.1-54
3.94 G.M. Hood, R.J. Schultz: J. Nucl. Mater. *69-70*, 607 (1978)
3.95 P.D. Fellows: Private Communication
3.96 S.M. Kim, W.J.L. Buyers: Phys. Lett. A *49*, 181 (1974)
3.97 R.P. Gupta, R.W. Siegel: Phys. Rev. Lett. *39*, 1212 (1977)
3.98 S.M. Kim, W.J.L. Buyers, P. Martel, G.M. Hood: J. Phys. F *4*, 343 (1974)

3.99 K.P. Singh, G.S. Goodbody, R.N. West: Phys. Lett. A *55*, 237 (1975)
3.100 Sp. Dedoussis, Stef. Charalambous, M. Chardalas: Phys. Lett. A *62*, 359 (1977)
3.101 P. Rice-Evans, Tin. Hlaing, D.B. Rees: J. Phys. F *6*, 1079 (1976)
3.102 J.L. Campbell, C.W. Schulte, J.A. Jackman: J. Phys. F *7*, 1985 (1977)
3.103 C.W. Schulte, P.C. Lichtenberger, R.R. Gingerich, J.L. Campbell: Phys. Lett. A *41*, 305 (1972)
3.104 J.L. Campbell, T.E. Jackman, I.K. MacKenzie, C.W. Schulte, C.G. White: Nucl. Instrum. Methods *116*, 369 (1974)
3.105 J.L. Campbell, C.W. Schulte, D.K. Dieterly: Appl. Phys. *6*, 327 (1975)
3.106 C.W. Schulte, J.L. Campbell, J.A. Jackman: Appl. Phys. *16*, 29 (1978)
3.107 S.M. Kim, W.J.L. Buyers: J. Phys. F *8*, L103 (1978)
3.108 A. Seeger: Cryst. Lattice Defects *4*, 221 (1973)
3.109 C. Dauwe, M. Dorikens, L. Dorikens-Vanpraet: Appl. Phys. *5*, 45 (1974)
3.110 D.P. Kerr, D.J. Sullivan, P.D. Fellows, R.N. West: To be published
3.111 G.M. Hood: Private Communication
3.112 A.C. Damask, G.J. Dienes: *Point Defects in Metals* (Gordon and Breach, London 1963)
3.113 G.M. Hood, B.T.A. McKee: J. Phys. F *8*, 1457 (1978)
3.114 K. Kuribayashi, S. Tanigawa, S. Nanao, M. Doyama: Solid State Commun. *12*, 1179 (1973)
3.115 I.K. MacKenzie, P.C. Lichtenberger: Appl. Phys. *9*, 3311 (1976)
3.116 J.E. Dickman, R.N. Jeffrey, D.R. Gustafson: Phys. Rev. B *16*, 3334 (1977)
3.117 J.E. Dickman, R.N. Jeffrey, D.R. Gustafson: J. Nucl. Mater. *69-70*, 604 (1978)
3.118 I.Ya. Dekhtyar, D.A. Levina, V.S. Mikhalenkov: Dokl. Akad. Nauk SSSR *156*, 795 (1964). Sov. Phys.-Doklady *G*, 492 (1964)
3.119 S. Berko, J.S. Erskine: Phys. Rev. Lett. *19*, 307 (1967)
3.120 J.C. Grosskreutz, W.E. Millet: Phys. Lett. A *28*, 621 (1969)
3.121 P. Hautojärvi, A. Tamminen, P. Jauho: Phys. Rev. Lett. *24*, 459 (1970)
3.122 R.M.J. Cotterill, K. Petersen, G. Trumpy, J. Träff: J. Phys. F *2*, 459 (1972)
3.123 P. Hautojärvi, P. Jauho: Acta Polytech. Scand. Phys. *98*, 1 (1973)
3.124 K. Hinode, S. Tanigawa, M. Doyama: J. Phys. Soc. Jpn. *39*, 545 (1975)
3.125 J.H. Kusmiss, C.D. Esseltine, C.L. Snead, A.N. Goland: Phys. Lett. A *32*, 175 (1970)
3.126 C.L. Snead, A.N. Goland, J.H. Kusmiss, H.C. Huang, R. Meade: Phys. Rev. B *3*, 275 (1971)
3.127 K. Kuribayashi, S. Tanigawa, S. Nanao, M. Doyama: Phys. Lett. A *40*, 27 (1972)
3.128 D.C. Connors, V.H.C. Crisp, R.N. West: Phys. Lett. A *33*, 180 (1970)
3.129 C. Dauwe, M. Dorikens, L. Dorikens-Vanpraet, D. Segers: Appl. Phys. *5*, 117 (1974)
3.130 B.T.A. McKee, S. Saimoto, A.T. Stewart, M.J. Stott: Can. J. Phys. *52*, 759 (1974)
3.131 G. Dlubek, G. Brauer, O. Brümmer, W. Andrejtscheff, P. Manfrass: Phys. Status Solidi (a) *30*, K37 (1975)
3.132 J. Baram, M. Rosen: Phys. Status Solidi (a) *16*, 263 (1973)
3.133 G. Dlubek, O. Brümmer: Krist. Tech. *12*, 489 (1977)
3.134 P. Hautojärvi: Solid State Commun. *11*, 1049 (1972)
3.135 S. Mantl. W. Trifthäuser: Phys. Rev. B *17*, 1645 (1978)
3.136 I.K. MacKenzie: Private Communication
3.137 I.K. MacKenzie, T.E. Jackman, P.C. Lichtenberger: Appl. Phys. *9*, 259 (1976)
3.138 I.K. MacKenzie, J.A. Eady, R.R. Gingerich: Phys. Lett. A *33*, 279 (1970)
3.139 S. Saimoto, B.T.A. McKee, A.T. Stewart: Phys. Status Solidi (a) *21*, 623 (1974)
3.140 K.G. Lynn, R. Ure, J.G. Byrne: Acta Metall. *22*, 1075 (1974)
3.141 W. Brandt, R. Paulin, C. Dauwe: Phys. Lett. A *48*, 480 (1974)
3.142 W.H. Holt, M.F. Rose, S.Y. Chuang, S.J. Tao: Phys. Lett. A *32*, 422 (1970)
3.143 K.G. Lynn, C.M. Wan, R.W. Ure, J.G. Byrne: Phys. Status Solidi (a) *22*, 731 (1974)
3.144 C.F. Coleman, A.E. Hughes: "Positron Annihilation", in *Research Techniques in Non-destructive Testing*, ed. by R.S. Sharpe (Academic Press, New York 1977)
3.145 A.E. Hughes: Private Communication
3.146 H.P. Leighly: Appl. Phys. *12*, 217 (1977)

3.147 J.P. Pendrys, B.T.A. McKee, A.T. Stewart: Appl. Phys. *4*, 333 (1974)
3.148 I.K. MacKenzie, T.E. Jackman, C.G. White, C.W. Schulte, P.C. Lichtenberger: Appl. Phys. *7*, 141 (1975)
3.149 P. Rice-Evans, Tin Hlaing: J. Phys. F *7*, 821 (1977)
3.150 P. Rice-Evans, Tin. Hlaing, I. Chaglar: Phys. Rev. Lett. *37*, 1315 (1976)
3.151 I.K. MacKenzie: Phys. Rev. B *16*, 4705 (1977)
3.152 W.B. Gauster, S.R. Dolce: Solid State Commun. *16*, 867 (1975)
3.153 P. Hautojärvi, A. Vehanen, V.S. Mikhalenkov: Appl. Phys. *11*, 191 (1976)
3.154 B. Nielsen, N. Storgaard, K. Petersen: To be published
3.155 R. Myllylä, M. Karras, T.Miettinen: Appl. Phys. *13*, 137 (1977)
3.156 V.S. Mikhalenkov: Phys. Status Solidi (a) *24*, K111 (1974)
3.157 S. Mantl, W. Trifthäuser: Phys. Rev. Lett. *34*, 1654 (1975)
3.158 G. Dlubek, O. Brümmer, E. Hensel: Phys. Status Solidi (a) *34*, 737 (1976)
3.159 G. Dlubek, O. Brümmer, P. Sickert: Phys. Status Solidi (a) *34*, 401 (1977)
3.160 G. Dlubek, O. Brümmer: Phys. Lett. A *58*, 417 (1976)
3.161 W. Brandt, M. Oremland: Phys. Lett. A *57*, 387 (1976)
3.162 P. Hautojärvi, A. Vehanen, V.S. Mikhalenkov: Solid State Commun. *19*, 309 (1976)
3.163 K. Petersen, N. Thrane, R.M.J. Cotterill: Philos. Mag. *29*, 1 (1974)
3.164 K. Petersen, M. Knudsen, R.M.J. Cotterill: Philos. Mag. *32*, 417 (1975)
3.165 K. Petersen, J.H. Evans, R.M.J. Cotterill: Philos. Mag. *32*, 427 (1975)
3.166 M. Eldrup, O.E. Mogensen, J.H. Evans: J. Phys. F *4*, 499 (1976)
3.167 K. Hinode, S. Tanigawa, M. Doyama: Radiat. Eff. *32*, 73 (1977)
3.168 W. Schilling, G. Burger, K. Isebeck, H. Wenzl: In *Vacancies and Interstitials in Metals*, ed. by A. Seeger, D. Schumacher, W. Schilling, J. Diel (North-Holland, Amsterdam 1970) pp.255-362
3.169 W.R. Wampler, W.B. Gauster: J. Phys. F *8*, L1 (1978)
3.170 O. Mogensen, K. Petersen, R.M.J. Cotterill, B. Hudson: Nature *239*, 98 (1972)
3.171 R.M.J. Cotterill, I.K. MacKenzie, L. Smedskjaer, G. Trumpy, J.H.O.L. Träff: Nature *239*, 101 (1972)
3.172 D.R. Gustafson, G.T. Barens: J. Nucl. Mater. *48*, 79 (1973)
3.173 W. Trifthäuser, J.D. McGervey, R.W. Hendricks: Phys. Rev. B *9*, 3321 (1974)
3.174 K. Petersen, N. Thrane, G. Trumpy, R.W. Hendricks: Appl. Phys. *10*, 85 (1976)
3.175 V.W. Lindberg, J.D. McGervey, R.W. Hendricks, W. Trifthäuser: Philos. Mag. *36*, 117 (1977)
3.176 S. Nanao, K. Kuribayashi, S. Tanigawa, M. Doyama: Mater. Sci. Eng. *18*, 285 (1975)
3.177 M. Hasegawa, T. Suzuki: Radiat. Eff. *21*, 201 (1974)
3.178 S.M. Kim, W.J.L. Buyers, P. Martel, G.J.C. Carpenter: Can. J. Phys. *52*, 278 (1974)
3.179 C.L. Snead, A.N. Goland: Phys. Lett. A *55*, 189 (1975)
3.180 P. Jena, A.K. Gupta, K.S. Singwi: Solid State Commun. *21*, 293 (1977)
3.181 P. Hautojärvi, J. Heiniö, M. Manninen, R. Nieminen: Philos. Mag. *35*, 973 (1977)
3.182 C.H. Hodges, M.J. Stott: Solid State Commun. *12*, 1153 (1973)
3.183 N. Thrane, K. Petersen, J.H. Evans: Appl. Phys. *12*, 187 (1977)
3.184 N. Thrane, J.H. Evans: Appl. Phys. *12*, 183 (1977)
3.185 R. Grynszpan, K.G. Lynn, C.L. Snead, A.N. Goland: Phys. Lett. A *62*, 459 (1977)
3.186 R.M. Nieminen, J. Laakkonen, P. Hautojärvi, A. Vehanen: Phys. Rev. B *19* (February 1979)
3.187 W. Brandt: Adv. Chem. Ser. *158*, 219 (1976)
3.188 I.Ya. Dekhtyar: Phys. Lett. A *32*, 246 (1970)
3.189 I.Ya. Dekhtyar, M.M. Nistchenko, V. Pomashko: Phys. Status Solidi (b) *48*, K51 (1971)
3.190 V.L. Sedov, V.A. Teimurazova, K. Berndt: Phys. Lett. A *33*, 319 (1970)
3.191 M.J. Stott, A.T. Stewart, P. Kubica: Appl. Phys. *4*, 213 (1974)
3.192 P. Kubica, B.T.A. McKee, A.T. Stewart, M.J. Stott: Phys. Rev. B *11*, 11 (1975)

3.193 D.G. Lock, R.N. West: J. Phys. F *4*, 2179 (1974)
3.194 C. Koenig: Phys. Status Solidi (b) *88*, 569 (1978)
3.195 M. Doyama, S. Nanao, K. Kuribayashi, S. Tanigawa: J. Phys. F *3*, L125 (1973)
3.196 S. Tanigawa, S. Nanao, K. Kuribayashi, M. Doyama: Solid State Commun. *10*, 1025 (1972)
3.197 W.M. Lomer: *Vacancies and Other Point Defects in Metals and Alloys* (London, Inst. of Metals 1958) pp.79-98
3.198 M. Doyama, S. Tanigawa, K. Kuribayashi, S. Nanao: Cryst. Lattice Defects *4*, 255 (1973)
3.199 M. Doyama, K. Kuribayashi, S. Nanao, S. Tanigawa: Appl. Phys. *4*, 153 (1974)
3.200 S.M. Kim, W.J.L. Buyers, P. Martel, G.M. Hood: J. Phys. F *4*, 343 (1974)
3.201 C.L. Snead, T.M. Hall, A.N. Goland: Phys. Rev. Lett. *29*, 62 (1972)
3.202 S. Tanigawa, S. Nanao, K. Kuribayashi, M. Doyama: J. Phys. F *2*, L65 (1972)
3.203 M. Doyama, J.S. Koehler: In *The Properties of Liquid Metals*, ed. by S. Takeuchi (Taylor and Francis, London 1973) p.629
3.204 M. Doyama: Proc. 4th Intern. Conf. on Positron Annihilation (Helsingør 1976) unpublished
3.205 H. Fukushima, M. Doyama: J. Phys. F *6*, 677 (1976)
3.206 O. Sueoka: J. Phys. Soc. Jpn. *39*, 969 (1975)
3.207 P. Schultz, T.E. Jackman, I.K. MacKenzie, J.R. MacDonald: Proc. 4th Intern. Conf. on Positron Annihilation (Helsingør 1976) unpublished
3.208 P.J. Schultz, T.E. Jackman, J. Fabian, E.A. Williams, J.R. MacDonald, I.K. MacKenzie: Can. J. Phys. *56*, 1077 (1978)
3.209 I.Ya. Dekhtyar, S.G. Litovchenko, V.S. Mikhalenkov: Sov. Phys. Doklady *7*, 1135 (1963)
3.210 H. Morinaga: Phys. Lett. A *34*, 384 (1971)
3.211 M. Doyama, K. Kuribayashi, S. Tanigawa, S. Nanao: J. Phys. Soc. Jpn. *36*, 1706 (1974)
3.212 K. Kuribayashi, S. Tanigawa, S. Nanao, M. Doyama: Solid State Commun. *17*, 143-145 (1975)
3.213 C. Sykes, F.W. Jones: Proc. R. Soc. A *157*, 213 (1936)
3.214 H. Fukushima, J. Sugiura, M. Doyama: J. Phys. F *6*, 1845 (1976)
3.215 K. Kuribayashi, S. Tanigawa, S. Nanao, M. Doyama: Scr. Metall. *9*, 423 (1975)
3.216 T. Kojima, K. Kuribayashi, M. Doyama: Scr. Metall. *9*, 1071 (1975)
3.217 T. Kojima, K. Kuribayashi, M. Doyama: Appl. Phys. *12*, 179 (1977)
3.218 T. Troev, K. Hinode, S. Tanigawa, M. Doyama: Appl. Phys. *13*, 105 (1977)
3.219 M. Meurtin, P. Lesbats: Proc. 4th Intern. Conf. on Positron Annihilation (Helsingør 1976) unpublished
3.220 D.G. Lock, R.N. West: Phys. Lett. A *49*, 137 (1974)
3.221 D.R. Gustafson, A.R. Mackintosh: Phys. Lett. *5*, 234 (1963)
3.222 D.R. Gustafson, A.R. Mackintosh, D.J. Zaffarano: Phys. Rev. *130*, 1455 (1963)
3.223 R.N. West, R.E. Borland, J.R.A. Cooper, N.E. Cusack: Proc. Phys. Soc. *92*, 195 (1967)
3.224 O.E. Mogensen, G. Trumpy: Phys. Rev. *188*, 639 (1969)
3.225 L.E. Ballentine: Can. J. Phys. *44*, 2533 (1966)
3.226 J.D. McGervey: In *Positron Annihilation*, ed. by L.O. Roellig, A.T. Stewart (Academic Press, New York 1967) p.305
3.227 W. Brandt, H.F. Waung: Phys. Lett. A *27*, 700 (1968)
3.228 F. Itoh, M. Kuroha, K. Kai, S. Takeuchi: J. Phys. Soc. Jpn. *32*, 567 (1972)
3.229 J.A. Arias-Limonta, P.G. Varlashkin: Phys. Rev. B *1*, 142 (1970)
3.230 W. Brandt, J. Reinheimer: Phys. Lett. A *35*, 109 (1971)
3.231 F. Itoh: J. Phys. Soc. Jpn. *41*, 824 (1976)
3.232 T.E. Faber: *An Introduction to the Theory of Liquid Metals* (Cambridge Univ. Press, Cambridge 1972) p.288
3.233 M. Doyama, S. Tanigawa, K. Kuribayashi, H. Fukushima, K. Hinode, F. Saito: J. Phys. F *5*, L230 (1975)
3.234 H. Weisberg, S. Berko: Phys. Rev. *154*, 249 (1967)
3.235 M.V. Chu, C.J. Jan, P.K. Tseng, W.F. Huang: Phys. Lett. A *43*, 423 (1973)
3.236 R.N. West, V.H.C. Crisp, G. De Blonde, B.G. Hogg: Phys. Lett. A *45*, 441 (1973)

3.237 F. Itoh, K. Kai, M. Kuroha, S. Takeuchi: Phys. Status Solidi (b) *75*, 559 (1976)
3.238 H.S. Chen, S.Y. Chuang: Phys. Status Solidi (a) *25*, 581 (1974)
3.239 S.Y. Chuang, S.J. Tao, H.S. Chen: J. Phys. F *5*, 1681 (1975)
3.240 K. Tsuji, H. Endo, Y. Kita, M. Ueda, Z. Morita: In *Liquid Metals 1976*, ed. by R. Evans, D.A. Greenwood (Inst. of Physics, Bristol and London 1977) pp.367-71

4. Positrons in Imperfect Solids: Theory

R. M. Nieminen and M. J. Manninen

With 16 Figures

The physics of imperfect crystalline solids involves a vast class of phenomena. In the wide sense of the word, imperfections in solids include point defects (vacancies, impurities) and their aggregates, various types of dislocations, stacking faults, grain boundaries as well as surfaces. Besides being physically interesting and challenging in their own right, imperfections in crystals are important in technological materials science, greatly influencing the macroscopic physical world around us. There exists an enormous literature covering the many facets of the broad topic of defected solids. Recent reviews, which emphasize an atomistic approach include [4.1-6].

Following the experimental discovery of the sensitivity of positrons to lattice defects, the application of the positron annihilation method to investigate defect properties of solids has increased dramatically during the last decade, and the technique has developed into a standard tool of solid-state science, capable of yielding unique information of concentration, configuration and internal structure of defects in condensed matter. In this chapter we will try to elucidate the basic physical theory behind the process of positron trapping and annihilation at lattice imperfections. The aim of the paper is twofold: I) to understand how and why positrons are attracted to defects and II) to interprete the information conveyed by experimental results in terms of the atomic and electronic structure of the defect involved. We will discuss the positron distribution in solids prior to trapping, its tendency to seek out regions in the crystal with lower than average ion density, the factors contributing to defect-positron interaction and the characteristics of the trapped state. The selection of the topics is not exhaustive, but will reflect our own interests and prejudices. The emphasis will be on metals, with only short excursions to insulating solids. Besides other articles in this volume, the reader is referred to a number of texts [4.7-14] of relevance to the present subject. A particularly useful reference is WEST's review [4.14], which should be consulted for a clear and compact presentation of basic positron physics.

The organization of the paper is the following: In Sect.4.1, we discuss the positron mobility, distribution and annihilation in perfect crystals, and the effect of temperature on them. The trapping model is introduced, and the temperature dependence of the trapping process is analyzed. Sect.4.2 covers defects in metals. The electronic structure of defects in metals is discussed, with emphasis on the density functional approach. Positron-defect interaction and annihilation characteristics of the trapped positrons are described, with applications to a number of specific cases. Sect.4.3 is a short discussion of positrons in defects of nonmetals, and in Sect.4.4 we conclude.

4.1 Positron Distribution, Mobility, and Trapping

4.1.1 Positron Implantation, Slowing Down, and Thermalization

Positrons are injected into a solid from a radioactive source, either external or internal, which emits a beta-energy spectrum of positrons with a maximum energy around 1 MeV. The positrons lose their energy via ionizing collisions, plasmon and electron-hole generation, and finally by phonon scattering coming to thermal equilibrium with the host matter. The positron implantation profiles have been studied by BRANDT and PAULIN [4.15]. They found that the absorption coefficient is in a wide variety of materials given by

$$\alpha_+ = R_+^{-1} = (16 \pm 1)\,\frac{[d\ \mathrm{g\,cm^{-3}}]}{E_{max}^{1.43}[\mathrm{MeV}]}\ [\mathrm{cm^{-1}}] \tag{4.1}$$

where d is the sample mass density and E_{max} the maximum kinetic energy of emitted positrons. Thus the typical implantation range R_+ is a fraction of millimeter in solids, and is little influenced by channelling conditions.

The slowing down and thermalization times are usually assumed to be short compared to mean positron lifetimes in solids, and the high resolution angular correlation experiments by KUBICA and STEWART [4.16] do show that positrons in metals thermalize down to nearly liquid helium temperatures before annihilation. This indicates that phonon scattering must play a major role in the final stages of thermalization as was argued theoretically by PERKINS and CARBOTTE [4.17].

The apparent effective mass m^* of thermalized positrons in metals can be deduced from the smearing of the angular correlation curve near the Fermi cutoff. The effective mass has contributions arising from the static positron-lattice interaction (band mass), from positron-phonon scattering and from positron-electron interactions (many-body mass). The theoretical estimates for these effects combined yield values of $m^* = 1.2\ m$ in alkalis [4.17-19], while the early experiments [4.20] indicated somewhat higher masses. Improvements in the experimental resolution [4.16] have alleviated this discrepancy. Also the notion of a thermal effective mass is not quite clear, as was first noted by MIKESKA [4.21]. He found that the lifetime broadening of a positron quasi-particle due to the phonon scattering leads to a broader momentum distribution than the Boltzmann distribution, which would result in a 15-30% increase in the apparent effective mass in a Boltzmann-like momentum distribution [4.22]; see also the discussion by BERGERSEN and PAJANNE in [4.23].

4.1.2 Mobility and Diffusion

The motion of a thermalized positron in solids is limited by positron-electron interaction, described by electron-hole pair generation, by positron lattice interaction and by scattering off impurities. An understanding of the mobility and its temperature dependence is important for a proper interpretation of the trapping process at defects, as will be discussed in the next section.

Owing to experimental difficulties, there exist few direct measurements of the positron mobility in condensed matter. MILLS and PFEIFFER [4.24] determined the mobility of positrons in Ge, observing the Doppler shift of the annihilation gamma as a function of the bias electric field across an intrinsic Ge detector. A similar technique was employed to measure the mobility in Si [4.25]. Their results, and those of SUEOKA and KOIDE [4.26], and BRANDT and PAULIN [4.15], are summarized in Table 4.1, where we give values of the mobility μ_+ and the diffusion constant, defined via the Einstein relation

$$D_+ = \mu_+ \frac{kT}{e} \quad . \tag{4.2}$$

Apart from the estimates of [4.15], based on the field dependence of the implantation profile, the mobilities in insulators seem to be a few percent of the electron mobilities at comparable temperatures. The results of BRANDT and PAULIN [4.15] may be influenced by the effect of the electric field on the kinetic mobility prior to thermalization. In semiconductors, the positron mobility is limited by acoustic phonons, and is in the deformation potential approximation [4.27]

$$\mu_+ = \frac{2\hbar^4 s}{3E_d^2} \left(\frac{2\pi}{m^{*5}}\right)^{1/2} \left(\frac{1}{kT}\right)^{3/2} \quad , \tag{4.3}$$

where s is the sound velocity and E_d the deformation potential constant. The results of MILLS and PFEIFFER [4.24] show that in Ge E_d for positrons is about 19 eV, but the fit to (4.3) is not perfect indicating optical phonon generation at low temperatures.

$\hbar = h/2\pi$ (normalized Planck's constant)

Table 4.1. Experimental estimates for positron mobility μ_+. D_+ is the diffusion constant defined via Einstein relation $D_+ = kT\mu_+/e$

	T [K]	μ_+ $[cm^2V^{-1}s^{-1}]$	D_+ $[cm^2s^{-1}]$	Ref.
Ge	36 ± 5	350 ± 17	1.09	[4.24]
Ge	93 ± 5	124 ± 10	0.99	[4.24]
Si	80 ± 2	460 ± 20	3.24	[4.25]
Si	184 ± 1	173 ± 15	2.74	[4.25]
Si	300	430 ± 100	10 ± 3	[4.15]
C	300	< 20	< 0.52	[4.26]
C	300	215 ± 100	5 ± 3	[4.15]
SiO_2	300	< 15	< 0.39	[4.26]
$BaTiO_3$	300	< 80	< 2.07	[4.26]

A few workers have tried to estimate the positron mobility and diffusion constant indirectly from defect trapping. However, care should be taken when using such an approach, because the results depend crucially on what specific trapping model is used. If the mobility of positrons is high, the trapping process is not diffusion limited and the extraction of diffusion-related parameters is not meaningful. This will be discussed in more detail in Sect.4.1.5.

In metals the measurement of positron mobility is obviously prohibitively difficult, and one has to rely on theoretical estimates. BERGERSEN et al. [4.28] have calculated the various contributions in a number of simple metals. The inverse relaxation time due to scattering off conduction electrons near the Fermi surface is

$$\tau_e^{-1} = \frac{\pi}{4}\frac{m^*}{m}\frac{(kT)^2}{\hbar\varepsilon_F} , \qquad (4.4)$$

where ε_F is the Fermi energy. Acoustic phonon scattering contributes a relaxation time τ_{ph}, which is obtained from (4.3) by multiplying by m^*/e, and the relaxation time due to impurity scattering is proportional to $(kT)^{-1/2}$. BERGERSEN et al. [4.28] calculated the deformation potential constants

$$E_d = \Omega \frac{\partial E(\Omega)}{\partial \Omega} , \qquad (4.5)$$

where E is the energy of the lowest positron state in the crystal of volume Ω, incorporating the three major contributions to E as will be discussed in Sect.4.2.4e.

Table 4.2 summarizes the results of BERGERSEN et al. [4.28] in pure metals at T = 300 K, and it also lists the mean free path

$$l = \left(\frac{3kT}{m^*}\right)^{1/2}\left(\frac{1}{\tau_e} + \frac{1}{\tau_{ph}}\right)^{-1} = \frac{3D_+}{v_+} , \qquad (4.6)$$

Table 4.2. Mobility-related quantities for positrons in metals at T = 300 K [4.28]. τ_e, τ_{ph} are the relaxation times against electron and phonon scattering, μ_+ is the mobility and D_+ the diffusion constant. The mean free path is l, while the average diffusion length before annihilation is L

Metal	τ_e $[10^{-22}s]$	τ_{ph} $[10^{-14}s]$	μ_+ $[cm^2V^{-1}s^{-1}]$	D_+ $[cm^2s^{-1}]$	l [Å]	L [Å]
Li	3.3	2.7	26	0.7	23	3500
Na	2.2	2.6	26	0.7	23	3800
K	1.5	1.6	14	0.4	14	3100
Mg	5.0	2.1	21	0.5	18	2600
Zn	6.6	2.1	21	0.5	18	2300
Al	8.2	1.5	14	0.4	13	2100
Ga	7.3	1.4	14	0.4	12	2100
In	6.0	1.4	14	0.4	12	2300
Tl	5.6	1.3	13	0.3	11	2100

and the average diffusion length before annihilation

$$L = \sqrt{6D_+\tau} \quad . \tag{4.7}$$

Above, $v_+ = (3kT/m^*)^{1/2}$ is the thermal velocity and τ the lifetime against annihilation. It is seen that the mobility is phonon limited ($\tau_{ph} << \tau_e$) at all practical temperatures, with leading temperature dependence $D_+ \propto T^{-1/2}$. The order of magnitude of the diffusion constants in Table 4.2 is the same as that of the experimental results in insulators [4.24,25], which is expected in view of the smallness of the electronic contribution. In polar materials, the optical phonons determine the mobility; even there, however, there are good reasons to expect [4.29] that typical diffusion constants are of the order $D_+ \sim 0.1$-$1.0\ cm^2s^{-1}$ around room temperature.

4.1.3. Positron Distribution in Solids

Consider a perfect rigid lattice of ions and the related electronic configuration. A positron in such a system will be in its lowest energy Bloch state with nearly zero thermal momentum. As it is a light-mass positively charged particle it is strongly repelled from the ion cores and consequently its wave-function amplitude is small in these regions, increasing rapidly to become largest in the interstitial space between the ions. The spatial distribution of the thermalized positron has somewhat of a "Swiss cheese" character, the density distribution being relatively uniform apart from the "holes" around each ion due to the strong repulsion. This presents difficulties for a "nearly free-positron" plane-wave expansion [4.30,31] to converge near the ions. This problem can be overcome using Stott's positron pseudopotential theory [4.32,33], which allows one to calculate accurate positron wave functions and energies with modest numerical labor. This approach can readily be generalized to solids with defects, as will be discussed in Sect.4.2.2. The physical arguments, which lead to this theory are quite different from those familiar for valence electrons in simple metals, but the motivation is similar. We will concentrate on applying the method in metals, although it is equally feasible in insulators.

The construction of a single-particle potential for a positron in a metal is somewhat simpler than for electrons. There is no exchange repulsion, and calculations based on a positron in a uniform electron gas [4.34] indicate that the positron-electron correlation potential is a slowly varying function of density, and will be swamped out by the electrostatic potential near ions. To a good approximation the positron in a perfect metal moves in the Hartree potential of the ions and conduction electrons. In defected systems, the electron-positron correlation contribution may have to be included: in many cases a local density approximation will suffice (see Sect.4.2.2). The positron wave function for states near the bottom of the lowest energy band is separated into two factors. One reflects the strong repulsion of the positron from the ion core, and is insensitive to the positron energy or the environment of the core. The other is a smooth envelope, which is energy dependent, sensitive to the environment and reflects the positron distribution in the interstitial regions and/or between atomic cells in the crystal. Figure 4.1 schematizes this division for a nonzero $\underline{k}$ corresponding to a temperature of about 1000 K. The envelope satisfies a Schrödinger-like equation with a relatively weak potential term which, e.g., in perfect metals can be attacked with low-order perturbation theory. Thus the philosophy of using the term "pseudopotential".

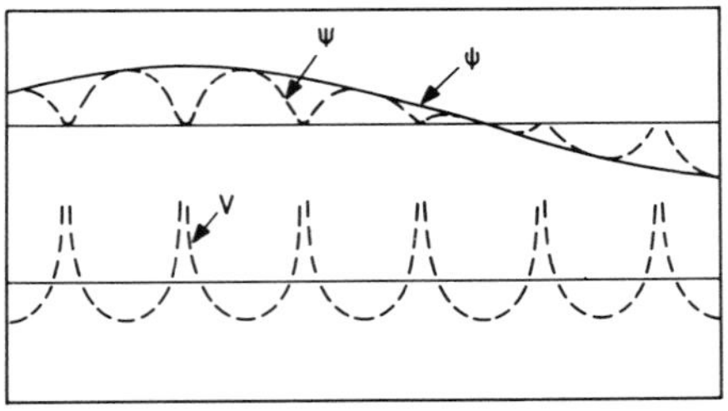

Fig.4.1. Schematic positron wave function and potential in a perfect metal. V is the full Hartree potential and Ψ the corresponding full wave function. ψ is the pseudo wave function corresponding to the nearly constant pseudopotential

The positron wave function for a wave vector $\underline{k} \approx 0$ is factorized as

$$\Psi_{\underline{k}}(\underline{r}) = U(\underline{r} - \underline{R})\psi_{\underline{k}}(\underline{r}) \quad , \tag{4.8}$$

where $\underline{r}$ lies in an atomic cell centered on $\underline{R}$, and $U(\underline{r})$ is chosen to satisfy the Schrödinger equation with the spherically symmetric single-ion potential $V_a(r)$ with Wigner-Seitz boundary conditions imposed at a conveniently chosen muffin-tin radius. In the spherical cell approximation thus

$$\left[-\frac{\hbar^2}{2m}\frac{\partial^2 r}{\partial r^2}+V_a(r)\right]U(r)=E_{WS}U(r) \quad , \tag{4.9}$$

$$\left.\frac{\partial U}{\partial r}\right|_{r=R_0}=0 \quad , \tag{4.10}$$

where R_0 is the muffin-tin radius. Between the muffin-tin spheres, the core factor $U(r)$ can be chosen to be a constant $U(r) = U(R_0)$. Substituting (4.8) with this kind of $U(r)$ into the full Schrödinger equation yields an equation for the pseudo wave function $\psi_{\underline{k}}(r)$

$$\left[-\frac{\hbar^2}{2m}\nabla^2+W(\underline{r})\right]\psi_{\underline{k}}(\underline{r})=E_0(\underline{k})\psi_{\underline{k}}(\underline{r}) \quad , \tag{4.11}$$

where

$$W(\underline{r})=\begin{cases} E_{WS}+V(\underline{r})-V_a(\underline{r}-\underline{R}_n)-\dfrac{\hbar^2}{2m}\dfrac{\nabla U(\underline{r})}{U(\underline{r})}\cdot\nabla & \\ & \text{within the } n^{th} \text{ muffin-tin} \\ V(\underline{r}) & \text{outside a muffin-tin} \end{cases} \tag{4.12}$$

and $V(\underline{r})$ is the full positron potential. It is easy to see that the pseudopotential $W(\underline{r})$ is much weaker than $V(\underline{r})$ in the core regions, and that the gradient part of it is generally small, since $\psi_{\underline{k}}(\underline{r})$ is a smooth function because the rapidly varying part of the full $\Psi_{\underline{k}}$ has been factorized out. Therefore (4.11,12) form a good starting point either for an expansion in Bloch waves or perturbation theory for any potential less singular than $1/r^2$ near the origin.

The method has been used to calculate positron ground-state wave functions, energies and band masses in simple metals [4.32,33,35] and ionic crystals [4.29], the pressure dependence of positron states in alkalis [4.36] as well as for the positron distribution in disordered alloys [4.37,38]. It also underlies the defect calculations described in Sect.4.2. In practice, the Wigner-Seitz energy E_{WS} is close to E_0 and the distinction between the two is often omitted. Values of E_0 are given in Table 4.6 for simple metals, and for transition and noble metals in Table 4.7. In *pure* metals at finite temperatures, most of the deviation of the positron energy dispersion relation $E = E_0(\underline{k})$ from the free-particle-like behavior is due to the gradient part of the pseudopotential. As an example of this, Table 4.3 lists the average scalar effective masses for a number of metals, obtained from a second-order perturbation expansion around $\underline{k} = 0$ [4.32].

Table 4.3. Positron k = 0 effective masses calculated to second-order in the positron pseudopotential [4.32]

	Li	Na	K	Be	Mg	Al
m^*/m	1.03	1.07	1.11	1.04	1.10	1.12

4.1.4 Annihilation Characteristics and Electron-Positron Correlation in Pure Metals

A first-principles description of the positron annihilation process is a complicated many-particle problem even for a perfect metal. Fortunately lifetime and angular correlation experiments give evidence that, at least in simple metals, the annihilation parameters are close to those in a homogeneous electron gas corresponding to the valence electron density of the metal. We are not going to describe how the annihilation characteristics for an electron gas are calculated (see [4.14] for a general discussion) but give approximations how the electron gas data can be used in estimating the annihilation characteristics in real metals and in defects.

In a homogeneous electron gas of density n the positron annihilation rate can be written as

$$\lambda(n) = \lambda_0(n)\gamma(n) = \pi r_0^2 cn\gamma(n) \quad , \tag{4.13}$$

where λ_0 is the Sommerfeld free-electron formula, r_0 the classical electron radius and c the velocity of light. γ is a density dependent enhancement factor caused by the strong electron-positron correlation which increases the electron density at the site of the positron. BRANDT and REINHEIMER [4.39,40] have suggested a practical interpolation formula for the enhancement factor and give

$$\begin{aligned}\lambda(n) &= \frac{12}{r_s^3}\left(1 + \frac{10+r_s^3}{6}\right) \times (ns)^{-1} \\ &= (2 + 134n) \times (ns)^{-1} \quad ,\end{aligned} \tag{4.14}$$

where we have introduced the usual density parameter

$$r_s = \left(\frac{3}{4\pi n}\right)^{1/3} \tag{4.15}$$

given in units of a_0, the Bohr radius of 0.529 Å. This reproduces within a few percent the results of most many-body calculations [4.41-43] for $2 < r_s < 6$, i.e., for the metallic density range. However, the formula fails at the high-density limit $r_s < 1$, where the leading term is

$$\lambda(n) = \frac{12}{r_s^3}(1 + 1.23\, r_s) \times (ns)^{-1} \quad . \tag{4.16}$$

In real metals a positron has a nonvanishing overlap with the more tightly bound "core" electrons, and consequently a faster annihilation rate than what would be predicted by the conduction electron gas alone. Especially large is the core contribution in transition and noble metals, which have an extrusive d shell.

WEST [4.45] has suggested that the annihilation with the core electrons can be calculated using the electron gas theories by renormalizing the valence electron density n according to the description

$$n_{eff} = n\left(1 + \frac{\Gamma_c}{\Gamma_v}\right) \quad , \tag{4.17}$$

where the Γ_c and Γ_v are the partial annihilation rates with the "core" and "valence" electrons, respectively. The ratio Γ_c/Γ_v can be estimated from the angular correlation curves, which consist of a clearly separable Gaussian core electron part and a free electron-like parabola. This separation is best done in simple metals, whereas it may be impossible in some transition metals. The ratio Γ_c/Γ_v is simply A_c/A_v where A_c and A_v are the areas under the Gaussian and the parabolic parts of the angular correlation curve. Experimental estimates for the core fraction are given in Table 4.6. Because of the uncertainty as to higher momentum components (Umklapp) from valence electron annihilations and the way in which the broad component should be extrapolated back to the origin, WEST [4.45] arbitrarily took A_c to be 80% of the measured area of the Gaussian portion of the angular distribution. This method produces quite well the experimental lifetimes in most simple metals as seen from Fig.4.2, which plots the theoretical electron gas result [4.43] for $\lambda = \lambda(n)$ together with experimental estimates using (4.17). However, there remains slight uncertainty in the theoretical annihilation rates for the homogeneous electron gas. Most calculations [4.41-43] give annihilation rates smaller than the experimental values in metals, but more recent computations [4.46] give values larger than the experimental ones. In the latter case WEST's renormalization technique would make, in simple metals, the agreement between experimental and theoretical values a little poorer.

The angular correlation and Doppler line-shape experiments measure the momentum distribution of the annihilating electron-positron pair. If the electrons and the positron are treated as independent particles the many-body wave function is a Slater determinant and the momentum distribution of the annihilation quanta is

$$\Gamma_0(\underline{p}) = \frac{\pi r_0^2 c}{(2\pi)^3} \sum_i \left| \int d\underline{r}\, \exp(-i\underline{p} \cdot \underline{r})\psi_i(\underline{r})\Psi_+(\underline{r}) \right|^2 \quad , \tag{4.18}$$

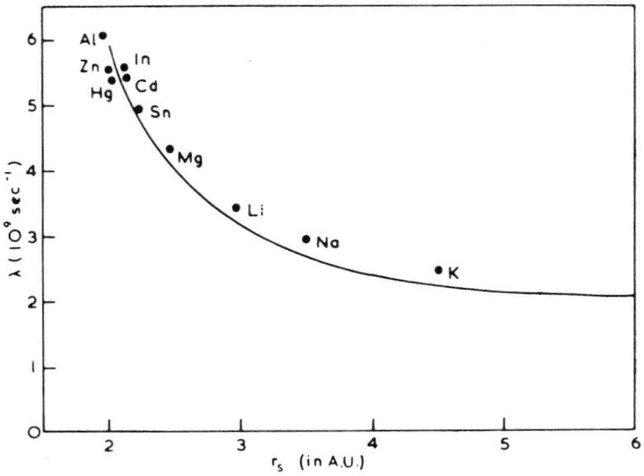

Fig.4.2. Experimental and theoretical annihilation rates. The theoretical curve is from [4.43]. The experimental points correspond to r_s values renormalized according to (4.17)

where Ψ_+ is the ground-state positron wave function and where the sum goes over the occupied electron states ψ_i. The conventional long-slit angular correlation apparatus (see Chap.1) measures only one component, p_z, of the momentum distribution, so that the angular correlation curve is

$$I(p_z) = \int dp_x \int dp_y \Gamma_0(\underline{p}) \quad , \tag{4.19}$$

which reduces in the isotropic case to

$$I(p_z) = 2\pi \int_{p_z}^{\infty} dp \, p\Gamma_0(p) \quad . \tag{4.20}$$

In a two-dimensional angular correlation measurement the average is of course taken only over one momentum component.

For a homogeneous electron gas the independent particle approximation (4.18) gives the 2γ-angular correlation curve

$$I(p_z) = \frac{r_0^2 c}{2\pi} (p_F^2 - p_z^2) \, (p_F - |p_z|) \quad , \tag{4.21}$$

which is an inverted parabola the width of which is proportional to the Fermi momentum p_F. This describes surprisingly well the observed angular correlation curves of simple metals if one neglects the small broad Gaussian part due to the core electrons and higher momentum components of the valence electrons. The independent-particle approximation does not take into account the strong electron-positron correlation and thus by integrating (4.21) over p_z one recovers the Sommerfeld formula for the total annihilation rate. The effect of the electron-positron correlations on the momentum distribution in the homogeneous electron gas can be included through a momentum dependent enhancement factor [4.14,41] $\varepsilon(p)$, so that

$$\Gamma(\underline{p}) = \varepsilon(p)\Gamma_0(\underline{p}) \quad . \tag{4.22}$$

$\varepsilon(p)$ can be written in the form

$$\varepsilon(p) = a + b\,(p/p_F)^2 + c\,(p/p_F)^4 \quad , \tag{4.23}$$

where a, b and c depend on the electron density [4.41], b and c being in metallic densities much smaller than a. This enhancement factor has a relative small effect on the doubly integrated long-slit angular correlation distribution, resulting in the experimentally verified [4.47] slight "swelling" of the free-electron parabola. The success of the independent-particle approximation is based just on the fact that the momentum dependence of the enhancement factor $\varepsilon(p)$ is weak. There is also strong evidence that the independent-particle approximation describes well also the core contribution of the angular correlation curve, calculated by using atomic states for the electrons and the Wigner-Seitz wave function for the positron [4.48,49]. However, the independent-particle approximation may not reproduce the full angular correlation curve, since the neglect of enhancement effects would probably mean wrong relative weights for the core and valence parts.

The electron-positron correlation energy is in the high-density limit given by the RPA expression [4.44]

$$E_{corr}(n) \xrightarrow[r_s \to 0]{} - \frac{1.56}{r_s^{1/2}} + 0.051(\ln r_s)^2 - 0.081 \ln r_s + 1.14 \text{ [Ry]} \tag{4.24}$$

while the low-density limit is the energy of a positronium ion (Ps^-) [4.14]

$$E_{corr}(n) \xrightarrow[r_s \to \infty]{} - 0.524\text{[Ry]} \quad . \tag{4.25}$$

In the intermediate metallic density region $2 < r_s < 6$, E_{corr} has been calculated by many-body techniques [4.34,50]. Generally, the correlation energy can be written via the Hellmann-Feynman theorem as

$$\begin{aligned} E_{corr}(n) &= \int_0^1 dZ\, \varepsilon_{corr}(Z) \\ &= - \frac{3}{r_s^3} \int_0^1 dZ \int_0^\infty dr\, r\Delta g(r,Z) \quad , \end{aligned} \tag{4.26}$$

where $\Delta g = \Delta n/n$ is the charge-density enhancement in the screening cloud around a positron with a fractional charge Z. Assume that the polarization cloud at full interaction strength has a universal shape, i.e., is proportional to a universal, decreasing function $f(\gamma r)$, where γ is some scaling parameter. Noting that the polarization cloud contains exactly one electron

$$\frac{3}{r_s^3}\int_0^\infty dr\, r^2 \Delta g(r,Z = 1) = 1 \tag{4.27}$$

it follows that

$$\begin{aligned}\varepsilon_{corr}(Z = 1) &\propto \left[\frac{3}{4\pi r_s^3}\,\Delta g(r = 0,\, Z = 1)\right]^{1/3} \\ &\propto (\Delta\lambda)^{1/3} \quad ,\end{aligned} \tag{4.28}$$

where $\Delta\lambda$ is the polarization (enhancement) contribution to the annihilation rate. For large densities linear response to the perturbing charge is a good approximation, i.e., $\Delta g(r,Z)$ is proportional to Z. The coupling constant integration (4.26) then gives $E_{corr} = 0.5\varepsilon_{corr}(Z = 1)$. On the other hand, the same relation also holds for the low-density Ps limit and is probably valid at all electron densities to a good approximation. Thus we may assert that

$$E_{corr}(n) \propto (\lambda - \lambda_0)^{1/3} \quad , \tag{4.29}$$

where λ is the actual annihilation rate and $\lambda_0 \propto n$ is the unperturbed Sommerfeld rate. This relation is indeed well satisfied for electron gas calculations [4.34, 43,50]. Using an interpolation formula for $\lambda(r_s)$ suggested by BRANDT and REINHEIMER [4.39] we get for the electron gas

$$E_{corr}(n) = -\,0.524\left(1 + \frac{10}{r_s^3}\right)^{1/3} \text{[Ry]} \tag{4.30}$$

or

$$E_{corr}(n) = -\,0.416(\Delta\lambda)^{1/3}\text{[Ry]} \quad , \tag{4.31}$$

where $\Delta\lambda$ is in units of $10^9\ s^{-1}$. The form (4.31) can be used to evaluate E_{corr} in any bulk metal using the experimental data, provided some reasonable estimates for λ and λ_0 can be found. Tables 4.6,7 show the correlation energies for a number of metals; in simple metals, the electron gas results [4.34] have been used, whereas in transition and noble metals use has been made of the experimental lifetime data [4.51] and (4.31). A complementary way to estimate the effect of the more tightly bound electrons on the correlation energy follows the idea of WEST [4.45] to use a renormalized conduction electron density in the interpolation formula (4.30), in a way discussed in the context of the annihilation rates.

4.1.5 Effect of Temperature on Annihilation Characteristics

Above we have argued that positrons in perfect crystals are typically in free-particle-like, delocalized Bloch states with a mobility essentially limited by lattice vibrations. Here we discuss the temperature dependence of the annihilation rate of such states in metals, following STOTT and WEST [4.35]. An understanding of this thermal dependence is crucial for a proper description of equilibrium defect measurements at high temperatures.

Using the positron pseudopotential formulation outlined above, STOTT and WEST [4.35] have expressed the positron density in a solid in its instantaneous ionic configuration. As a convenient label they use the core contribution of the annihilation rate, resulting from the overlap of the positron wave function with the tightly bound, nearly atomic core electrons. Within the independent-particle model, it can be written as

$$\lambda_c^0 \propto \frac{1}{\Omega - \chi(0)} \left[1 - \frac{1}{\Omega} \sum_{\underline{k} \neq 0} \langle |S(\underline{k})|^2 \rangle F(k)\right] \quad , \tag{4.32}$$

where Ω is the volume per atom, $\langle |S(\underline{k})|^2 \rangle$ the usual structure factor, and

$$\chi(\underline{k}) = \int d\underline{r} \, \exp(-i\underline{k} \cdot \underline{r}) \left\{1 - \left[\frac{U(r)}{U(R_0)}\right]^2\right\} \quad , \tag{4.33}$$

with the integral over the muffin-tin region $r < R_0$ [see (4.9,10)], measures the strength of the positron exclusion from the core. $F(k)$ is the "positron core rate wave-number characteristic", which is practically volume independent and carries all the details of the positron-ion interaction and the positron core electron overlap. The structural information is thus solely conveyed by the structure factor. The temperature dependence of the core rate is contained in Ω and $\langle |S(\underline{k})|^2 \rangle$. As temperature is raised from zero three effects lead to changes in the positron distribution and thus λ_c^0.

I) Anharmonic thermal expansion of the crystal leads to a uniform decrease of the positron amplitude and thus diminishes also the core rate. The leading factor $[\Omega - \chi(0)]^{-1}$, reflecting the free volume available for the positron, accounts for this.

II) Anharmonic thermal expansion leads also to a redistribution of the positron density from the core to the interstitial regions because of the larger interstitial space and thus to a further decrease in the core overlap.

III) Lattice vibrations cause an increase in positron density in any interstitial region which is momentarily enlarged by the thermal motion of the ions around their equilibrium positions. The effect enters λ_c through modifications to the structure factor. The peaks at the reciprocal lattice vectors are damped by the Debye-Waller factor, with a compensating diffuse background growing as the temperature rises.

STOTT and WEST performed quantitative calculations of these effects on the core rate, and their results are reproduced in Table 4.4. To make contact with experiments it is more appropriate to look at the changes in the total annihilation rate rather than the core contribution alone, as the separation of these two may not be too simple. Electron gas theories (see, e.g., [4.14]) express the total annihilation rate $\lambda(n)$ as a function of the conduction electron density n. WEST [4.45] has argued that in real metals the total rate can be described by the same theory through a renormalized electron density, see (4.17)

$$\lambda = \lambda[n(1 + A_c/A_v)] \quad , \tag{4.34}$$

where A_c and A_v are the fractional areas of the core (broad) and valence (parabola) parts of an observed long-slit angular correlation curve. Assuming that the temperature-induced relative changes in the many-body enhancement factors for the two groups of electrons are similar, the change in the total annihilation rate can be written as

$$\frac{\Delta\lambda}{\lambda} = \frac{\lambda-\lambda_p}{\lambda}\left(A_v \frac{\Delta\lambda_v^0}{\lambda_v^0} + A_c \frac{\Delta\lambda_c^0}{\lambda_v^0}\right) \quad , \tag{4.35}$$

where λ_p is the spin-averaged positronium annihilation rate and $\Delta\lambda_v^0$, $\Delta\lambda_c^0$ are the changes in the valence and core annihilation rates within the independent-particle model. The former may be incorporated solely by volume increase, i.e., $\Delta\lambda_v^0/\lambda_v^0 = -\Delta\Omega/\Omega$.

The predicted changes in Table 4.4 are in good agreement with experiment. The three contributions to the change in the core annihilation rate are of comparable magnitude, the effect of the lattice vibrations being the largest.

This thermal dependence of the annihilation rate in the perfect lattice in non-negligible. It will also be present in the high-temperature region where thermally activated defects may act as positron traps, and it has to be properly subtracted if one wants to focus on the changes in annihilation characteristics induced by an equilibrium defect concentration.

4.1.6 Trapping at Defects

The sensivity of positrons to various types of lattice defects in solids has a firm experimental basis, and the many important applications of it are discussed at length by WEST in this volume. The information conveyed by annihilation radiation from traps of the electron structure of defects will be discussed in Section 4.2. We shall now briefly discuss the basic dynamics of the trapping process.

Table 4.4. Calculated independent-particle core rate changes, $-\Delta\lambda_c^0/\lambda_c^0$ [%] due to I) volume, II) lattice, and III) lattice vibration effects. $-\Delta\lambda_c^0/\lambda_c^0$ is the valence rate change, and $-\Delta\lambda/\lambda|_{th}$ the total annihilation rate change due to temperatures increase. The last column gives the experimental value. [4.35]

	Volume I)	Lattice II)	Vibr. III)	$-\frac{\Delta\lambda_c^0}{\lambda_c^0}$	$-\frac{\Delta\lambda_v^0}{\lambda_v^0}$	$-\frac{\Delta\lambda}{\lambda}\Big\vert_{th}$	$-\frac{\Delta\lambda}{\lambda}\Big\vert_{exp}$
100 K → 300 K							
Na	4.3	2.6	5.1	12.0	3.7	1.2	0.9 ± 0.9
K	5.1	3.2	7.1	15.4	4.2	0.6	1.3 ± 0.8
Mg	1.6	1.3	4.2	7.1	1.3	1.1	?
Zn	2.0	1.8	6.1	9.9	1.7	3.3	2.5 ± 0.5
Cd	2.1	3.4	11.4	16.9	1.7	6.3	7.5 ± 0.5
Al	1.3	2.3	5.1	8.7	1.1	1.5	0.5 ± 0.4
In	1.9	1.9	15.8	19.6	1.5	5.2	4.2 ± 0.5
300 K → 800 K							
Cu	3.3	1.8	7.9	13.0	2.7	7.3	6.8 ± 0.8
Ag	3.9	2.4	10.0	16.3	3.1	9.8	9.7 ± 1.0
Au	2.9	2.0	8.5	13.4	2.2	8.7	5.1 ± 1.5

The current theories of defect trapping are based on the rate equation approach [Ref.4.8, pp.179-181]; see also [4.52,53]). The simplest "two-state" model can be presented as

$$\dot{n}_f = -\lambda_f n_f - \nu c_t n_f$$

$$\dot{n}_t = -\lambda_t n_t + \nu c_t n_f \quad , \qquad (4.36)$$

where n_t, n_f are the probabilities that the positron is trapped and free at time t, λ_t and λ_f are the annihilation rates in the two states, ν is the specific trapping rate for unit defect concentration, and c_t is the concentration of the traps. The model can readily be generalized to include more than one kind of traps or thermal detrapping [4.54,55], or even prethermalization trapping [4.56].

In Table 4.5 we present examples of the experimentally determined trapping rates ν for various types of defects. As the absolute defect concentrations are difficult to obtain accurately, there is scatter in the experimental values. The differences are linked to uncertainties in, e.g., the formation energies and entropies for thermally generated defects. On the other hand, there has been considerable experimental and theoretical effort to determine the correct temperature dependence of the specific trapping rate, since it has an important bearing on the determination of defect activation energies and concentrations from positron annihilation data. The fraction of positrons that get trapped gives a measure of the defect concentration, and its temperature dependence can be used to find the activation energy.

Table 4.5. Experimental estimates for the trapping rate ν for various types of defects. For vacancies and voids, ν is given per unit concentration; S_v and E_v are the vacancy formation entropy and energy, respectively. For dislocations, ν is given per unit dislocation density

Monovacancies	$\nu \exp(S_v/k)$ [10^{14} s^{-1}]	E_v [eV]	Ref.
Al	12	0.66 ± 0.04	[4.57]
	4.3	0.62 ± 0.02	[4.58][a]
	25.40	0.76	[4.59][a]
	5 ± 2	0.62 ± 0.02	[4.60]
	6 ± 0.3	0.66 ± 0.01	[4.61]
In	150	0.48 ± 0.01	[4.61]
Pb	160 ± 30	0.54 ± 0.02	[4.61]
	20...180	0.58...0.65	[4.62]
Cd	1.9	0.39 ± 0.04	[4.57]
Zn	36	0.54 ± 0.02	[4.57]
Cu	130 ± 30	1.29 ± 0.02	[4.61]
	100	1.26	[4.63]
Ag	174 ± 35	1.16 ± 0.02	[4.61]
Au	12 ± 2	0.97 ± 0.01	[4.61]
	15 ± 5	0.98 ± 0.01	[4.60]
Mo	2.6		[4.64]
Dislocations	ν [cm^2 s^{-1}]	T [K]	
Al	0.066	77	[4.59]
Cu	1.5	300	[4.65]
Ni	0.3	77 - 300	[4.161]
Voids	ν [10^{16} s^{-1}]	Void radius [Å]	
Mo	8...40	15	[4.64]

[a] Quenching experiment

The models of the trapping process have been reviewed by BERGERSEN and PAJANNE [4.23], and HODGES [4.67] has tried to reconcile the various viewpoints, classical and quantum mechanical. The key questions are, what aspects of positron motion before it is trapped, if any, are important in determining the trapping rate, and how do they affect it?

A quantum mechanical approach to the trapping rate was first proposed by HODGES [4.68] using the Golden Rule transition rate

$$\nu = \frac{2\pi}{\hbar} \sum_{i,f,t} P_i |M_{i,f,t}|^2 \, \delta(\varepsilon_i - \varepsilon_f - \varepsilon_t) \quad , \tag{4.37}$$

where P_i is the occupation propability of an initial positron state $|i\rangle$, ε denotes energy, and the subscripts i, f and t denote the initial and final states of the medium and the trapped positron states, respectively. The matrix element $M_{i,f,t}$ involves the overlap between the initial and final states as well as factors describing the specific energy absorption mechanics. Usually a binding energy of the order of 1 eV of the positron to the trap (see Sect.4.2) will be liberated and consumed by elementary excitations of the solid. These are either electron-hole excitations or lattice vibrations. In metals, the former is expected to be the dominant one, and inserted into (4.37) with a Boltzmann-type distribution P_i of initial positron states leads to an essentially temperature-independent trapping rate for small traps, i.e., for fairly localized trapped states with a spatial extent much less than the positron de Broglie wavelength. Thus for a monovacancy or a small vacancy cluster in a metal this approach predicts little or no temperature dependence for the trapping rate ν.

However, there is a possibility that the mobility of the untrapped positron sets a limit to the trapping rate, shifting the process from the *transition-limited* regime discussed above to the *diffusion-limited* one. Since the trapping rate is proportional to some average of the positron density surrounding the trap, it may be decreased by local depletion in that density produced by the presence of the trap. In other words, the positron current may be insufficient to maintain the Golden Rule rate, a depletion of the positron density around the trap ensues and leads to a diminished trapping rate ν.

The simplest approach to study this effect is to apply diffusion theory, which, however, is only valid over distances large compared to mean free path. It is crucial that proper and physical boundary conditions are imposed. For example, the trapping rate ν is linearly proportional to the diffusion constant D_+ *only* in the extreme case of diffusion-limited trapping, where the trap acts as a perfect absorber and the positron flux actually vanishes at the trap boundary. This is not normally the case, and thus reported values of the positron diffusion constant based on this idea ($\nu \approx 4\pi r_d D_+/\Omega$, where r_d is the trap radius) may underestimate D_+ by several orders of magnitude.

BERGERSEN and McMULLEN [4.69] have applied the diffusion theory to look at the temperature and density dependence of trapping in a network of straight dislocations. Considering only electron-hole mediated trapping, they found a temperature-independent trapping rate at low temperature or if the trap potential was shallow. At high temperatures, the diffusion-induced decrease in the trapping rate became important if the trap potential was deep. The phonon contribution to the trapping rate, which would increase with T, was not included in the calculation as its effect would seem to be small.

McMULLEN [4.70] has formulated a quantum-mechanical perturbation approach to the effect of the positron mobility, described in terms of positron-phonon scattering, on the trapping rate at vacancies in metals and, in accord with BERGERSEN and TAYLOR [4.71], and McMULLEN and HEDE [4.72], concludes that the effect is small, even if the trapping is unphysically strong. Thus there is little reason to assume any temperature dependence for vacancy trapping in metals. McMullen's formulation is yet to be applied to more extended defects or to systems with phonon-mediated trapping.

For very large metallic voids with a radius greater than positron thermal wavelength, BRANDT's [4.40] semiclassical arguments would suggest leading T^{+1}-dependence of the trapping rate in the transition regime, determined by the absorption of the incident positron wave at the void boundary. This dependence would eventually saturate as one moves over to the diffusion-limited regime at high temperatures, where the strong trapping starts to deplete the positron flux in the vicinity of the void.

In conclusion, we stress that the process of positron capture at defects is basically a quantum-mechanical one, and classical diffusion ideas must be applied with extreme care. While the temperature dependence of trapping rates in vacancies and dislocations in metals is expected to be very small, that is not necessarily the case for voids or surfaces.

4.1.7 Self-trapping

As will be discussed in more detail below (Sect.4.2), any region of lower than average ion density is attractive to the positron and is a possible trap. Thus it may happen in principle that the positron itself deforms the lattice enough to produce a trap in a way analogous to a polaron in polar materials. Formation of such an entity would imply reduced positron mobility and reduced annihilation rate and perhaps temperature dependence of these quantities.

The idea of positron self-trapping was first introduced and studied for ionic crystals by GOLDANSKII and PROKOPEV [4.73]. In these materials the coupling to optical phonons is the important mechanism. Later theoretical and experimental [4.29,74] work has brought more evidence of the occurrence of such states, reminiscent of large polarons with a relatively modest lattice deformation extending to large distances. The self-energies of such states are believed to be small, around 20 meV, and the effective polaron masses are estimated to be 20 - 30 times the free mass. This kind of behavior could be called extended self-trapping, where the positron is accompanied by a weakly distorted lattice that has retained much of its periodic character.

Self-trapping in metals was first suggested by STOTT [4.75], whose idea has been followed by LICHTENBERGER et al. [4.76], SEEGER [4.77], and others, often in the context of any "anomalous" temperature dependence of annihilation characteristics. SEEGER [4.77] used an elastic continuum model with the positron coupled to the lattice dilation by a deformation potential, and suggested that a metastable self-trapped positron state would account for the observed behavior in Cd [4.76].

The idea of positron self-trapping in metals has been examined in more detail by HODGES and TRINKAUS [4.78], and LEUNG et al. [4.79]. The latter use a variational method to minimize the total ground-state energy of the positron in a deformable continuum system. If $\Psi(\underline{r})$ denotes the positron wave function, this energy is

$$E[\Psi] = -\frac{\hbar^2}{2m^*}\int d\underline{r}\Psi^*(\underline{r})\nabla^2\Psi(\underline{r})$$

$$-\frac{1}{2}\int d\underline{r}\int d\underline{r}'|\Psi(\underline{r}')|^2\Phi(\underline{r}-\underline{r}')|\Psi(\underline{r})|^2 \quad , \tag{4.38}$$

where the effective "elastic potential" is

$$\Phi(\underline{r}) = \frac{E_d^2}{2\pi^2 B}\left(\frac{\sin(q_D r)}{r^3} - q_D\frac{\cos(q_D r)}{r^2}\right) \quad . \tag{4.39}$$

E_d is the deformation potential constant and B the bulk modulus of the medium. The discrete nature is taken into account by introducing the Debye cutoff wave vector q_D to the elastic waves. The coupling constant characterizing the strength of the positron-lattice interaction is E_d^2/B. For a typical value of $q_D \sim 1.5$ Å^{-1} LEUNG et al. [4.79] found that self-trapping becomes energetically favorable ($E[\Psi] < 0$) for values of $E_d^2/B > 80$, corresponding to a localized self-trapped state of spatial extent of 2-3 Å. The lattice dilation that would accompany such a state is rather localized, with a mean width of about twice the Wigner-Seitz radius R_{WS}, with maximum local displacements around $0.2R_{WS}$. This kind of state should be termed compact self-trapping as it is formed by gross ionic rearrangement that destroys the periodic character. It also follows that the harmonic approximation that underlies the calculation would not give a particularly accurate description of such states.

However, the values of the coupling constant E_d^2/B in real metals lie well below the self-trapping threshold, typically $E_d^2/B \sim 20$-45. As the model of LEUNG et al. [4.79] is equivalent to a discrete ion model using a Debye phonon spectrum along with a Coulomb positron-ion potential, screened in the $q \to 0$ limit and with no Umklapp processes, it actually overestimates the positron-phonon coupling and thus provides a lower bound for the self-trapping threshold. More accurate screening and inclusion of Umklapp would push the threshold to even higher couplings. The conclusion to be drawn from the calculation of LEUNG et al. [4.79] is that positron self-trapping is unlikely to occur in metals. This is substantiated by the discrete lattice calculations of HODGES and TRINKAUS [4.78].

The extreme limit of compact self-trapping would be the case of a positron in a vacancy. Since the binding energy E_b of a positron in a vacancy in many cases exceeds the vacancy formation energy E_v, it certainly would be energetically favorable for a positron to generate a vacancy and be trapped in it. However, this kind of process cannot occur within the positron lifetime of $\sim 10^{-10}$ s, since the ion rearrangements require considerably longer times than this. The formation and migration energies of interstitials and vacancies are of the order of 0.5 eV in most metals, yielding self-diffusion constants typically less than 10^{-15} cm^2s^{-1} around room temperature. Thus their mobility is so low that there is no time for the rearrangements and trapping to occur; an atom typically takes a time of the order of a second to move a few lattice constants. At high temperatures, of course, an equilibrium nonzero vacancy concentration is present and trapping at these becomes the important process.

4.2 Defects in Metals

4.2.1 Electronic Structure of Defects

The defect properties of metals are markedly different from those of insulating solids. Metals are electrical conductors, and any charge perturbations produced by local rearrangements of ions at the defect region are effectively screened out within a length of the order of the unit cell radius. The electronic structure of defects is not easy to describe in terms of local wave functions, since the charge transferred to or from the defect region has contributions of all symmetries and energies in the conduction band. Furthermore, especially if one is interested in calculating defect formation energies, the calculations should be carried out self-consistently. This task is enormously complicated by the loss of three-dimensional lattice symmetry at the vicinity of the defect.

Nevertheless considerable effort has been put into calculating various electronic defect properties of metals. Excluding the whole realm of magnetic and nonmagnetic impurity problems, the theoretical attack has mostly been on static properties, e.g., formation energies and volumes, including lattice relaxations near the defect, or on the effect of lattice defects on transport related properties (electrical resistivity, thermopower, specific heat and spin susceptibility, etc.), or on the mechanical properties of defective solids. All these problems are related to important materials science problems, e.g., the recovery of radiation-induced damage in metals [4.80].

The theoretical approaches to defect electronic structure are numerous. The applications to vacancies in metals have recently been reviewed by EVANS [4.81]. There are many current texts about surfaces (see, e.g., [4.82,83]). One line of approach has been to apply tight-binding ideas in various forms [4.84-86] or related methods based on generalized Wannier functions [4.87]. Although this class of methods is both physically intuitive and in principle powerful for systems with narrow d or f bands, difficulties arise with proper inclusion of sp-type electrons and achievement of true self-consistency, and so far no really quantitatively meaningful results have been obtained for metals.

A widely used method in defect calculations is based on the use of empirical pairwise potentials between metallic atoms [4.88]. Defect formation energies and equilibrium atomic positions can be obtained via minimizing total energies calculated by straightforward summations. The method is simple in principle and can be quite useful in predicting qualitative trends or in, for example, simulating radiation damage production together with the molecular dynamics technique. The quantitative results are, however, sensitive to the choice of the interatomic potential and the method gives no information about the redistribution of electrons around the defect. Somewhat more sophisticated is the pseudopotential approach [4.89], which has proved quite successful in explaining the bulk properties of sp-bonded simple metals. In that method one constructs the effective interatomic potential by linearly screening a "bare" pseudopotential, which is designed to have the correct scattering properties for conduction electrons in a way that the orthogonality to core states is "pseudized" out. The screening dielectric function is obtained from the theory of the uniform electron gas. Since the pseudopotential is weak, perturbation theory is expected to work, and in most cases one evaluates energies to second order in the pseudopotential. In the defect case, due to the linearity assumption, this amounts to just evaluating the lattice sums with a structure factor modified for the defect geometry. There is no guarantee, however, that the linear response is valid at the defect region and one has to be careful in assessing the apparent merits of this method [4.90].

The pseudopotential method can also be incorporated in a nonlinear, self-consistent calculation suitable for simple metals. In that one takes the bare atomic pseudopotentials and calculates the conduction electron charge density in their presence, with appropriate approximation for exchange and correlation. The lack of symmetry in the defect region makes such a calculation very difficult as one has to numerically solve the Schrödinger equation in awkward geometry. Such calculations have been carried out for some simple metal surfaces, see [4.91], by matching the numerical solutions to bulk band states a few atomic layers inside the metal.

Another possible approach would be to use a large supercell containing the defect and impose periodic boundary conditions. Thus one could make use of the fast and effective schemes developed for band-structure calculations and would not be limited to simple metals. Apart from surface calculations simulated by thin films [4.92], such calculations in a self-consistent form have not been attempted so far. LOUIE et al. [4.93] have used related techniques for a vacancy in Si. Calculations with finite clusters of atoms could in some cases also have relevance to defect studies.

A great simplification in the problem can be achieved if the discrete lattice structure is smeared out to a more symmetric background potential so that conduction electron distribution can be solved self-consistently with only modest numerical labor. Hopefully, the lattice effects can be reintroduced via perturbation theory. This is best done for simple metals, which have nearly free-electron type Fermi surfaces. The exchange and correlation effects are most often included as local density dependent potentials; the formal foundations of the so-called density-functional method were laid by HOHENBERG and co-workers [4.94,95]. LANG and KOHN [4.96] were the first to apply these ideas in a fully self-consistent study of metal surfaces, and the method has subsequently been applied to the vacancy problem [4.97-99], to adhesion [4.100,101], and to various impurity problems [4.102, 103]. It has also been refined to take better account of the lattice potential [4.104,105]. Since the following chapters will mostly make use of electron structure results of defects obtained by the KOHN-SHAM method [4.95], we will give here a short description. For more details, the reader is referred to the recent article of GUNNARSSON and LUNDQVIST [4.106] and to the reviews by LANG [4.107], and EVANS [4.81].

The ground-state energy of a system of electrons interacting with static ions via a local pseudopotential can be written as a functional of the electron number density $n(\underline{r})$

$$E[n] = T_s[n] + E_{xc}[n] + \frac{e^2}{2} \int d\underline{r} \int d\underline{r}' \frac{n(\underline{r})n(\underline{r}')}{|\underline{r} - \underline{r}'|}$$

$$+ \int d\underline{r} \sum_{\underline{R}} w(|\underline{r} - \underline{R}|)n(\underline{r}) + \frac{1}{2} \sum_{\substack{\underline{R}\underline{R}' \\ \underline{R}\neq\underline{R}'}} \frac{(Ze)^2}{|\underline{R} - \underline{R}'|} \quad . \tag{4.40}$$

The first terms of (4.40) are the noninteracting kinetic energy, exchange-correlation energy and Hartree electrostatic energy, respectively. The ionic potential w has a long-range attractive Coulomb part and a short-range repulsive part w_R due to core orthogonalization

$$w(\underline{r}) = -\frac{Ze^2}{r} + w_R(r) \quad . \tag{4.41}$$

It is customary and convenient to add and substract a fictitious neutralizing positive background (ionic density distribution) $n_+(\underline{r})$ to (4.40). Then the auxiliary one-electron KOHN-SHAM wave functions ψ_i of the density-functional theory satisfy the self-consistent Schrödinger equation

$$\left[-\frac{\hbar^2}{2m}\nabla^2 + \mu_{xc}(n;\underline{r}) + \phi(n;\underline{r}) + \delta V(\underline{r})\right]\psi_i(\underline{r}) = \varepsilon_i\psi_i(\underline{r}) \quad . \tag{4.42}$$

Above, $\mu_{xc}(n;\underline{r}) = \delta E_{xc}[n]/\delta n(\underline{r})$ is the exchange-correlation potential,

$$\phi(n;\underline{r}) = e^2 \int d\underline{r}' \frac{n(\underline{r}') - n_+(\underline{r}')}{|\underline{r} - \underline{r}'|} \tag{4.43}$$

is the full electrostatic potential and

$$\delta V(\underline{r}) = \sum_{\underline{R}} w(|\underline{r} - \underline{R}|) + e^2 \int d\underline{r}' \frac{n_+(\underline{r}')}{|\underline{r} - \underline{r}'|} \quad . \tag{4.44}$$

The electron density is constructed by summing over the occupied states below the Fermi energy ε_F

$$n(\underline{r}) = \sum_{\varepsilon_i < \varepsilon_F} |\psi_i(\underline{r})|^2 \quad . \tag{4.45}$$

The exchange and correlation energy $E_{xc}[n]$ is not known exactly, and one must in practice resort to some approximation. In most cases, a local density approximation to $\mu_{xc}(n;\underline{r})$ works surprisingly well [4.106].

In a simplified defect problem $n_+(\underline{r})$ is specified to mimic the ionic configuration, with enough symmetry to facilitate the numerical solution; $\delta V(\underline{r})$ is then added via perturbation theory. For example, a three-dimensional vacancy cluster can be approximated by

$$n_+(\underline{r}) = n_0\theta(r - R) \tag{4.46}$$

where n_0 is the constant number density of the ions and conduction electrons in the metal and R is the radius of the void, namely

$$R = \sqrt[3]{N}\, R_{WS} \tag{4.47}$$

where N gives the number of vacancies contained in the void. In this continuum or "jellium" model the ionic charges can be thought to have been smeared to a uniform background density with a spherical hole in it. Figure 4.3 shows the electron density profiles at vacancy clusters in aluminum, calculated by this method [4.108, 109].

The density-functional approach, augmented with the jellium-based viewpoint, is a powerful and practical method to calculate electron redistribution and potentials at defects, and it provides input for the positron-state calculations to be

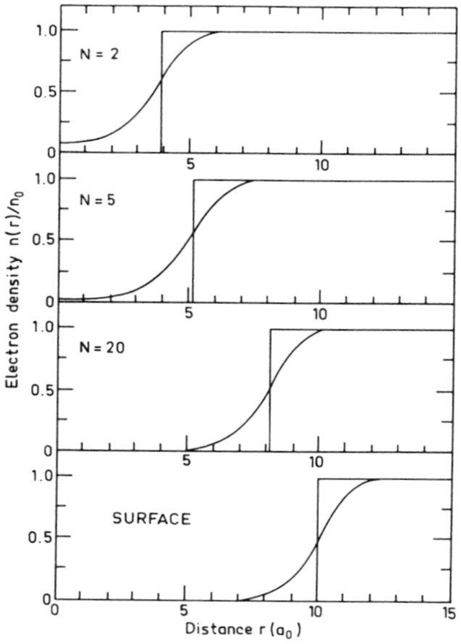

Fig.4.3. Self-consistent electron density profiles as a function of the microvoid size in aluminum. N is the number of vacancies in the void and the sharp edge in the background density denotes the radius of the void. The lowest profile corresponds to a planar surface placed arbitrarily at r = 10 a.u.

described below by giving the Hartree potential $\phi(\underline{r})$ and the electron density $n(\underline{r})$. With these, a local positron potential can be constructed within the positron pseudopotential scheme (see Sect.4.1.3). It may also be noted that the density-functional approach can in principle be applied to the nonlinear screening problem of the positron itself; in fact, an approximate solution has been presented by LEUNG et al. [4.110] by scaling arguments from the corresponding proton screening problem [4.102].

4.2.2 Positron-Defect Interaction

As argued in Sect.4.1.3 a good approximation for the positron potential in a perfect metal is provided by the Hartree potential of the ions and conduction electrons. The electron-positron correlation energy only affects the absolute energy value; as far as the positron distribution is concerned it can be neglected since it is a slowly varying function of the electron density which, on the other hand, is almost constant in the interstitial regions between the ions. This is not generally the case in the vicinity of a lattice defect, where the variations in the conduction electron density may be quite drastic, as in the extreme case of a surface where the electron density eventually goes to zero. Thus in a defective metal the positron potential has to include besides the full Hartree potential of the *defective* metal also contributions from the electron-positron correlation.

In calculating the positron wave function and the positron defect interaction, the positron pseudopotential picture introduced in Sect.4.1.3 is again the most practical. Figure 4.4 gives a schematic picture of the positron full wave function

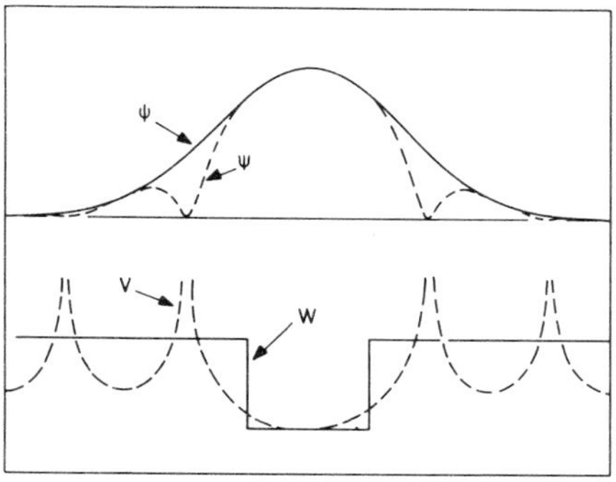

Fig.4.4. Schematic positron wave function and potential in a metal vacancy. Ψ and V are the full wave function and potential whereas ψ and W are the pseudo wave function and pseudopotential

and pseudo wave function in a localized state at a metal vacancy. Now it is convenient to write the full positron potential in a system consisting of different types of atoms A_i in sites R_i as

$$V(\underline{r}) = \sum_i V_{ref}(\underline{r} - \underline{R}_i) + \Delta V(\underline{r}) + V_{corr}(\underline{r}) \quad , \tag{4.48}$$

where

$$V_{ref}(\underline{r}) = \begin{cases} V_{A_i}(\underline{r}) \quad , & \underline{r} \text{ in the atomic cell centered in } \underline{R}_i \\ \\ 0 \quad , & \underline{r} \text{ outside the atomic cell} \end{cases} \tag{4.49}$$

and $V_{A_i}(\underline{r})$ is the Hartree potential in a perfect reference crystal consisting solely of type-A_i atoms, evaluated with the same cell size as in the actual crystal. If r in (4.48) is in a vacant cell, $V_{ref}(\underline{r}) = 0$. $\Delta V(\underline{r})$ is the difference between the full electrostatic potential and $\sum_i V_{ref}(|\underline{r} - \underline{R}_i|)$ and is due to the charge transfer between different cells in the defective crystal. $V_{corr}(\underline{r})$ is the potential term due to electron-positron correlation. In factorizing the positron wave function (4.8) we now define $U(\underline{r})$ so that in each cell it is the Wigner-Seitz wave function $U_i(\underline{r})$ calculated for the potential $V_{A_i}(\underline{r})$. Thus in a perfect, pure metal $U(\underline{r})$ coincides with the ground-state Wigner-Seitz wave function. The positron pseudopotential is thus in the ith cell

$$W(\underline{r}) = \Delta V(\underline{r}) + V_{corr}(\underline{r}) + E_{0i} - \frac{\nabla U_i(\underline{r})}{U_i(\underline{r})} \cdot \nabla \tag{4.50}$$

where E_{0i} is the Wigner-Seitz energy for a perfect metal of atoms A_i. Thus, apart from the gradient term in (4.50), the construction of the positron pseudopotential in the imperfect solid boils down to evaluating the E_{0i} for the constituent cells and the change in the electron density $n(\underline{r})$ between the real crystal and the reference one. The latter determines the electrostatic contribution $\Delta V(\underline{r})$ and, in the local density approximation, also the electron-positron correlation potential $V_{corr}(\underline{r})$.

The calculation of the charge transfer and the electrostatic defect potential $\Delta V(\underline{r})$ proceeds via the methods discussed in the preceding section. A particularly useful approach is the jellium-based density-functional theory, in which $\Delta V(\underline{r})$ for a positron is equal to the negative of the potential $\phi(\underline{r})$ defined in (4.43). This and some other ways of estimating $\Delta V(\underline{r})$ are discussed in the applications below.

The electron-positron correlation energy $E_{corr}(n)$ in a homogeneous electron gas is a slowly varying function of the electron density n [4.34,50]. Thus for electron densities varying not too rapidly at the defect region a local density approximation for the correlation potential will suffice,

$$V_{corr}(\underline{r}) = E_{corr}[n(\underline{r})] \quad . \tag{4.51}$$

The evaluation of the correlation energy in real metals has been discussed in Sect.4.1.4 in some detail, and bulk values are given in Tables 4.6 and 4.7. For defected metals, the spatial dependence of the "real" metal correlation energy may be significant. In model calculations based on the continuum approximation for the ionic charges the core effects can be estimated as follows. Assuming that the core electrons are tightly bound and consequently do not relax appreciably around a defect, the effective electron density can be written as

$$n_{eff}(\underline{r}) = n(\underline{r}) + Zn_{+}(\underline{r}) \frac{\Gamma_c}{\Gamma_v} \quad , \tag{4.52}$$

where $n(\underline{r})$ is the valence electron density, Z the valency of the metal and n_+ the ion number density. The latter term represents the effective core electron density estimated from the unrelaxed valence electron profile $Zn_+(\underline{r})$. This effective density can again be inserted in (4.51). In simple metals where the core fraction is small the correlation energy can to good accuracy be estimated from the conduction electron density $n(\underline{r})$ alone.

Outside a metal surface the electron density goes to zero and the local approximation for the correlation potential approaches the positronium value -0.52 Ry. However, well outside the metal the positron is not screened and the polarization of the surface leads to an image charge interaction. Thus in the case of a metal surface the local approximation is no more valid and one has to find an approximation which gives the image potential as an asymptotic limit. The positron surface interaction is described in more detail in Sect.4.2.4.

Table 4.6. Values for the positron zero-point energy E_0 [4.34], the positron-electron correlation energy E_{corr} [4.112] and the experimental core contribution in the angular correlation curve $A_c/(A_c+A_v)$ in simple metals; for references see [4.97]

Metal	E_0 [eV]	E_{corr} [eV]	$A_c/(A_c+A_v)$ [%]
Li	1.8	-7.3	17
Na	1.8	-7.1	19
K	1.4	-6.9	25
Rb	1.2	-6.9	28
Cs	1.1	-6.9	35
Mg	3.1	-7.9	18
Zn	4.6	-8.4	49
Cd	4.4	-8.0	57
Hg	4.2	-7.9	71
Al	4.8	-8.8	17
Ga	4.5	-8.6	39
In	4.2	-8.2	41
Tl	4.1	-8.0	38
Sn	4.6	-8.6	32
Pb	4.4	-8.4	42

Table 4.7. Values for the positron zero-point energy E_0 and the electron-positron correlation energy E_{corr} in transition metals [4.111]

Metal	E_0 [eV]	E_{corr} [eV]
Sc	3.7	-8.2
Ti	4.8	-9.2
V	5.7	-9.4
Cr	6.3	-9.8
Fe	5.7	-10.1
Co	5.7	-9.9
Ni	5.3	-9.9
Cu	4.9	-9.5
Y	3.3	-8.0
Zr	4.3	-9.0
Nb	5.4	-9.6
Mo	6.0	-9.8
Rh	5.2	-9.8
Pd	5.4	-9.6
Ag	4.5	-9.5
Ta	5.3	-10.1
W	6.3	-10.2
Ir	6.7	-10.1
Pt	5.6	-9.9
Au	5.3	-9.9

The last two terms of the pseudopotential (4.50) can loosely be interpreted as the kinetic energy contributions. E_{0i} is the ground state Wigner-Seitz energy ("zero point energy") of the positron in a metal consisting of i-type atoms. E_0 values for most of the simple [4.34] and transition [4.111] metals are given in Tables 4.6,7. They are roughly proportional to the square root of the electron density at the Wigner-Seitz cell boundary [4.112,113], a result readily understood from an analogy with the harmonic oscillator: the curvature of the positron potential around its minima in the interstices is proportional to the interstitial electron density as seen from the Poisson equation.

In a defective metal or in an alloy the cell size is not necessarily the same as in a pure metal. We can avoid the calculation of E_0 for different sized cells by approximating the dilation effect as suggested by ARPONEN et al. [4.114] taking

$$E_0(\Omega) = E_0\left(\frac{\Omega_0}{\Omega}\right)^{2/3} \tag{4.53}$$

where Ω_0 is the volume of the unit cell used to calculate E_0 and Ω is the cell size in the defective metal. Using this approximation in a continuum model the local part of the kinetic energy in the positron pseudopotential is

$$E_0(\underline{r}) = E_0 \left[\frac{n_+(\underline{r})}{n_{+0}(\underline{r})}\right]^{2/3} \tag{4.54}$$

where n_{+0} and n_+ are the ion densities of the perfect and defective lattice, respectively. This formula offers a practical way to estimate the positron kinetic energy in complicated defect structures.

The final contribution to the positron pseudopotential is the gradient term, which depends on the pseudo wave function and, in principle, requires a self-consistent calculation. In the case of a defect-localized positron state the gradient term is more significant than in a pure metal, where it represents only a minor correction. It should therefore be carefully considered whether the gradient term can be neglected in comparison with the other terms of the pseudopotential. We return to this question in Sect.4.2.4a.

4.2.3 Annihilation Characteristics

In an inhomogeneous electron system the annihilation rate is given by

$$\lambda = \tau^{-1} = \pi r_0^2 c \int d\underline{r} |\Psi(\underline{r})|^2 \tilde{n}(\underline{r}) \tag{4.55}$$

where $|\Psi|^2$ is the positron density and $\tilde{n}(\underline{r})$ is the electron density at the positron site. Using (4.13), we obtain the local density formula [4.115]

$$\lambda = \int d\underline{r} |\Psi(\underline{r})|^2 \lambda[n(\underline{r})] \quad . \tag{4.56}$$

In real metals positrons annihilate in addition to the nearly free valence electrons also with the core electrons of the ions. In simple metals the annihilation probability with a core electron is about 20%, as seen from Table 4.6, whereas in transition metals it may be up to 90%. The core electron annihilations cannot be treated through (4.13) and (4.56) since as tightly bound electrons they have a smaller enhancement factor than that in (4.14) and their density variations may be too rapid for the local approximation to work.

Instead, West's approach (4.17) can in the defect region be adopted into (4.56), in the continuum model, using the effective density (4.52) in λ. This allows us to take into account all electrons. However, there is no firm evidence of the applicability of this approach to metals where the core annihilation probability is large due to the d electrons. Around the defect the outermost d shell may relax significantly resulting in a different core annihilation probability than in bulk metal.

The angular correlation curve measures the momentum distribution of the annihilating electron-positron pair. At the vicinity of a lattice defect the electron density distribution deviates from that of the perfect lattice; for a positron localized at the defect the overlap with the core electrons is diminished as ions are missing from the defect. This brings about narrowing in the angular correlation curve. On the other hand, the conduction electron density is also depleted at the defect region, with a concomittant narrowing in momentum distribution. Thus the angular correlation curve for a trapped positron is always more peaked at small momenta than for a free Bloch-like positron. On the other hand, the localization increases the positron momentum which is reflected as a slightly pronounced tail at large momenta. In simple metals like the alkalis, Al or Mg, the small core contribution can be neglected altogether and one can study how the free-electron parabola is changed for a positron trapped by a defect.

The independent-particle model (4.18) can, of course, be directly used in calculating the angular correlation curve of a trapped positron. Approximating the angular correlation curve of the real metal by an inverted parabola we have already assumed the positron pseudopotential picture and the continuum model for the metal. We have to use the same model also for the defect. The electron states ψ_i in (4.18) around the defect have to be calculated in the continuum model and for Ψ_+ we have to use the wave function of the trapped positron. The independent-particle approximation is not, however, always practical since a single-particle Hartree-Fock representation for the valence electron distribution has to be known. If the electron density is calculated by the density-functional theory one can use the KOHN-SHAM wave functions. These are not, in principle, the proper electron wave functions, but the error due to this can be expected to be small [4.97]. The KOHN-SHAM wave functions for a homogeneous electron gas are just the plane waves producing the free-electron parabola.

Two approximations have been presented which try to estimate the momentum distribution directly from the electron density. BRANDT [4.115] has suggested a local formula

$$I(p_z) = \frac{r_0^2 c}{2\pi} \int d\underline{r} |\Psi(\underline{r})|^2 \lambda[n(\underline{r})][p_F(\underline{r})^2 - p_z^2]\theta(p_F(\underline{r})^2 - |p_z|^2) \quad , \tag{4.57}$$

which means simply that at each point $\underline{r}$ the positron annihilates as in a uniform electron gas and produces a free-electron parabola the width of which is determined by the local density $n(\underline{r})$. ARPONEN et al. [4.114] have proposed an approximation which partly takes into account the nonlocal character of the momentum distribution. The partial annihilation rate at total momentum $\underline{p}$ in this so-called mixed-density approximation is

$$\begin{aligned} \Gamma(\underline{p}) \; &\propto \int d\underline{r} \int d\underline{r}' \exp[i\underline{p} \cdot (\underline{r} - \underline{r}')]\Psi^*(\underline{r})\Psi(\underline{r}') \\ &\times g[p_F(\underline{R})|\underline{r} - \underline{r}'|]\sqrt{\lambda[r_s(\underline{r})]\lambda[r_s(\underline{r}')]} \quad , \end{aligned} \tag{4.58}$$

where $\underline{R} = (\underline{r} + \underline{r}')/2$ and g is a function related to the electron-electron pair correlation function,

$$g(z) = \frac{3}{z^3} [\sin(z) - z \cos(z)] \quad . \tag{4.59}$$

The mixed density approximation takes into account also the momentum associated to the localized positron state, which is ignored in the local model. If the positron wave function and the electron density vary slowly compared to g, one can substitute $\Psi^*(\underline{r})\Psi(\underline{r}') \approx |\Psi(\underline{R})|^2$ and $\sqrt{\lambda[r_s(\underline{r})]\lambda[r_s(\underline{r}')]} \approx \lambda[r_s(\underline{R})]$ and the mixed-density formula reduces to the local formula (4.57).

As already mentioned, the independent-particle model (4.18) does not account for the enhancement in the absolute annihilation rate, which is, on the other hand, included in both the local and the mixed-density approximation. Because of the nonlocal character of the momentum distribution, the enhancement does not play as important a role as in the case of the total rate and the independent-particle model works quite well as discussed earlier.

4.2.4 Applications

In this section we review the results of calculations carried out by using different types of models for the defects and positron-defect interaction. Most of the calculations are based on ideas presented in the previous sections, but also some supplementary models are described. The aim of the calculations is to evaluate the measurable quantities, the lifetime, and the angular correlation curve of the trapped positron. Also the trapping rate can be measured, but with a much poorer accuracy (see Sect.4.1.5). Except for thermal detrapping at very high temperatures [4.116] there is no direct method to measure the binding energy of the trapped positron and thus very little can be said about the accuracy of the binding energies predicted by different models.

Most of the calculations have been done for aluminum, which seems to be the "best" metal for positron studies. Aluminum has an almost spherical Fermi surface and thus the conduction band can be approximated by nearly free electrons. The positron annihilation probability with core electrons is small and can to a first approximation be neglected. The positron-defect interaction in Al is strong. The positron gets trapped in vacancies, dislocations and voids, and in each type of defect the annihilation characteristics differ markedly from those of free positrons and also from each other. In alkali metals, which are perhaps the simplest metals, no positron vacancy trapping has been found. The explanation of this provides another crucial test for the theories.

In the following we will discuss separately different types of defects: vacancies, dislocations, impurities, alloys, vacancy clusters, and surfaces.

a) Vacancies

Several modifications of the positron pseudopotential picture have been used to calculate the positron trap states in vacancies of simple metals [4.68,97,114,117]. MANNINEN et al. [4.97] used the jellium model for the metal and calculated the electron density around the vacancy by the density-functional methods as discussed in Sect.4.2.1. They used the positron pseudopotential (4.50) omitting the gradient term and calculating the correlation potential locally from the valence electron density. The resulting trapping potential and the positron pseudo wave function are shown in Fig.4.5 for aluminum. The three different contributions to the trapping potential are also shown. The positron binding energy and the lifetime, calculated using (4.14,52,56), are given in Table 4.8, where also results of other calculations for aluminum are collected.

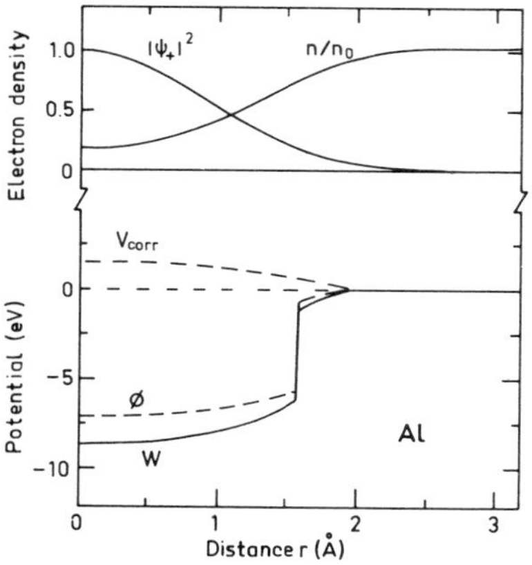

Fig.4.5. Electron and positron densities, n and $|\psi_+|^2$, and positron trapping potential V_+ in an aluminum vacancy. V_{corr} and ϕ are the correlation and electrostatic parts of the trapping potential

Table 4.8. Calculated positron binding energies E_b and lifetimes τ_v in an aluminum vacancy

	τ_v [ps]	E_b [eV]
HODGES [4.68]		3.81, 2.04
ARPONEN et al. [4.114]	243	2.62
MANNINEN et al. [4.97]	237	1.75
BRANDT [4.40]	241	6
MORI [4.117]	273	0.8
GUPTA and SIEGEL [4.119]	231	3.31

HODGES [4.68] used the pseudoatom picture in estimating the electrostatic part of the positron potential. His two values for the binding energy E_b in Table 4.8 are calculated with and without including electron-electron correlations into the screening dielectric function. ARPONEN et al. [4.114] used the jellium model and the nonlinear Thomas-Fermi approximation in calculating the electron density and the electrostatic potential in the defect. MORI [4.117,118] also used the Thomas-Fermi model but tried to take into account the discrete lattice structure around the vacancy. BRANDT [4.40] used quite a different model for the trapping potential. In his model the electron density in the jellium vacancy is calculated in a Thomas-Fermi approximation, but includes also the trapped positron as a stationary external charge distribution contributing to the self-consistent electrostatic potential. The electron density found in this way, together with the positive background, generate an electrostatic potential, which is used as the trapping potential for the positron. This model is not in the spirit of the positron pseudopotential idea, and it clearly overestimates the strength of the attraction to the vacancy, since it ignores the fact that the positron is accompanied by a screening cloud also in the bulk of the metal. Recently, GUPTA and SIEGEL [4.119] have performed an augmented plane-wave calculation for a vacancy in Al, and have estimated the positron binding energy and lifetime in it; they also noted that the temperature dependence of the trapped-state lifetime is considerable weaker than that of the bulk lifetime. Their calculation indicated somewhat lower electron densities inside the vacancy. Unfortunately, their calculation was not self-consistent and thus it is not possible to draw definite conclusions. Also the electron-positron correlation potential was not included. The lifetime estimates of various models are rather close to the experimental value, which differs clearly from the bulk rate of 160-170 ps [4.59,60]. This shows that the lifetime is rather insensitive to the model, whereas the binding energy is not.

The angular correlation curve for positrons annihilating in vacancies in aluminum have been calculated by using the approximative methods of Sect.4.2.3. ARPONEN et al. [4.114] proposed the mixed density approximation. Their results are shown in Fig.4.6 together with those obtained from the local formula (4.57) and the experimental points of KUSMISS and STEWART [4.120]. The core contribution has been subtracted from their data. The mixed density approximation is in a fairly good agreement with the experimental results, whereas the local approximation clearly overestimates the narrowing of the curve from the free-electron parabola, also shown in Fig.4.6. MANNINEN et al. [4.97] calculated the angular correlation curve in the independent-particle approximation by using the KOHN-SHAM wave functions. The result, also shown in Fig.4.6, is in agreement with both the experimental and the mixed-density results. Fig.4.7 shows the derivatives of the angular correlation curves in various approximations. Both the independent-particle and the mixed-density approximation seem to overestimate the portion of annihilations at large momentum values. This large angle tail is caused by the modifications of the positron momentum distribution due to the localization. Thus the experimental results seem to indicate a somewhat weaker localization and consequently a smaller binding energy for the positron. The local approximation for the angular correlation

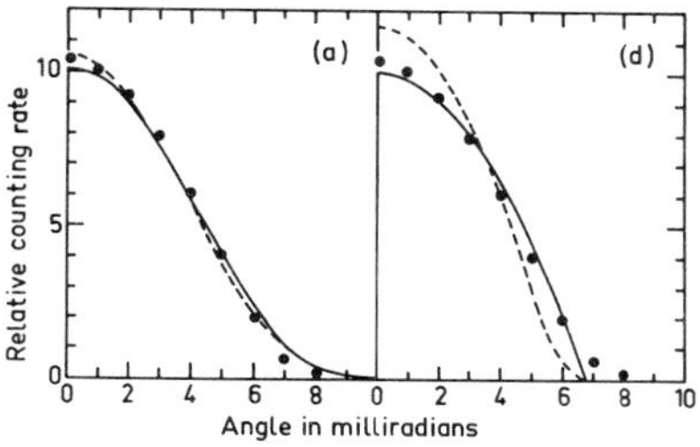

Fig.4.6a and b. Angular correlation curves for valence electrons in an aluminum vacancy: a) independent particle (solid curve) and mixed density (dashed line) approximations, b) local approximation (dashed line) and the free-electron parabola (solid curve). The experimental points are from [4.120]

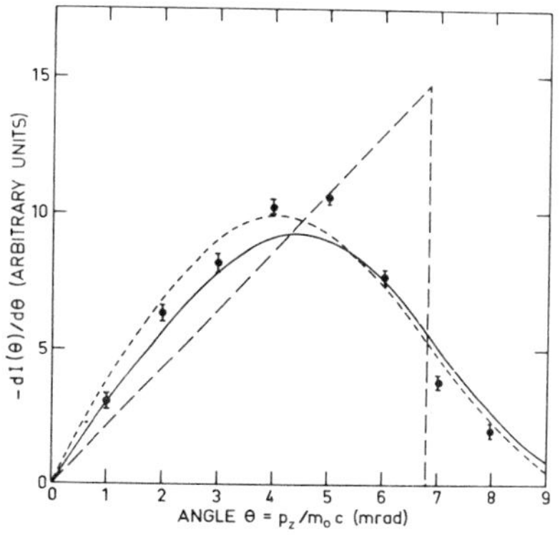

Fig.4.7. Derivatives of angular correlation curves of positrons trapped at vacancies in aluminum. The solid and the dotted lines are the results of the independent-particle and mixed-density approximations, respectively. The dashed line corresponds to the free-electron parabola. The experimental points are from [4.120]

curve seems not to be useful, when density variations are as rapid as in the case of a vacancy.

The calculations in metals other than aluminum have concentrated on estimating the binding energy and the lifetime of the trapped positron [4.68,40,97]. The only calculation of the angular correlation curve was made by ITOH [4.121] in liquid gallium using the mixed density formula, but his trapping potential was constructed as by BRANDT [4.40] and is therefore probably too strong. In Table 4.9 we give positron binding energies and lifetimes in monovacancies calculated via the positron pseudopotential picture [4.97]. The results of the lifetimes are rather close to the experimental values in most metals. In the alkali metals the binding energies are so small that the trapping does not seem probable, and even if it occurs, the lifetime of the trapped state differs so little from the bulk value that it is difficult to observe experimentally. This is in agreement with the experimental evidence that no trapping occurs in alkali metals [4.122]. On the other hand, also thermal detrapping operates most effectively in the alkalis.

No really quantitative calculations of the trapping rate ν have been made. HODGES [4.68] used the Golden Rule formula and, including electron-hole excitations, estimated the trapping rate per unit concentration to be of the order of $10^{15}\ s^{-1}$, which is of the correct order of magnitude (see Table 4.5).

The theories of the positron vacancy states in simple metals produce rather correctly the lifetime and (in Al) the angular correction results. Yet in all

Table 4.9. Positron binding energies E_b and lifetimes τ_v at metal vacancies. τ_b is the bulk lifetime. For references of the experimental lifetimes see [4.97]

Metal	E_b [eV]	τ_b[ps] Theory	τ_b[ps] Experiment	τ_v[ps] Theory	τ_v[ps] Experiment
Li	0	325	291	325	b
Na	0.02	381	338	391	b
K	0.02	426	397	436	b
Rb	0.26 [a]	436	406	458	b
Cs	0.01	445	418	452	b
Be	0.80	142	213	177	
Mg	0.59	249	235	307	255
Zn	1.70 [a]	161	179	233	240
Cd	1.77 [a]	185	175 [c]	276	250 [c]
Hg	1.74 [a]	169	138	266	165
Al	1.75	162	161	237	243
Ga	2.65 [a]	156	190	248	260
In	2.55 [a]	185	182	289	240
Tl	1.72	199	210	302	230
Sn	2.64	167	202	282	
Pb	3.44 [a]	168	195 [d]	291	280 [d]
Cu	0.44 [a]	163	132	205	
Ag	0.73 [a]	164		227	
Au	0.93 [a]	143	125 [c]	212	210 [c]

[a] Calculated from the Thomas-Fermi electron density
[b] Experiments seem to show that no trapping occurs in these metals [4.122]
[c] [4.158]
[d] [4.62]

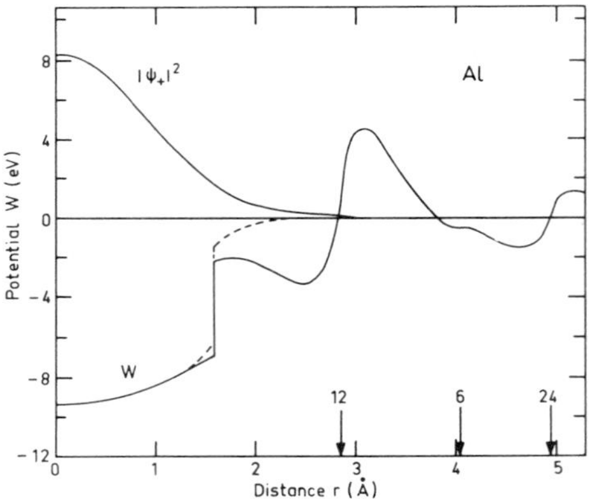

Fig.4.8. The positron pseudopotential W in aluminum vacancy including the gradient term. The dashed line denotes the pseudopotential without the gradient term. $|\psi_+|^2$ is the positron pseudodensity. The arrows indicate the positions of neighboring atomic cells, with the number of ions in them

calculations so far the gradient term of the positron pseudopotential (4.50) has been neglected. Outside the vacancy the gradient of the pseudo wave function is approximatively $|\nabla\psi/\psi| \sim \sqrt{2E_b} + 1/r$ which does not die away at long r values and is rather large at the site of the nearest neighbors even if the binding energy E_b approaches zero. Thus the next step should be a self-consistent calculation with the full positron pseudopotential. Figure 4.8 shows a result of such a calculation [4.123] for Al, where the gradient part of (4.50) has been spherically symmetrized around the vacancy, and the results of STOTT and KUBICA [Ref.4.32, Fig.1] have been used for $\nabla U(r)/U(r)$. The electron density profile was obtained as in MANNINEN et al. [4.97]. In comparison with [4.97], the binding energy of a positron in Al vacancy is found to increase from 1.8 to 2.4 eV. A similar calculation [4.123] for Na increases the binding energy from 0.02 to 0.17 eV. It may be argued that the spherical symmetrization overestimates the true effect of the gradient part of the pseudopotential. Yet the changes in vacancy lifetimes are small, less than 3%. Apart from including the gradient part of the pseudopotential one should try to go beyond the jellium model in producing the electron density and the electrostatic potential in the vacancy.

b) Dislocations

Dislocations, which are technologically the most important defects, are both experimentally and theoretically much more difficult to study than vacancies. Dislocations are usually produced by plastic deformation [4.124-126]. Unfortunately deformation generates, besides many types of dislocations, also other defects, grain boundaries, vacancies and perhaps small vacancy clusters. COTTERILL et al. [4.59] produced dislocation loops in aluminum by annealing of quenched specimens, but the dislocation density was too low to give accurate annihilation parameters for the trapped state. Theoretically a major problem is to find a realistic model for the atomic configuration around the dislocation core. Knowing this, the conduction electron density and the positron state can be solved using the methods described in the previous sections.

Two calculations have been carried out for positron trapped states at edge dislocations in aluminum [4.127,114]. MARTIN and PAETSCH [4.127] described the positron potential at the dislocation by a two-dimensional square-well potential. The dimensions of this square-well were estimated via a pairwise positron-ion interaction which was taken to be the screened Coulomb potential. The resulting estimates for the positron binding energy were 0.02-0.1 eV. As an approximation for the valence electron density MARTIN and PAETSCH [4.127] used the linear screening results calculated by BROWN [4.128]. In this model the reduction of the electron density at the dislocation core is only 15% from the bulk value, which together with the small binding energy gives the lifetime of the trapped state within 1% of the bulk lifetime 166 ps [4.59], whereas the experimental estimate for the lifetime at a dislocation in Al is about 230 ps [4.126]. MARTIN and PAETSCH estimated the shape of the angular correlation curve using a two-dimensional Gaussian as a trial wave function for the positron and unperturbed plane waves for the electrons in the independent-particle approximation. Comparing the results with the experimental curve [4.129] they estimated that the width of the positron wave function (deviation of the Gaussian) in a dislocation is about 2.5 Å.

ARPONEN et al. [4.114] used a somewhat different model for the edge dislocation in Al. For the ion density the continuum theory of the elasticity was used outside a core radius R_b, inside which the ion density was set to zero (this removes the unphysical singularity of classical elasticity theory at the origin). The conduction electron density was evaluated in the nonlinear Thomas-Fermi approximation. The positron trapping potential was constructed in the pseudopotential picture neglecting the gradient term. The core radius R_b = 0.85 Å was found to reproduce the experimental lifetime [4.129] of the trapped positron and the resulting positron wave function and electron density profile were used to calculate the angular correlation curve in the mixed-density approximation. In Fig.4.9 the ion and electron densities around the dislocation and the electrostatic part of the positron pseudopotential are shown. The electron density inside the dislocation core is less than one half of the bulk density and consequently the trapping potential is much stronger than in the model of MARTIN and PAETSCH [4.127] resulting in a higher binding energy of 2.8 eV. The calculated angular correlation curve together with the experimental points is shown in Fig.4.10. As in the case of a vacancy the calculated curve overestimates annihilations at large momenta, indicating a slightly too large binding energy. In fact HAUTOJÄRVI [4.129] has estimated from his angular correlation data that the binding energy is of the order of 0.6 eV.

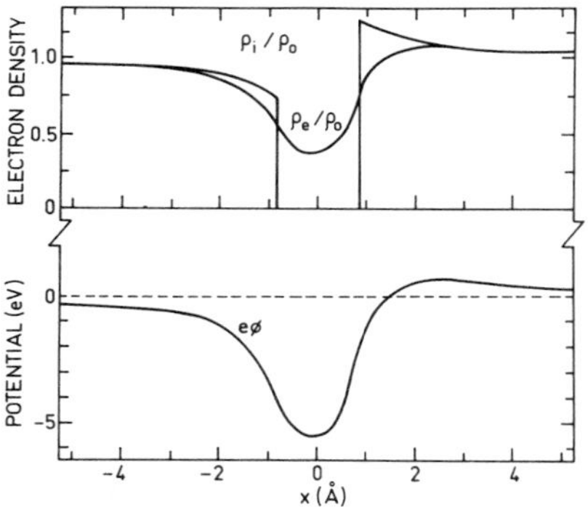

Fig.4.9. The ionic charge distribution ρ_i and the corresponding electron density ρ_e at an edge dislocation, plotted along the axis perpendicular to the glide plane. The resulting electrostatic potential is ϕ. [4.114]

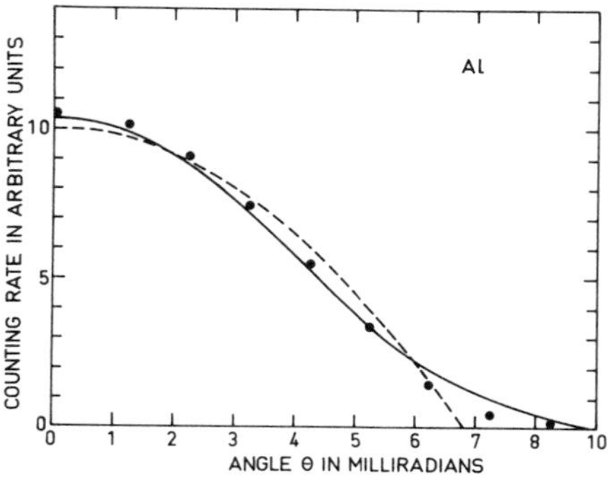

Fig.4.10. Angular correlation curve for positrons annihilating in dislocations of aluminum [4.114]. The solid curve is the result of the mixed-density approximation. The experimental points are from [4.129]. The dashed line is the free-electron parabola

The positron trapping rate at dislocations has been studied by BERGERSEN and McMULLEN [4.69]. They used the same model as HODGES [4.68] in the case of vacancies, but took into account besides the electron-hole production also the phonon excitations. The resulting trapping rate is almost proportional to the positron binding energy and dominated by the electron-hole excitations. For the binding energy of ~ 1 eV the calculated trapping rate in Al, 0.3 $cm^2\ s^{-1}$, is somewhat larger than the experimental estimate 0.07 $cm^2\ s^{-1}$ [4.59]. It is presumable that the orthogonality of the initial positron strate to the trapped state, which was not taken into account in [4.69], would reduce the theoretical estimate.

In all calculations so far infinitely long straight dislocations have been studied, where the positron is free to move along the dislocation line. However, there is no experimental evidence that this is the actual trapped state in a sample containing dislocations. There is also a possibility that the positron gets trapped at jogs, kinks or corners of the dislocations and is bound in all three dimensions. Only a clear anisotropy [4.114] in the angular correlation curve measured for oriented dislocation networks would prove that the positron is free along the dislocation line.

c) Impurities and Alloys

Small amounts of interstitial impurities have no effect on positron annihilation, since the positron is repelled from the region of high ion density. The substitutional impurities may attract or repel the positron depending as a first estimate on the impurity size: the smaller impurity the more free space is left for the positron. The only experimental suggestion for actual positron trapping by substitutional impurities is for helium in aluminum [4.130]. However, positron trapping at lattice defects, vacancies, dislocations and voids, may be strongly affected by the precence of impurities which are also attracted into these defects [4.131-133]. In the following we are concentrating only on substitutional impurities and alloys for which the pseudopotential picture can directly be applied [4.32] and discuss the affinities of positrons to the components of the alloy.

Assuming an average atomic cell radius R_a we can calculate the Wigner-Seitz wave functions $U_x(r)$ for different types x of cells. For simplicity we assume only two kinds of atoms A and B. Since the positron wave function must be continuous the Wigner-Seitz wave functions have to be adjusted to give the same amplitude at the interstices, so at the cell radius $U_A(R_a) = U_B(R_a)$. This means that the Wigner-Seitz wave function does not have the same normalization in different cells. In order to calculate the positron pseudo wave function we have to find approximations to the charge transfer between the cells, which gives the electrostatic and correlation parts of the pseudopotential. STOTT and KUBICA [4.32] estimated the electrostatic potential via the internal Fermi energy E_F^x [4.134] in the cell of type x. This approximation overestimates the charge transfer and gives an upper limit for the actual electrostatic potential. If the charge transfer is small the correlation potential in each cell can be approximated with that of the pure metal of the cell type. Neglecting the gradient term one can define the pseudopotential difference between different type of cells, i.e., $V_0 = E_0^A - E_0^B + E_F^A - E_F^B + E_{corr}^A - E_{corr}^B$ [4.32]. In Table 4.10 the pseudopotential differences and the three components of it are listed for several binary alloys with 50 at.% composition [4.32].

Table 4.10. Mean atomic radius, three contributions to the positron pseudopotential difference, and the net pseudopotential difference V_0 for a number of 50 at.% alloys. A negative V_0 in column 6 indicates a positron affinity for A over B in the alloy. The square-well strength parameter is also listed [4.32]

A	B	R_a [a.u.]	$E_0^A - E_0^B$ [Ry]	$E_{corr}^A - E_{corr}^B$ [Ry]	$E_f^A - E_f^B$ [Ry]	V_0 [Ry]	ξ
Al	Zn	2.94	0.04	-0.04	0.08	0.08	0.8
Mg	Al	3.09	-0.08	0.03	0.08	0.03	0.5
Mg	Zn	3.06	-0.03	0.00	0.13	0.10	1.0
Li	Mg	3.26	-0.10	0.04	-0.02	-0.08	0.9
Mg	Cd	3.29	-0.05	0.00	0.07	0.02	0.5
In	Cd	3.35	0.06	-0.03	0.06	0.09	1.0
Ag	Au	3.01	-0.03	0.00	-0.02	-0.05	0.7

Using the V_0 values of Table 4.10 one can roughly estimate if a single impurity atom is able to localize the positron. An isolated spherical potential well of depth V_0 and radius R_a will have at least one bound state when $\xi = R_0(2\pi V_0/h)^{1/2}$ is greater than $\pi/2$. The values for the parameter ξ are given in Table 4.10. In no cases does a bound state occur, but a large ξ indicates an enhancement of the positron wave function at the site of the impurity. Since the parameter ξ is proportional to the well radius R_a it is evident that large isolated impurity clusters are able to trap the positron. In an alloy the situation is somewhat more complicated. In a random alloy there exist, of course, clusters of one type of atoms but the surrounding medium is not a pure metal made of the other type atoms but an alloy and thus the trapping potential is weaker than the parameter ξ indicates.

STOTT and KUBICA [4.32] have shown that the positron affinities for different sorts of atoms in alloys can be determined by measuring the positron annihilation rate with the core electrons. The core fraction of the total annihilation rate can be separated using the angular correlation curve. STOTT and KUBICA [4.32] define for an alloy a reduced core rate

$$\Delta\lambda_c = \frac{\lambda_c - \lambda_c^A}{\lambda_c^A - \lambda_c^B} \qquad (4.60)$$

where λ_c, λ_c^A and λ_c^B are, respectively, the core rates for the alloy and the pure metals A and B. For the pure metal of type A $\Delta\lambda_c$ is zero and for the metal B it is equal to one. In the pseudopotential picture, assuming the pseudo wave function to be slowly varying and the core electron states well localized inside the Wigner-Seitz sphere, $\Delta\lambda_c$ can be expressed in the form

$$\Delta\lambda_c = \frac{1 - \sigma^A}{1 + \left(\frac{\alpha^A}{\alpha^B} - 1\right)\sigma^A} \tag{4.61}$$

where

$$\alpha^x = \int_\Omega d\underline{r}\,|U_x(\underline{r})|^2 \tag{4.62}$$

and

$$\sigma^A = \frac{\sum_{\underline{R}_A} \int_\Omega d\underline{r}\,|\psi(\underline{r} - \underline{R}_A)|^2}{\int d\underline{r}\,|\psi(\underline{r})|^2} \quad . \tag{4.63}$$

The advantage of (4.61) is that detailed computations of the overlap of the positron with the core electrons are unnecessary and the difficult problem of estimating the enhancement of the annihilation rate due to positron-electron correlations is avoided. If the valencies and the sizes of the ion cores of the constituents of the alloy are similar, the ratio α^A/α^B is very nearly unity and the denominator in (4.61) is also nearly unity. Furthermore, if the positron pseudopotential difference V_0 is small in magnitude the positron pseudodensity is nearly uniform throughout the alloy. Under these circumstances the reduced core rate would decrease linearly from the value 1 for pure B to 0 for pure A as the alloy composition is varied assuming that there is no volume change in alloying. Any deviation from a straight line of the reduced core rate vs concentration plot would indicate a preference of the positron for either A sites or B sites. For a concave curve the positron pseudo wave functions would be larger at A sites than B sites as a result of $V_0 > 0$. In contrast a convex curve would indicate a positron affinity for B over A with $V_0 > 0$.

KUBICA et al. [4.37] have measured the reduced core rate for LiMg alloys of different concentrations. Their result is shown in Fig.4.11. A clear departure from the linear dependence, shown with a dashed line, indicates positron preference for Li. The dotted line is calculated from (4.61) by using first-order perturbation theory to produce the positron pseudodensity in the random alloy. The solid lines in Fig.4.11 are calculated by using the following approach to estimate the positron pseudo wave function in dilute alloys. The pseudo wave function was calculated with periodical boundary conditions in a large spherical cell (radius $R = R_a/C^{1/3}$, C is the Li concentration) containing one impurity atom. The potential inside the impuri-

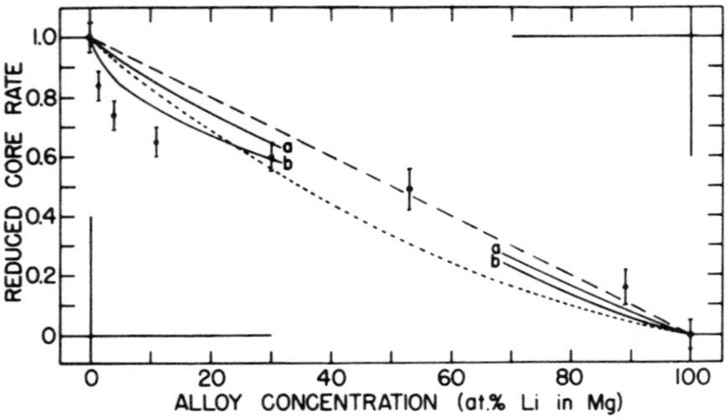

Fig.4.11. Reduced core annihilation rate as a function of the concentration in a LiMg alloy [4.37]. The dashed line is straight. The two solid lines and the dotted line are results of the pseudopotential model calculations described in the text

ty cell ($r < R_a$) was the pseudopotential difference V_0 and zero outside. The lines a and b in Fig.4.11 correspond to two different values 0.08 Ry and 0.2 Ry for V_0, respectively. The experimental points indicate much stronger positron affinity for Li than the theoretical result. One possible reason for this discrepancy is that lithium clusters are present at small lithium concentrations and do localize the positron.

LOCK and WEST [4.38] have made a detailed study of positron density enhancement effects in the alloy systems PbIn, MgIn, CuAl and CuNi. The experimental results were given in terms of a parameter indicating the peak height of the angular correlation curve. The measurements showed a strong affinity to Pb atoms in PbIn and a somewhat weaker preference of Cu over Ni and Al over Cu. LOCK and WEST presented an analysis of their results based on the pseudopotential concept, and showed that an assumption of a uniform positron density throughout the alloy fails for PbIn, CuAl and CuNi, but is fairly good for MgIn. This is in general agreement with the positron pseudopotential predictions.

HONG and CARBOTTE [4.135] have used the coherent potential approximation (CPA) [4.136] to calculate the positron distribution and the angular correlation curve in binary random alloys. Using the single-site approximation to CPA and omitting vertex corrections, they obtain the smearing in both the positron momentum distribution and in the electron Fermi momentum. These two are found to have an effect of the same order of magnitude on the angular correlation curve. The fraction of positrons f_A found in A sites is in their model

$$f_A = \frac{C_B\sigma_0^2}{C_B\sigma_0^2 + C_A(\sigma_0 - V_0)^2} \qquad (4.64)$$

where C_A, C_B are the atomic percentages of A and B atoms, respectively, and σ_0 is the coherent potential at the positron band edge. For small pseudopotential differences ($|V_0| \simeq 1$ eV) their results for the positron spatial distribution (4.63), are close to those of KUBICA et al. [4.37]. This seems to indicate that in most actual cases (see Table 4.10) the simple perturbation treatment of KUBICA et al. is rather good, but special attention must be paid to the possible existence of isolated precipitates in the actual alloy samples. It is also clear that a more careful treatment of the parameter V_0, including charge transfer, is in order.

Although isolated impurities have far weaker effects on positrons than, e.g., monovacancies, there are many interesting and important alloy phenomena to be explored by the positron technique, and more studies, both experimental and theoretical, should be concentrated on this area.

d) Vacancy Clusters

The most drastic changes in the annihilation parameters are brought about by vacancy clusters and surfaces [4.137]. Small spherical cavities, "microvoids", can be treated in the positron pseudopotential picture in a way similar to vacancies. HAUTOJÄRVI et al. [4.108], and JENA et al. [4.109] have calculated the electron density inside small voids in Al using the density-functional method (see Fig.4.3) and calculated the positron state in the pseudopotential model. The resulting lifetime of the trapped positron is shown in Fig.4.12 as a function of the void radius. When the void size increases the lifetime approaches rapidly its saturation value 500 ps, which is the low-density limit of (4.14), showing that the positron wave function is concentrated in the central part of the void which is practically empty of electrons. The model probably overestimates the lifetimes in di- and trivacancies, which in reality are far from spherical.

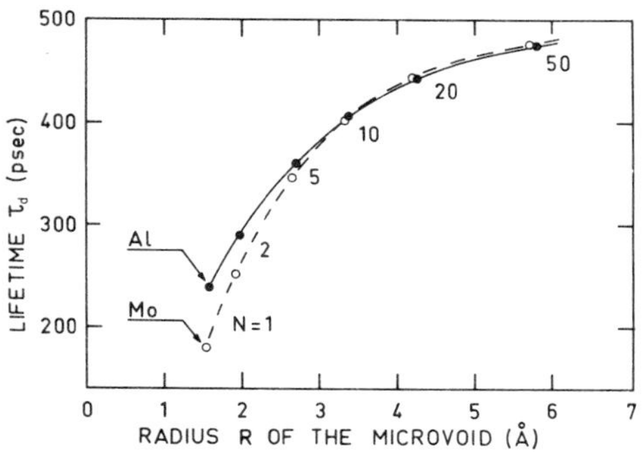

Fig.4.12. Positron lifetimes in microvoids of aluminum and molybdenum as a function of the void radius R [4.108]. N is the number of vacancies in the void

From a density-functional solution for the electron density profile the angular correlation curve can be calculated both in the independent-particle and mixed-density approximations. JENA et al. [4.109] have used the independent-particle model.

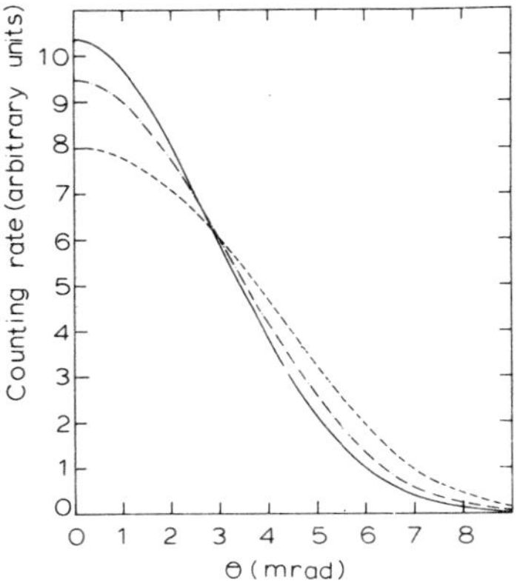

Fig.4.13. Angular correlation curves for microvoids in aluminum calculated in the independent-particle approximation [4.109]: --- for monovacancy; -.-.- for 8-atom void and —— for 27-atom void

Their results for a monovacancy, 8-atom void and 27-atom void in Al are shown in Fig.4.13. The curves for voids are broader than those of the mixed-density approximation [4.108], which tends to overestimate the narrowing in the case of large voids.

An unambiguous comparison between the theoretical and experimental annihilation characteristics is not possible in the case of microvoids since there is no independent method to measure the size of the voids (the resolution of the electron microscope is a few tens of Å). HAUTOJÄRVI et al. [4.108] have used the results of Fig.4.12 to estimate the dependence of the apparent microvoid size on the annealing temperature in electron irradiated molybdenum. In Fig.4.14 the lifetime of the trapped positron, measured by ELDRUP et al. [4.64], has been plotted as a function of the annealing temperature. The corresponding microvoid size, taken from Fig.4.11, is shown on the vertical scale on the right. LINDBERG et al. [4.138] have made annealing studies in neutron-irradiated aluminum single crystals. Their lifetime value for large voids is a little bit over 500 ps and for defects, tentatively identified as vacancy clusters bound to transmutation-produced Si precipitates, it varies from 250 to 400 ps depending on the annealing temperature. LINDBERG et al. also

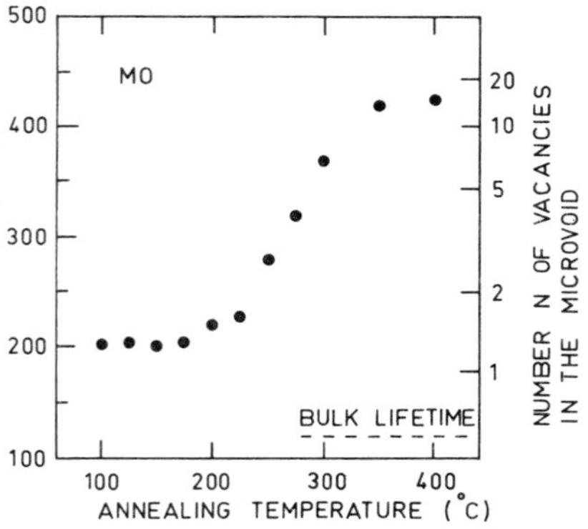

Fig.4.14. The lifetimes of the trapped positrons reflecting the apparent size of the microvoid as a function of the annealing temperature in molybdenum [4.108]. The experimental lifetimes are from [4.64] and the number of vacancies in the void is estimated from Fig.4.12

measured the angular correlation curve of the trapped state corresponding to the lifetime value 400 ps. The full width at half maximum of the curve is about 2.5 mrad. This together with the corresponding lifetime value of 400 ps is in agreement with the theoretical estimates of HAUTOJÄRVI et al. [4.108] for a spherical cluster consisting of 8-10 vacancies, whereas the angular correlation curves of the independent-particle model, Fig.4.13, seem to be too broad to be consistent with the lifetime value.

It is obvious that the model discussed above may break down for large clusters, where the idea of positron states extending through the void is not necessarily realistic. The correlation potential inside the void approaches the image potential limit, and the electrostatic part of the pseudopotential goes over to the surface dipole layer. In large voids a surface state may be energetically more favorable (see below). The transition from defect-centered to surface states is schematized in Fig.4.15, where a model potential

$$\begin{aligned} V(r) &= -\frac{e^2}{4(R-r_0-r)}\left[1-\frac{R-r_0-r}{2(R-r_0)}\right]^{-1} , \quad r < R \\ &= E_0 + E_{corr} + \Delta V , \quad r > R \end{aligned} \tag{4.65}$$

with a constraint $V(r) \geqq E_{corr}$ has been used to mimic the dependence of the positron-defect interaction on the microvoid radius R in Al. Equation (4.65) for $r < R$ is just the static image potential inside a hole of radius R in a perfect conductor; r_0 is the shift of the effective image plane with respect to the jellium edge ($r_0 = 0.8$ Å for an Al planar surface according to the self-consistent calculations of LANG and KOHN [4.139]. The quantity ΔV is the surface dipole potential for positrons (see below). We see that the positron density maximum shifts from the center of the cluster toward the surface for voids consisting of 20 - 40 vacancies. Bigger clusters should be considered as essentially macroscopic surfaces.

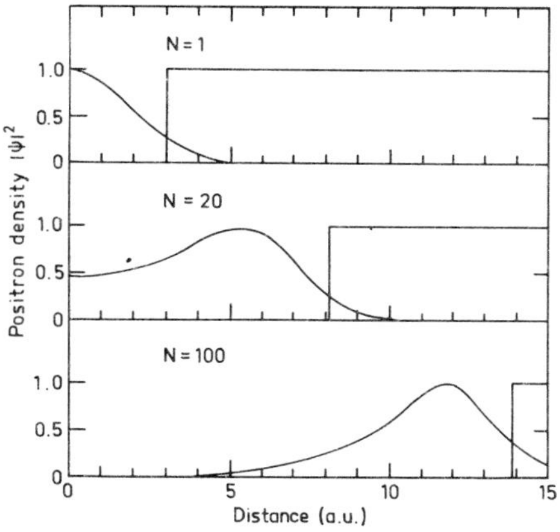

Fig.4.15. Schematic picture of the transition from an extended positron state to surface state as the void size increases

e) Surfaces

The areas of positron physics with possible relevance to surface science have recently been reviewed by BRANDT [4.137]. HODGES and STOTT [4.140] were the first to suggest that the extremely large changes in the annihilation characteristic observed in void-containing samples [4.141,142] could be accounted for by a positron surface state at the void internal surface. This state would be localized in the direction perpendicular to the surface, while the positron would be free to move parallel to the surface. The key ingredient in this localization is supposedly the long-range attraction between a charged particle and a semi-infinite medium: the image force between a positron in the vacuum and its screening cloud in the metal. Figure 4.16 illustrates schematically the positron potential at the surface region. Within the positron pseudopotential philosophy, the strictly one-particle contribution to this potential is given by $E_0 + D$, with E_0 the bulk band energy (Tables 4.6,7) and D the electrostatic surface dipole (generally $D > 0$ for electrons) originating from the spill-over of the conduction electrons into the vacuum. In addition, there is the electron-positron correlation potential V_{corr}, which in the surface case becomes energy dependent and nonlocal. Far inside the metal, however, it approaches the constant bulk value E_{corr}. Thus in the bulk

$$V(z) \xrightarrow[z \to -\infty]{} -\phi_p = D + E_0 + E_{corr} \tag{4.66}$$

which defines the positron work function ϕ_p. HODGES and STOTT [4.34] have estimated the positron work function in simple metals employing the surface dipole values from the work of HEINE and HODGES [4.112], they are displayed in column 2 of Table 4.11. HODGES and STOTT [4.140] took (4.66) to hold up to the surface $z = 0$, and in the vacuum side used the classical image potential for a static particle

$$V(z) = \frac{e^2}{4z} \quad , \tag{4.67}$$

imposing a minimum cutoff at the positronium energy of -6.8 eV, and found surface-bound states in a number of metals. NIEMINEN and MANNINEN [4.143] later refined this model by estimating the velocity-dependent corrections to the static result (4.67), and obtained estimates of the annihilation characteristics. NIEMINEN and HODGES [4.144] carried out a more detailed analysis of the dynamic image potential. They expressed the electron-positron interaction in terms ov virtual excitations of surface and bulk plasmons, and applied variational methods to the resulting polaron-type hamiltonian. The nonlocality of the image interaction is preserved in this approach, and the variational principle provides a lower limit to the positron binding on the surface. The binding energies E_b relative to the vacuum are given in Table 4.11. A stable surface state must fulfil $E_b > \phi_p$ and $E_b > 0$.

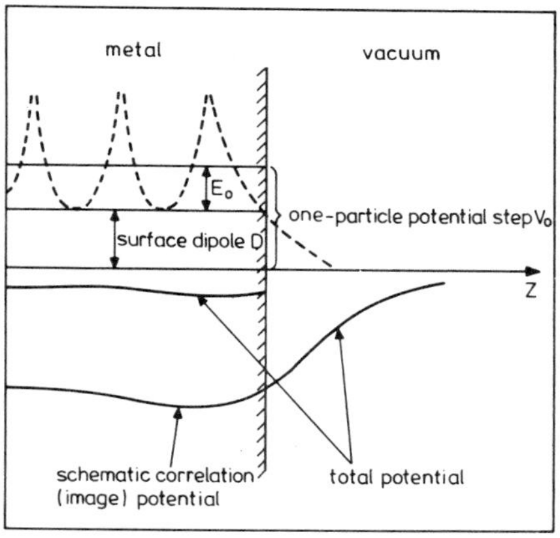

Fig.4.16. A schematic presentation of the positron potential near a metal-vacuum interface. $V_0 = E_0 + D$ is the one-particle potential step at the surface. Note that the image potential in reality is nonlocal

Table 4.11. Positron surface-state paramters [4.144]. D is the surface dipole [4.112] and V_0 the Hartree potential step (see Fig.4.15) E_b and ϕ_p are the positron binding energy and the positron work function, respectively. The last column gives two estimates for the ratio of the surface state and the bulk lifetimes

Metal	D [eV]	V_0 [eV]	E_b [eV]	ϕ_p [eV]	Stable surface states	Lifetime enhancement
Li	1.2	3.0	2.8	4.4	no	
Na	0.7	2.4	2.7	4.6	no	
K	0.2	1.8	2.6	5.2	no	
Rb	0.1	1.5	2.7	5.4	no	
Cs	-0.06	1.2	2.9	5.7	no	
Mg	2.0	5.2	2.3	2.7	no	
Zn	2.9	7.5	2.0	0.9	yes	2.3-3.0
Cd	1.9	6.3	2.1	1.8	yes	1.9-2.3
Hg	0.8	5.0	2.4	2.9	no	
Al	3.4	8.2	2.1	0.7	yes	1.8-3.2
Ga	2.0	6.5	2.3	2.0	yes	2.0-2.5
In	1.4	5.6	2.4	2.6	no	(2.2-2.7)
Tl	0.4	4.5	2.7	3.5	no	
Sn	1.2	5.8	2.5	2.7	no	(1.8-2.1)
Pb	0.3	4.6	2.8	3.8	no	

The results of NIEMINEN and HODGES [4.144], some of which are summarized in Table 4.11, confirm that stable surface states indeed are energetically favorable in some metals. The estimates of the lifetime enhancement show that the surface state has a lifetime, which is typically 2-3 times the bulk lifetime. A systematic experimental survey of the simple metals would be highly desirable; so far there are data only for Al [4.138,145], which are in rough agreement with the theoretical predictions [4.144]. The role of the defects on surfaces, which could presumably have a large effect on the positron localization, is yet to be explored.

The positron work function ϕ_p is of interest also in the context of the observed emission of low-energy positrons from some moderator surfaces [4.146]. Just as electrons must overcome their work function to emerge from a solid surface, the spontaneous emission of positrons can be associated with a *negative* positron work function. A work function negative enough could give a positron sufficient kinetic energy to avoid trapping by the image potential and eventually emerge free in the vacuum. The estimates of the positron work function in simple metals (Table 4.11) are all positive and thus do not allow positron emission. Those simple metals with the lowest work function (Zn, Cd, Al, Ga) are expected to show surface trapping, while in the others positrons will mainly reside in the internal regions of the sample.

NIEMINEN and HODGES [4.111] have calculated the positron work functions in transition and noble metals. The correlation and band-structure contributions E_{corr} and E_0 have been discussed in Sect.4.2.2 and are listed in Table 4.7. The surface dipole terms were obtained by NIEMINEN and HODGES [4.111] via a theorem [4.113] that relates the absolute electron band positions to the interstitial electrostatic potential at zero pressure. The surface dipoles and positron work functions for transition and noble metals are given in Table 4.12. In a number of cases, a negative work function is obtained. It is remarkable that the three metals with the most negative work function, namely Cr, Mo and W also show a sharp symmetrical peak in the energy distribution of positrons emerging from the solid into the vacuum [4.146]. This peak appears at an energy which roughly corresponds to the work function estimates. The second, broad and nondistinct peaks may have other, possibly instrumental origins not linked to material-characteristic properties of the solid.

Table 4.12. Values of the electron surface dipole D and the positron work function ϕ_p for transition and noble metals [4.111]

Metal	D [eV]	ϕ_p [eV]
Sc	2.8	1.7
Ti	4.3	0.1
V	5.4	-1.7
Cr	5.7	-2.2
Fe	5.2	-0.8
Co	5.0	-0.8
Ni	5.0	-0.4
Cu	3.8	0.8
Y	2.6	2.1
Zr	4.3	0.4
Nb	5.4	-1.2
Mo	5.8	-2.0
Rh	4.6	0.0
Pd	3.8	0.4
Ag	2.6	2.4
Ta	4.8	0.0
W	6.0	-2.1
Ir	4.8	-1.4
Pt	4.5	-0.2
Au	3.5	1.1

Judging from the simple metal calculations [4.144], one may expect that a number of transition metals and noble metals (some of which support fairly well defined surface and bulk plasmons) are likely to have positron surface states. This comes about because those metals have generally low or even negative positron work functions.

We finish this section by noting that the experiments [4.147] on positron diffusion through metal-metal interfaces are in accord with the theoretical work function estimates. The intermetallic junction presents a potential barrier for the positron, the height of which is to a first approximation equal to the difference of the positron work functions of the constituent metals. Thus it acts as a rectifier for thermalized positrons which can be detected as a preferential annihilation on one side of the junction.

4.3 Nonmetals

Positron behavior in ionic solids is discussed at length by DUPASQUIER in Chap.5. The theory of positron states localized at F-color centers in ionic crystals has been elaborated by BEREZIN [4.148], HAUTOJÄRVI et al. [4.149], and FARAZDEL and CADE [4.150]. The occurence of other kinds of defect trapping in ionic solids is a complicated and somewhat controversial subject, and no really quantitative theoretical studies have been put forward. See, however, the papers by GOLDANSKII et al. [4.151]. In semiconductors, observable changes in the annihilation characteristics are produced by irradiation [4.152,153] or plastic deformation [4.154, 155]. It is believed that under such circumstances trapping occurs at extended defects, e.g., vacancy clusters or line dislocations, whereas atomic size traps brought about by doping or by temperature seem to have very little effect. It is quite probable that even if trapping does occur at such defects, the trapped state is so extended in space due to the shallow defect potential and the large dielectric constant of the medium that the annihilation parameters do not change appreciably. Again, theoretical work in this area is virtually nonexisting.

Finally, we note in passing that the dynamics of positronium atoms in insulators has been the subject of a number of studies. For example, the diffusion of positronium atoms in insulators with defects has been discussed by BRANDT and PAULIN [4.156], and the annihilation of positronium in quartz including Umklapp processes and the role of defects in it has drawn some attention [4.157]. Positronium interaction with defects and impurities in ice has been investigated by MOGENSEN and collaborators [4.159,160].

4.4 Conclusions

The positron-defect interaction opens up new possibilities to investigate the electronic structure of defects as well as their nature and concentration. We have reviewed some of the basic features of positrons in solids with defects, emphasizing the applications to metal physics. We have adopted a pragmatic point of view and tried to simplify the complicated electron-positron many-body problem into a one-particle representation suitable for numerical calculations. Although we feel that it is fair to say that the trapping-related phenomena in metals are basically understood, much remains to be done in improving our ability to calculate reliable annihilation characteristics for positrons in imperfect solids. Progress, both experimental and theoretical, would be particularly valuable in understanding positron behavior in alloys and on surfaces. The developments in positron studies in imperfect crystalline matter will be tied to those in understanding electronic structure of different types of defects, alloys and surfaces, and we may expect continued and growing activity in these areas.

Acknowledgements. We wish to thank a number of friends and colleagues for useful discussions, comments and criticism, especially Jouko Arponen, Mike Finnis, Pekka Hautojärvi, Chris Hodges, Malcolm Stott, and Roy West.

References

4.1 C.P. Flynn: *Point Defects and Diffusion* (Clarendon Press, Oxford 1972)
4.2 *Theory of Imperfected Crystalline Solids*, Trieste Lectures 1970 (IAEA, Vienna 1971)
4.3 G. Caglioti (ed.): *Atomic Structure and Mechanical Properties of Metals* (North-Holland, Amsterdam 1976)
4.4 M. Stoneham: *Theory of Defects in Solids* (Clarendon Press, Oxford 1975)
4.5 J.H. Crawford, L.M. Slifkin (eds.): *Point Defects in Solids*, Vols.1-3 (Plenum Press, New York 1972-1977)
4.6 N.B. Hannay (ed.): *Treatise in Solid State Chemistry*, Vol.2 (Plenum Press, New York 1975)
4.7 P.R. Wallace: In *Solid State Physics*, Vol.10, ed. by F. Seitz, D. Turnbull (Academic Press, New York 1960) pp.1-46
4.8 A.T. Stewart, L.O. Roellig (eds.): *Positron Annihilation* (Academic Press, New York 1967)
4.9 P. Hautojärvi, A. Seeger (eds.): *Proceedings of The Third International Conference on Positron Annihilation* (Springer, Berlin, Heidelberg, New York 1974)
4.10 I.Ya. Dekhtyar: Phys. Rep. C *9*, 243 (1974)
4.11 A. Seeger: J. Phys. F *3*, 248-294 (1973)
4.12 M. Doyama, R. Hasiguti: Cryst. Lattice Defects *4*, 139 (1973)
4.13 W. Brandt: Sci. Am. *233*, 34 (1975)
4.14 R.N. West: Adv. Phys. *22*, 263 (1973)
4.15 W. Brandt, R. Paulin: Phys. Rev. B *15*, 2511 (1977)
4.16 P. Kubica, A.T. Stewart: Phys. Rev. Lett. *34*, 852 (1975)

4.17 A. Perkins, J.P. Carbotte: Phys. Rev. B *1*, 101 (1970)
4.18 C.K. Majumdar: Phys. Rev. *149*, 406 (1966)
4.19 D.R. Hamann: Phys. Rev. *146*, 277 (1966)
4.20 A.T. Stewart, J.B. Shand: Phys. Rev. Lett. *16*, 261 (1966)
4.21 H.J. Mikeska: Z. Phys. *232*, 159 (1970)
4.22 B. Bergersen, E. Pajanne: Phys. Rev. B *3*, 1588 (1971)
4.23 B. Bergersen, E. Pajanne: Appl. Phys. *4*, 25 (1974)
4.24 A.P. Mills, Jr., L. Pfeiffer: Phys. Rev. Lett. *36*, 1389 (1976)
4.25 A.P. Mills, Jr., L. Pfeiffer: Phys. Lett. A *63*, 118 (1977)
4.26 O. Sueoka, J. Koide: J. Phys. Soc. Jpn. *41*, 116 (1976)
4.27 J. Bardeen, W. Shockley: Phys. Rev. *80*, 72 (1950)
4.28 B. Bergersen, E. Pajanne, P. Kubica, M.J. Stott, C.H. Hodges: Solid State Commun. *15*, 1377 (1974)
4.29 R.M. Nieminen: J. Phys. C *8*, 2077 (1975)
4.30 D. Stroud, H. Ehrenreich: Phys. Rev. *171*, 399 (1968)
4.31 A.G. Gould, R.N. West, B.G. Hogg: Can. J. Phys. *50*, 2294 (1972)
4.32 M.J. Stott, P. Kubica: Phys. Rev. B *11*, 1 (1975)
4.33 P. Kubica, M.J. Stott: J. Phsy. F *4*, 1969 (1974)
4.34 C.H. Hodges, M.J. Stott: Phys. Rev. B *7*, 73 (1973)
4.35 M.J. Stott, R.N. West: J. Phys. F *8*, 635 (1978)
4.36 P. Kubica, M.J. Stott: Can. J. Phys. *53*, 450 (1975)
4.37 P. Kubica, B.T.A. McKee, A.T. Stewart, M.J. Stott: Phys. Rev. B *11*, 11 (1975)
4.38 D.G. Lock, R.N. West: J. Phys. F *4*, 2179 (1974)
4.39 W. Brandt, J. Reinheimer: Phys. Lett. A *35*, 109 (1971)
4.40 W. Brandt: Appl. Phys. *5*, 1 (1974)
4.41 S. Kahana: Phys. Rev. *129*, 1622 (1963)
4.42 A. Sjölander, M.J. Stott: Phys. Rev. B *5*, 2109 (1972)
4.43 P. Bhattacaryya, K.S. Singwi: Phys. Rev. Lett. *29*, 22 (1972)
4.44 J. Arponen: J. Phys. C *11*, L739 (1978)
4.45 R.N. West: Solid State Commun. *9*, 1417 (1971)
4.46 D.N. Lowy, A.D. Jackson: Phys. Rev. B *12*, 1689 (1975)
4.47 J.J. Donaghy, A.T. Stewart: Phys. Rev. *164*, 396 (1967)
4.48 S. Berko, J. Plaskett: Phys. Rev. *112*, 1877 (1958)
4.49 K.G. Lynn, J.R. MacDonald, R.A. Boie, L.C. Feldman, J.D. Gabbe, M.F. Robbins, E. Bonderup, J. Golovchenko: Phys. Rev. Lett. *38*, 241 (1977)
4.50 P. Bhattacaryya, K.S. Singwi: Phys. Lett. A *41*, 457 (1972)
4.51 I.K. MacKenzie, T.E. Jackman, N. Thrane: Phys. Rev. Lett. *34*, 512 (1975)
4.52 D.C. Connors, R.N. West: Phys. Lett. A *30*, 24 (1969)
4.53 B. Bergersen, M.J. Stott: Solid State Commun. *7*, 1203 (1969)
4.54 M. Doyama: J. Phys. Soc. Jpn. *33*, 1495 (1972)
4.55 A. Seeger: Appl. Phys. *4*, 183 (1974)
4.56 W.K. Warburton, M. Shulman: Phys. Lett. A *60*, 448 (1977)
4.57 B.T.A. McKee, W. Triftshäuser, A.T. Stewart: Phys. Rev. Lett. *28*, 358 (1972)
4.58 C.L. Snead, Jr., T.M. Hall, A.N. Goland: Phys. Rev. Lett. *29*, 62 (1972)
4.59 R.M.J. Cotterill, K. Petersen, G. Trumpy, J. Träff: J. Phys. F *2*, 459 (1972)
4.60 T.M. Hall, A.N. Goland, C.L. Snead, Jr.: Phys. Rev. B *10*, 3062 (1974)
4.61 W. Triftshäuser: Phys. Rev. B *12*, 4634 (1975)
4.62 S.C. Sharma, S. Berko, W.K. Warburton: Phys. Lett. A *58*, 405 (1976)
4.63 P. Rice-Evans, T. Hlaing, D.B. Rees: J. Phys. F *6*, 1079 (1976)
4.64 M. Eldrup, O.E. Mogensen, J.H. Evans: J. Phys. F *6*, 499 (1976)
4.65 B.T.A. McKee, S. Saimoto, A.T. Stewart, M.J. Stott: Can. J. Phys. *52*, 759 (1974)
4.66 K. Hinode, S. Tanigawa, M. Doyama: J. Phys. Soc. Jpn. *41*, 2037 (1976)
4.67 C.H. Hodges: J. Phys. F *4*, L230 (1974)
4.68 C.H. Hodges: Phys. Rev. Lett. *25*, 284 (1970)
4.69 B. Bergersen, T. McMullen: Solid State Commun. *24*, 421 (1977)
4.70 T.McMullen: J. Phys. F *7*, 3041 (1977); J. Phys. F *8*, 87 (1978)
4.71 B. Bergersen, D.W. Taylor: Can. J. Phys. *52*, 1594 (1974)
4.72 T. McMullen, B. Hede: J. Phys. F *5*, 669 (1975)
4.73 V.I. Goldanskii, E.P. Prokopev: Sov. Phys.-Solid State *6*, 2641 (1965)
4.74 K. Fujiwara, T. Hyodo, Y. Takakusa, O. Sueoka: J. Phys. Soc. Jpn. *39*, 403 (1975)

4.75 M.J. Stott: In the Abstracts of papers of the 3rd Intern. Conf. on Positron Annihilation (Helsinki 1973, unpublished)
4.76 P.C. Lichtenberger, C.W. Schulte, I.K. MacKenzie: Appl. Phys. *6*, 305 (1975)
4.77 A. Seeger: Appl. Phys. *7*, 85 (1975)
4.78 C.H. Hodges, H. Trinkaus: Solid State Commun. *18*, 857 (1976)
4.79 C.H. Leung, T. McMullen, M.J. Stott: J. Phys. F *6*, 1063 (1976)
4.80 F.L. Vook, H.K. Birnbaum, T.H. Blewitt, W.L. Brown, J.W. Corbett, J.H. Crawford, Jr., A.N. Goland, G.L. Kulcinski, M.T. Robinson, D.N. Seidman, F.W. Young, Jr., J. Bardeen, R.W. Balluffi, J.S. Koehler, G.H. Vineyard: Rev. Mod. Phys. *47*, Suppl.3, 51 (1975)
4.81 R. Evans: In *Vacancies 1976* (The Metals Society, London 1977) pp.30-44
4.82 R. Gomer (ed.): *Interactions on Metal Surfaces*, Topics in Applied Physics, Vol.4 (Springer, Berlin, Heidelberg, New York 1975)
4.83 E.C. Derouane, A. Lucas (eds.): *Electronic Structure and Reactivity of Metal Surfaces* (Plenum Press, New York 1977)
4.84 R. Haydock, V. Heine, M.J. Kelly: J. Phys. C *5*, 2845 (1972)
4.85 M. Lannoo: Phys. Rev. B *10*, 2544 (1974)
4.86 G. Allan, M. Lannoo: J. Phys. Chem. Sol. *37*, 699 (1976)
4.87 J.G. Gay, J.R. Smith: Phys. Rev. B *9*, 4151 (1974)
4.88 R.A. Johnson: J. Phys. F *3*, 295 (1973)
4.89 W.A. Harrison: *Pseudopotential Theory of Metals* (Benjamin, New York 1966)
4.90 R. Evans, M. Finnis: J. Phys. F *6*, 483 (1976)
4.91 J.A. Appelbaum, D.R. Hamann: Phys. Rev. B *6*, 2166 (1972)
4.92 J.G. Gay, J.R. Smith, F.J. Arlinghaus: Phys. Rev. Lett. *38*, 561 (1977)
4.93 S.G. Louie, M. Schlüter, J.R. Chelikowsky, M.L. Cohen: Phys. Rev. B *13*, 1654 (1976)
4.94 P. Hohenberg, W. Kohn: Phys. Rev. *136*, B 864 (1964)
4.95 W. Kohn, L.J. Sham: Phys. Rev. *140*, A1133 (1965)
4.96 N.D. Lang, W. Kohn: Phys. Rev. B *1*, 4555 (1970)
4.97 M. Manninen, R. Nieminen, P. Hautojärvi, J. Arponen: Phys. Rev. B *12*, 4012 (1975)
4.98 G.G. Robinson, P.F. de Chatel: J. Phys. F *5*, 1502 (1975)
4.99 M. Finnis, R. Nieminen: J. Phys. F *7*, 1999 (1977)
4.100 R. Nieminen: J. Phys. F *7*, 375 (1977)
4.101 J. Ferrante, J.R. Smith: Solid State Commun. *20*, 393 (1977)
4.102 Z.D. Popovic, M.J. Stott, J.P. Carbotte, G.R. Piercy: Phys. Rev. B *13*, 590 (1976)
4.103 M. Manninen, P. Hautojärvi, R. Nieminen: Solid State Commun. *23*, 795 (1977)
4.104 J.P. Perdew, R. Monnier: Phys. Rev. Lett. *37*, 1286 (1976); Phys. Rev. B *17*, 2595 (1978)
4.105 R. Nieminen: J. Nucl. Mater. *69-70*, 633 (1978)
4.106 O. Gunnarsson, B. Lundqvist: Phys. Rev. B *13*, 4274 (1976)
4.107 N.D. Lang: In *Solid State Physics*, Vol.28, ed. by H. Ehrenreich, F. Seitz, D. Turnbull (Academic Press, New York 1973) pp.225-300
4.108 P. Hautojärvi, J. Heiniö, M. Manninen, R. Nieminen: Philos. Mag. *35*, 973 (1977)
4.109 P. Jena, A.K. Gupta, K.S. Singwi: Solid State Commun. *21*, 293 (1977)
4.110 C.H. Leung, M.J. Stott, C.O. Almbladh: Phys. Lett. A *57*, 165 (1976)
4.111 R. Nieminen, C.H. Hodges: Solid State Commun. *18*, 1115 (1976)
4.112 V. Heine, C.H. Hodges: J. Phys. C *5*, 225 (1972)
4.113 R. Nieminen, C.H. Hodges: J. Phys. F *6*, 573 (1976)
4.114 J. Arponen, P. Hautojärvi, R. Nieminen, E. Pajanne: J. Phys. F *3*, 2092 (1973)
4.115 W. Brandt: In [Ref.38, p.176]
4.116 K. Maier, H. Metz, D. Herlach, H.E. Schaefer, A. Seeger: Phys. Rev. Lett. *39*, 484 (1977)
4.117 G. Mori: J. Phys. F *7*, L89 (1977)
4.118 G. Mori: J. Phys. F *7*, L7 (1977)
4.119 J.P. Gupta, R.W. Siegel: Phys. Rev. Lett. *39*, 1212 (1977)
4.120 J.H. Kusmiss, A.T. Stewart: Adv. Phys. *16*, 471 (1967)
4.121 F. Itoh: J. Phys. Soc. Jpn. *41*, 824 (1976)
4.122 I.K. MacKenzie, T.W. Craig, B.T.A. McKee: Phys. Lett. A *36*, 227 (1971)

4.123 M. Manninen, R. Nieminen: Unpublished
4.124 I.Ya. Dekhtyar, D.A. Levina, V.S. Mikhalenkov: Sov. Phys.-Dokl. *9*, 492 (1964)
4.125 S. Berko, J. Erskine: Phys. Rev. Lett. *19*, 307 (1967)
4.126 P. Hautojärvi, A. Tamminen, P. Jauho: Phys. Rev. Lett. *24*, 459 (1970)
4.127 J.W. Martin, R. Paetsch: J. Phys. F *2*, 997 (1972)
4.128 R.A. Brown: Phys. Rev. *141*, 568 (1966)
4.129 P. Hautojärvi: Solid State Commun. *11*, 1049 (1972)
4.130 C.L. Snead Jr., A.N. Goland, F.W. Wiffen: J. Nucl. Mater. *64*, 195 (1977)
4.131 P. Hautojärvi, A. Vehanen, V.S. Mikhalenkov: Solid State Commun. *19*, 309 (1976)
4.132 P. Hautojärvi, A. Vehanen, V.S. Mikhalenkov: Appl. Phys. *11*, 191 (1976)
4.133 N. Thrane, J.H. Evans: Appl. Phys. *12*, 183 (1977)
4.134 C.H. Hodges, M.J. Stott: Philos. Mag. *26*, 375 (1972)
4.135 K.M. Hong, J.P. Carbotte: Can. J. Phys. *55*, 1335 (1977)
4.136 R.J. Elliott, J.A. Krumhansl, P.L. Leath: Rev. Mod. Phys. *46*, 465 (1974)
4.137 W. Brandt: Adv. Chem. Ser. *158*, 219 (1976)
4.138 V.W. Lindberg, J.D. McGervey, R.W. Hendricks, W. Triftshäuser: Philos. Mag. *36*, 117 (1977)
4.139 N.D. Lang, W. Kohn: Phys. Rev. B *7*, 3541 (1973)
4.140 C.H. Hodges, M.J. Stott: Solid State Commun. *12*, 1153 (1973)
4.141 O. Mogensen, K. Petersen, R.M.J. Cotterill, B. Hudson: Nature *239*, 98 (1972)
4.142 R.M.J. Cotterill, I.K. MacKenzie, L. Smedskjaer, G. Trumpy, J.H.O.L. Träff: Nature *239*, 101 (1972)
4.143 R.M. Nieminen, M. Manninen: Solid State Commun. *15*, 403 (1974)
4.144 R.M. Nieminen, C.H. Hodges: Phys. Rev. B *18*, 2568 (1978)
4.145 K. Petersen, N. Thrane, G. Trumpy, R.W. Hendricks: Appl. Phys. *10*, 85 (1976)
4.146 S. Pendyala, D. Bartell, F.E. Girouard, J.W. McGowan: Phys. Rev. Lett.*33* , 1031 (1974)
4.147 W. Swiatkowski, B. Rozenfeld, H.B. Kolodziej, S. Szuszkiewicz: Acta Phys. Pol. A *47*, 79 (1974)
4.148 A.A. Berezin: Phys. Status Solidi (b) *50*, 71 (1972)
4.149 P. Hautojärvi, R. Nieminen, P. Jauho: Phys. Status Solidi (b) *57*, 115 (1973)
4.150 A. Farazdel, P.E. Cade: Phys. Rev. B *9*, 2036 (1974
4.151 V.I. Goldanskii, E.P. Prokopev: Sov. Phys.-Solid State *13*, 2481 (1972) and references therein
4.152 L.J. Cheng, C.K. Yeh, S.I. Ma, C.S. Su: Phys. Rev. B *8*, 2880 (1973)
4.153 W. Brandt, L.J. Cheng: Phys. Lett. A *50*, 439 (1975)
4.154 E. Kuramoto, S. Takeuchi, M. Noguchi, T. Chiba, N. Tsuda: J. Phys. Soc. Jpn. *34*, 103 (1973)
4.155 E. Kuramoto, S. Takeuchi, M. Noguchi, T. Chiba, N. Tsuda: Appl. Phys. *4*, 41 (1974)
4.156 W. Brandt, R. Paulin: Phys. Rev. B *5*, 2430 (1972)
4.157 C.H. Hodges, B.T.A. McKee, W. Triftshäuser, A.T. Stewart: Can. J. Phys. *50*, 103 (1972)
4.158 D. Herlach, H. Stoll, W. Trost, T.E. Jackman, K. Maier, H.E. Schaefer, A. Seeger: Appl. Phys. *12*, 59 (1977)
4.159 O.E. Mogensen, G. Kvajić, M. Eldrup, M. Milosević-Kvajić: Phys. Rev. B *4*, 71 (1971)
4.160 M. Eldrup, O.E. Mogensen, G. Trumpy: J. Chem. Phys. *57*, 495 (1972)
4.161 G. Dlubek, O. Brümmer, E. Hensel: Phys. Status Solidi (a) *34*, 737 (1976)

5. Positrons in Ionic Solids

A. Dupasquier

With 8 Figures

In 1963 BISI et al. [5.1] demonstrated that positrons implanted in ionic crystals end in annihilation through different channels. Five years later, BRANDT et al. [5.2] observed that annihilation characteristics are sensitive to lattice defects. These two discoveries have stimulated intensive activity in experimental and theoretical research. As a consequence, modern views on positron interaction with ionic media have developed in a direction completely unforeseen at the time when FERRELL [5.3]. WALLACE [5.4], and GREEN and LEE [5.5] were writing their reviews. Great attention has since been given to the formation of positron states bound to defects, as this phenomenon may provide the basis for an alternative to the classical methods of investigating microscopic imperfections in solids. Positron trapping by defects is not peculiar to ionic crystals, insofar as it is clearly apparent also in metals and semiconductors. However, no other type of material offers equal opportunities for a detailed study of the problem. The reasons are clear: I) the nature and the properties of lattice defects and color centers in ionic crystals (in particular, alkali halides) are generally well known; II) established methods of optical spectroscopy and magnetic resonance measurements allow a precise characterization of the defect structure of the sample; III) it is relatively easy to produce certain defects in controlled amounts; IV) annihilation characteristics for different states are sometimes so distinct that it is possible to attain complete separation of the associated spectral components.

The first step towards a correct understanding of processes leading to annihilation when defects are present in the sample is, of course, a knowledge of the behavior of positrons in perfect crystals. Ironically, for a long time this was the most controversial point; indeed, only very recently HYODO and TAKAKUSA [5.6] definitively confirmed the existence of a delocalized positroniumlike complex intrinsic to perfect lattices. Many points concerning the formation and the properties of positron states in ionic media still need clarification; nevertheless, the general aspects of the problem seem to be correctly understood, and a systematic presentation of the matter is now possible. Such is the aim of this chapter, which is an expanded and updated version of a review talk given by the author at the 4th International Positron Conference, Helsingør (1976). The same subject has recently been reviewed in part by BARTENEV et al. [5.7], also in monographs by VOROBEV [5.8], and by WEST [5.9]. In particular, this latter work has been taken as the starting point for this chapter.

The matter is presented here according to the following scheme: Section 5.1 is devoted to experimental techniques, with emphasis on special methods of interest in the field of our concern, and on difficulties that may be encountered in measurements and in data analysis. In Sect.5.2 the experimental results are systematically presented for the case of alkali halides, which are of course the most intensively studied ionic materials; to avoid any mixing of ideas with facts, in general we have chosen not to offer an interpretation of results unless completely unambiguous, although this procedure may render our report more laborious for the reader. On the other hand, in Sect.5.3, which is also devoted to alkali halides, we present models that have been developed for the various positron states, and so far as possible we link experimental results with theory. Finally, in Sect.5.4 we give a brief presentation of results concerning ionic and partially ionic insulators not of the alkali halide family.

5.1 Experimental Methods

5.1.1 Standard Experimental Techniques

Most of the experiments on positron annihilation in ionic solids have been performed in accordance with standard techniques, namely: measurements of lifetime spectra, angular correlation curves and Doppler-broadened annihilation line shapes. We shall not discuss here the principles underlying the above methods, nor describe typical setups, as this subject is treated in Chap.1.

It must be mentioned, however, that lifetime studies in the field of ionic media are quite demanding from the time-resolution viewpoint, because typical spectra may contain three or more components with mean lives all falling in the 100-100 ps range. The apparatuses used in modern experiments generally give ^{60}Co prompt curves with full width at half maximum (FWHM) of 250-300 ps or thereabouts, with energy-window setting as in the actual measuring conditions.

For angular correlation studies, angular resolution is not a problem; experiments are usually performed with long-slit equipment having resolutions in the 0.5-1 mrad range. An angular correlation system well suited for the investigation of ionic solids should have the highest photon collection efficiency in order to minimize the radiation dose received by the sample during the experiment.

The photon collection efficiency of a semiconductor spectrometer used for line-shape measurements is of course much better than that of an angular correlation apparatus. However, the annihilation line-shape method is limited by energy resolution; the resolution normally obtained (~ 1.5 keV) is equivalent to an angular resolution of about 6 mrad. This is hardly sufficient to reveal the structure of momentum distributions, and will be more useful for the study of gross effects.

A fourth method of investigation, the measurement of three-quantum production, has not been used in recent works. The correct application of this method to ionic crystals requires a direct experimental approach with the use of a triple coincidence apparatus, as made by BISI et al. [5.10]. In fact the three-quantum annihilation probability in nonpowder specimens is of the order of 3×10^{-3}, and this is too low to be inferred with real accuracy from the analysis of photon energy spectra [5.11,12].

5.1.2 Special Experimental Techniques

a) Two-Parameter Age-Momentum Measurements

The understanding of positron interaction with ionic solids would be greatly aided by the measurement of correlations between the age of the positron at the instant of annihilation and the momentum of the annihilating pair. Such measurement is, however, very difficult if the timing system is to operate in coincidence with an angular correlation apparatus, because of the low collection efficiency of this type of instrument. The experiment becomes more feasible if the timing system is coupled to a Ge(Li) spectrometer used as a momentum analyzer, as early suggested by MACKENZIE and MCKEE [5.13].

To our knowledge this technique has been applied only once in the field of ionic crystals, i.e., in an experiment on KCl performed by BRÜMMER et al. [5.14]. In the scheme adopted by the above authors the Ge(Li) detector was used not only to provide the energy information but also to give the stop signal to the timing system. The time resolution (FWHM of the ^{60}Co prompt curve) and the energy resolution (FWHM of the ^{85}Sr 514-keV line) of their setup were, respectively, 1.68 ns and 3.38 keV. These values compare very poorly with those obtainable with conventional apparatuses used for one-parameter measurements. Nevertheless their experiment was successful in showing up the existence of an age-momentum correlation.

Somewhat different is the scheme developed by MACKENZIE and MCKEE [5.15], and by MACKENZIE and SEN [5.16,17] for their studies on plastics and on neutron-irradiated molybdenum. In the setups of the above authors, one of the two annihilation photons is used as stop signal in a conventional timing system, while the other one provides the energy information. The energy and time resolutions quoted by MACKENZIE and SEN had the very good values of 1.45 keV and 290 ps, respectively. Of course, much could be done with this equipment also in the field of ionic solids.

b) Magnetic Quenching Measurements

It is well known that the application of a static magnetic field modifies the properties of the positronium atom by mixing the para state with that ortho state which has a zero spin component along the field direction (m = 0 ortho state) [5.18,19]. Two new eigenstates with m = 0 are thus formed, and they can both decay by two-photon annihilation, a process which is forbidden for pure ortho states. The effect becomes manifest through the reduction of the mean life and of the three-quantum annihilation probability; conversely, in the angular correlation curve an increase may be observed of the narrow component associated with self-annihilation in two photons.

The above phenomena, usually indicated as *magnetic quenching*, are helpful in demonstrating the formation of positronium. We call attention, however, to the point that in principle magnetic quenching may occur with positron-electron or positron-multielectron systems that are not positronium atoms [5.20]. The condition for the effectiveness of the magnetic field on a given system is that this system is formed in at least two states of different lifetime and/or momentum distribution, and that these states become mixed in a magnetic field. The simplest case occurs when the positron has a nonnegligible overlap probability with an unpaired-spin electron; in this case, similarly as for positronium, ortho and para states of different lifetime will exist. From now on, we shall refer to a system of this kind as *quasi-positronium* (qPs). On the contrary, when a positron is coupled to a multielectron system having full closed-shell symmetry, no magnetic quenching can occur. We shall refer to complexes of this kind as *closed-shell* systems. A simple example of a closed-shell system is a positron bound to a negative halogen ion in the $...ns^2np^6$ electronic configuration, while a positron bound to a negative halogen ion in the $...ns^2np^5(n+1)s$ configuration is a qPs system.

The interest of magnetic quenching measurements does not rest solely in the mere indication of the existence of a qPs system; it is also helpful in characterizing the properties of the qPs system, as we shall now briefly discuss. The quantitative treatment of the magnetic quenching effect on positronium, as given by HALPERN [5.19] remains valid for any qPs system, with modifications necessary when accounting for different singlet-triplet energy splitting and different annihilation rates. We observe now that the rate of annihilation with the unpaired-spin electron is proportional to the positron overlap probability with this electron. The energy splitting is also proportional to this overlap probability, if the relative wave function ψ of the interacting pair has s symmetry; this fact remains approximately true even if the above wave function contains a mixing of s and non-s terms [5.20, 21]. As far as magnetic quenching is concerned, the above would justify the consideration of almost any qPs system as a positronium atom with a different value of the wave function ψ at zero electron-positron distance. The result of the quantitative analysis of the magnetic quenching effect will give the value of $|\psi|^2$ at zero electron-positron distance.

The experimental procedure followed for studying magnetic quenching by the Milano group [5.22-24] consisted of measuring the fraction of annihilation events occurring in a suitably chosen lifetime interval, as a function of an externally applied static magnetic field. The explored interval of field intensity ranged up to 17 kG. The lifetime setup was a conventional fast-slow coincidence system, with FWHM of 280 ps for ^{60}Co prompt curves in normally operating conditions, and of 540 ps when light pipes were inserted between scintillators and photomultipliers. This indirect photomultiplier mounting, with careful magnetic shielding, is necessary to avoid disturbance by spurious effects. HERLACH and HEINRICH [5.25], HERLACH [5.26], and DANNEFAER and SMEDSKJAER [5.27,28] have, in turn, studied the magnetic quenching effect on momentum distributions with conventional long-slit angular correlation setups. In the experiments of [5.25,26] the positrons were injected into the sample from an external β source mounted on the axis of the field. Because of the helicity of positrons emitted in β decay, with this arrangement there exists a nonzero degree of spin polarization along the axis of the field. Thus the formation of singlet or of triplet states is favored depending on whether positron polarization is parallel or antiparallel to the field [5.29], and the effect of reversing the field becomes evident whenever Ps or qPs is actually formed inside the crystal. Of course, such an effect will not exist if a source-sample sandwich arrangement is adopted, as in the experiments of [5.27,28], no net average polarization remaining in this case along the field axis.

5.1.3 Experimental Difficulties

Discrepancies are often found in literature concerning annihilation in ionic crystals. It is our feeling that most of them originate in difficulties encountered in the collection and/or in the interpretation of experimental data. We list here a few of these problems.

a) Analysis of Multicomponent Lifetime Spectra

It is now an accepted fact that lifetime spectra in ionic crystals are complex. According to the interpretation of the data given by the authors of the most recent studies, even in ideal crystals at least three different positron states are formed, each of them contributing a component to the lifetime spectrum; furthermore, other states, and thus other components, are formed in the presence of lattice defects. Given this complicated situation, the reduction of raw experimental data to information with a direct physical meaning becomes a problem. To discuss this point, let us recall briefly certain facts. If one assumes that: I) the slowing down of the positron injected into the crystal is instantaneous; II) distinct positron states are possible; III) a positron entering one state may leave it either by annihilation or by transition to another state; IV) annihilation and transition rates are time-independent; then lifetime spectra are composed by the sum of exponentially decaying terms. The decay constants γ_i of the spectral components will depend on the annihilation and on the transition rates; the relative weights I_i of the exponential terms will depend on the above rates and also on the population of the states at the end of the slowing down (t = 0). Knowledge of I_i and γ_i parameters is the aim of the spectrum analysis, which is usually made by fitting a model curve to the raw experimental data. The finite resolution of the time spectrometer must also be taken into account, and therefore the model curve will have the mathematical form

$$s^*(t) = \int_0^\infty p(t - x)s(x)dx \tag{5.1}$$

where p(t) is the normalized instrumental response to simultaneous start-stop events (prompt curve), and s(t) is the true spectrum. According to the hypotheses outlined above, it is assumed that

$$s(t) = \sum_i I_i\gamma_i \exp(-\gamma_i t) \quad . \tag{5.2}$$

A number of computer programs have been developed to perform the fitting: for instance, a flexible program used by many groups is *POSITRONFIT* [5.30]. However, while it is normally possible to obtain with reasonable accuracy and good reproducibility the parameters of long-living components well separated from the rest of the spectrum, even sophisticated computer programs are of little help in resolving components with short and not too different lifetimes. It is a common experience that one can try fits with an increasing number of components without any real improvement in goodness-of-fit, until the program fails to converge. A laborious *poor-man* spectrum reduction, made with the help to a desk calculator and semilog paper, shows convincingly that it is impossible to extract from the data whatever is contained in a blurred and indistinct form. This is unfortunately the rule with the shortest-living components in the spectra for ionic crystals.

A number of conclusions may be drawn, as presented below.

I) In several cases, the values quoted in the literature as intensities and lifetimes of short-living components may acutally represent nothing more than the adjustable parameters of an interpolatory formula. While this is often sufficient to reveal, for instance, the effect of a given treatment on the sample, any more speculative interpretation may result in mistakes.

II) In consequence of I), it is impossible to reconcile the results of analysis made on the basis of a differing number of components.

III) The convergence of the computer program should be rendered as sharp as possible by fixing all the parameters (for instance, background and well-separated long-living components) obtained by preliminary analysis, or by means of independent measurements (characteristics of the prompt curve).

IV) Advantage should be taken of information given by other sources. For instance, DANNEFAER et al. [5.31] made a combined analysis of results of lifetime and line-shape measurements on the same samples.

V) The number of components and possible constraints or relations may be fixed on the basis of a theoretical model. Care must, however, be taken not to force the results into a predetermined scheme, thereby ignoring any unexpected behavior.

b) Analysis of Multicomponent Momentum Distributions

Different positron states give rise to different momentum distributions that overlap one another in complex angular correlation curves and Doppler-broadened energy profiles. In a few lucky cases the components are so distinct in shape that the separation can be established, so to speak, at first glance (an example as given in Sect.5.2.2, Fig.5.4). In other cases, some manipulation of the sample, or controlled modifications of the experimental conditions, may produce a new component, which can then be identified by difference with a *no-effect* standard (an example is to be found in [5.32]). When the above circumstances do not exist, authors attempt to analyze their data with a best-fit procedure on the basis of a model curve. Fits that are satisfactory from a numerical viewpoint are often obtained with model curves made up by a sum of Gaussians. While the distribution of the momentum component along one axis is actually represented by a Gaussian for thermalized positron-electron free pairs, there is no a priori reason why this should be so for pairs that interact with the lattice. To be on the safe side, in the absence of any concomitant evidence, the paramters obtained by such a procedure cannot be considered as directly related to individual states.

c) Source and Surface Contributions to Lifetime Spectra

Depending on the preparation of the source-sample assembly used for lifetime measurements, very long-living components may show up in the spectra. Their characteristics are rather erratic: it is typical that when the source is deposed on mica foils and sandwiched between single crystals, the decay time falls in the ns range and the intensity is of the order of 1% or less. It has also been observed that they undergo the magnetic quenching effect with the features of undistorted positronium [5.24].

An investigation of the origin of these components has never been carried out, but many authors agree in attributing their presence to positronium which has been formed at the source-sample interface and whose properties are strongly influenced by moisture and any other impurity collected on the surfaces exposed to positrons.

For studies concerning positron behavior in the crystal bulk, the appearance of these spurious components is a hindrance. The accurate manipulation of freshly cleaved crystals and of the source in a dry-box, and the sealing of the assembly in an hermetical container during measurement, may help in controlling the disturbance. Of course, when possible it is preferable that the positron-emitting isotope should be a normal constituent of the sample itself [5.33].

d) Radiation Damage Due to the Positron Source

The β and γ irradiation, to which the samples are necessarily exposed during an annihilation experiment, may produce: I) electron excitation followed by trapping in pre-existing lattice defects; II) ion displacement with creation of vacancy-interstitial pairs. Both processes can modify the annihilation characteristics that the experiment was intended to study. Actually, it is not immediately possible to define to what extent this effect is a disturbing one, since many factors have their influence: dose, dose rate, nature of the sample, impurity content, previous stress and thermal history, irradiation temperature, illumination conditions, etc. We limit ourselves to giving, for general guidance, values of the dose rate, K, absorbed in the region where positrons annihilate for two typical experimental arrangements. For instance, supposing that, as usual for lifetime measurements, a ^{22}Na source of 5 μCi is sandwiched between a pair of sample, the result is $K \simeq 5 \times 10^{14}$ eV cm^{-3}s^{-1}; supposing, on the contrary, that for an angular correlation measurement a ^{58}Co source of 50 mCi is placed at a distance of 0.7 cm from the sample, one obtains $K \simeq 10^{16}$ eV cm^{-3}s^{-1}. For comparison, we note that the dose rate of 10^{16} eV cm^{-3}s^{-1} is sufficient to build an F center concentration of $\sim 10^{17}$ cm^{-3} in NaCl natural crystal at room temperature in the time of 10^5 s [5.34]. This concentration is sufficient to produce observable effects on positron annihilation characteristics. We must therefore conclude that in many cases, particularly when the strong sources necessary for angular correlation measurements are used, allowance for defect creation by irradiation is mandatory. This point has been clearly brought out by DANNEFAER and SMEDSKJAER [5.27,28]. and by FUJIWARA et al. [5.35].

5.2 Annihilation Characteristics in Alkali Halides

5.2.1 Room Temperature Measurements on Crystals with Low Defect Concentration

a) Lifetime Spectra

The lifetime spectra of positrons implanted in alkali halides display a complex structure. This fact, demonstrated by measurements of BISI et al. in 1963 [5.1], is now firmly established. The shapes of the spectra are satisfactorily repre-

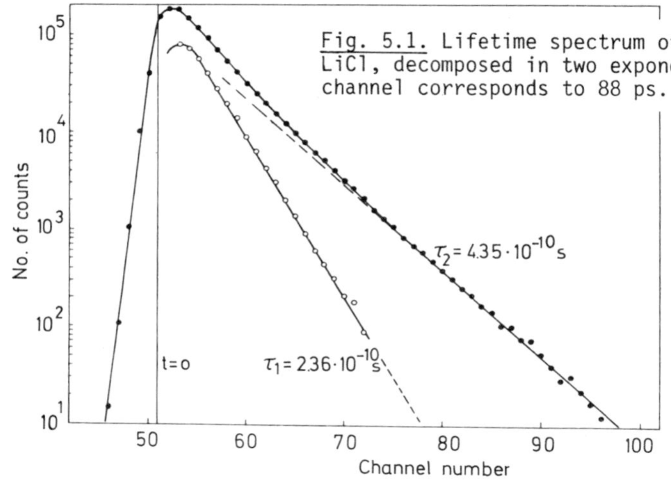

Fig. 5.1. Lifetime spectrum of positrons annihilating in LiCl, decomposed in two exponential decaying terms. One channel corresponds to 88 ps. [5.36]

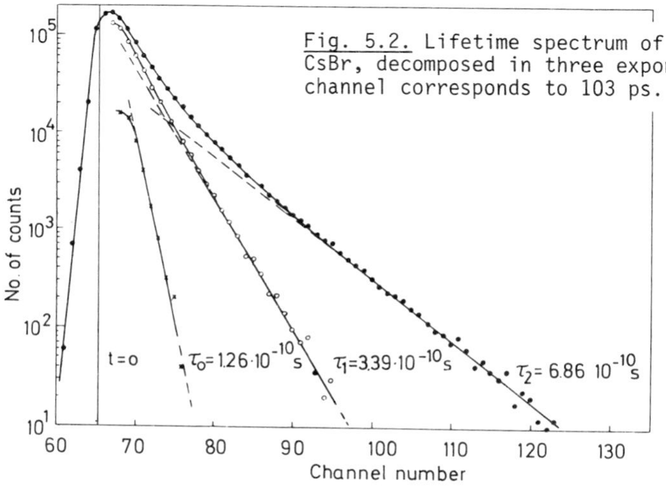

Fig. 5.2. Lifetime spectrum of positrons annihilating in CsBr, decomposed in three exponential decaying terms. One channel corresponds to 103 ps. [5.36]

sented by model curves calculated according to (5.1,2) (see Figs.5.1,2). Therefore, it is customary to describe a spectrum in terms of intensities I_i and lifetimes (to be precise, time-decay constants) $\tau_i = 1/\lambda_i$. Generally, it is found that these parameters are sensitive to the experimental conditions and to the state of the sample. This, and the experimental difficulties outlined in Section 5.1.3, explain discrepancies in the data of various authors. Nevertheless, the results of experiments made with adequate time resolution are not in marked disagreement when the same number of components is used for the numerical fit. Therefore we think that it is meaningful to define the general features of lifetime spectra giving here a homogeneous set of data concerning the whole family of alkali halides. In Table 5.1 we report the values obtained some years ago by BUSSOLATI et al.

[5.36]; the time resolution of the setup used for this work (280 ps FWHM for ^{60}Co prompt curves) was good even in comparison with more modern facilities. In our Table we retain the scheme used in the original paper for ordering lifetimes and intensities in series displaying a more or less regular variation within each group of salts. This ordering was tentatively explained by the above authors with a model that has been disproven by more recent evidence; though, it still seems reasonable to attribute the same physical origin to spectral components labeled with the same index. The parameters labeled *tail* refer to long-living components probably due to surface annihilation (see Sect.5.1.3).

The results reported in Table 5.1 are obtained with single crystals or large-grain powders of chemical grade purity; no thermal or mechanical treatment was performed to alter the original defect concentration of the samples. All the measurements were taken at room temperature. For some of the compounds listed in Table 5.1 the data of other researchers are also available [5.37-41]. The case of KCl has been reexamined recently, be DANNEFAER et al. [5.31], with a new approach. These authors have performed simultaneously lifetime and line-shape measurements, with time resolution of 330 ps (FWHM) and energy resolution of 1.5 keV (FWHM). To obtain internal consistency between the two sets of data, they analyze the time spectrum in three main components plus a tail. For untreated samples at room temperature, the main lifetimes are found to be $\tau_1 \approx 190$ ps, $\tau_2 = 464 \pm 30$ ps, $\tau_3 = 774 \pm 45$ ps. In a second work by the same group [5.42] the four-component fit was extended to eighteen more alkali halides. In general, this procedure does not, however, give a substantial improvement in the goodness-of-fit.

All the values quoted for untreated alkali halides are not necessarily representative of the positron behavior in perfect crystals. Since impurities and defects are certainly present, it may well be that some of the spectral components that have been observed owe their origin to such impurities and defects. On the other hand, it must be emphasized that even a careful anneal of high-purity single crystals cannot cancel the complex structure of the spectra.

b) Angular Correlation Curves

The angular correlation curves of photons from two-quantum positron annihilation are known for several alkali halides. They are given by the superimposition of at least two independent components having distinct properties. The existence of this complex structure was discovered by HERLACH and HEINRICH first for KCl colored by F centers [5.32,43], and later for KCl without coloration [5.26]. We may observe, however, that a hint regarding the presence of a narrow component superimposed on a broader one had already been given by the results obtained by STEWART and POPE for the family of chlorides and for sodium halides [5.44]. These authors, noting that the width of the correlation curves is essentially determined by the anion, attempt to fit their experimental data on the assumption that annihilation takes

Table 5.1. Lifetime spectra for alkali halides containing low defect concentration. Results of decomposition in two and three exponential terms [5.36]

Halide	τ_0 [10^{-10} s]	τ_1 [10^{-10} s]	τ_2 [10^{-10} s]	τ_3 [10^{-10} s]	τ_{tail} [10^{-10} s]	I_0 [%]	I_1 [%]	I_2 [%]	I_3 [%]	I_{tail} [%]
LIF	-	1.32±0.07	2.97±0.09	-	11.0; 18.7[a]	-	31±2	63.1±3.1	-	1.8; 0.4[a]
NaF	-	1.93±0.10	3.90±0.09	-	36.0; 14.4[a]	-	27±2	66.0±3.2	-	0.8; 1.3[a]
KF	-	1.80±0.09	4.24±0.13	10.0± 0.3	-	-	34±3	50.0±2.5	9.0±0.2	-
RbF	-	1.55±0.08	3.51±0.11	7.44±0.22	-	-	21±2	61.7±3.1	22.8±0.7	-
CaF	-	1.45±0.08	3.89±0.12	9.38±0.28	-	-	30±3	45.9±2.6	23.9±0.6	-
LiCl	-	2.36±0.12	4.35±0.13	-	17.5	-	51±4	49.1±2.4	-	0.6
NaCl	-	3.13±0.16	6.84±0.20	-	28.9	-	65±5	27.6±0.9	-	1.2
KCl	-	2.75±0.14	6.28±0.19	-	40.5; 24.3[a]	-	48±4	41.3±2.0	-	1.3; 1.6[a]
RbCl	-	1.99±0.10	4.56±0.14	10.0±0.3	-	-	27±3	59.3±3.0	7.0±0.3	-
CsCl	-	2.02±0.11	4.51±0.14	12.1±0.3	-	-	42±4	54.6±2.7	5.0±0.2	-
LiBr	-	2.97±0.15	5.56±0.17	-	-	-	79±6	17.1±0.8	-	-
NaBr	-	2.92±0.15	7.46±0.22	-	33.7	-	72±6	16.5±0.7	-	2.0
KBr	-	3.47±0.17	7.55±0.22	-	-	-	64±6	24.2±1.2	-	-
CsBr	1.26±0.09	3.39±0.17	6.86±0.21	-	-	26±3	55±5	22.7±1.1	-	-
LiI	-	3.27±0.17	6.60±0.20	-	-	-	85±7	12.9±0.6	-	-
NaI	-	3.28±0.18	7.76±0.23	-	42.2	-	82±7	10.4±0.5	-	0.8
KI	1.69±0.12	4.01±0.20	7.84±4.54	-	-	17±2	63±5	19.6±0.9	-	-
RbI	1.69±0.12	4.25±0.21	8.05±0.24	-	-	15±1	53±5	24.1±1.2	-	-
CsI	1.83±0.13	3.55±0.20	9.80±0.30	-	-	24±2	70±6	4.0±0.2	-	-

[a] The first number refers to powders, the second to single crystals

place in the outermost anion shell. In fact, they obtain a fair fit only at large angles, all their model curves tending to give too small values in the region $p \leqq 2\ mc \times 10^{-3}$. Similar results are obtained by HAUTOJÄRVI and NIEMINEN for KCl and KI using their own experimental data and a choice of positron wave functions [5.39]. It is worth emphasizing, however, that the evidence of the complex structure of the correlation curves is not based on the discrepancy between measurements and model calculations, but rather that a narrow component has been actually isolated in the case of KCl, by the observation of the enhancement given by a magnetic field [5.26]. The isolated narrow components, for uncolored KCl and for KCl containing F centers, are shown in Fig.5.3. DANNEFAER and SMEDSKJAER [5.27,28] perform a numerical decomposition of the correlation curves for KCl and NaCl. Their results with ascleaved specimens are collected in Table 5.2. Further convincing evidence of the complexity of the angular correlation curves, obtained by low temperature measurements, will be presented in Sect.5.2.2. It should be noted that the interpretation of angular correlation results is made difficult by possible radiation damage effects. The results of HERLACH's experiments on uncolored KCl [5.26] are, however, free from disturbing effects attributable to F center production, the samples being optically bleached continuously throughout the measurements.

For a complete description of complex angular correlation curves it would be necessary to have knowledge of the shape and intensity of each independent compo-

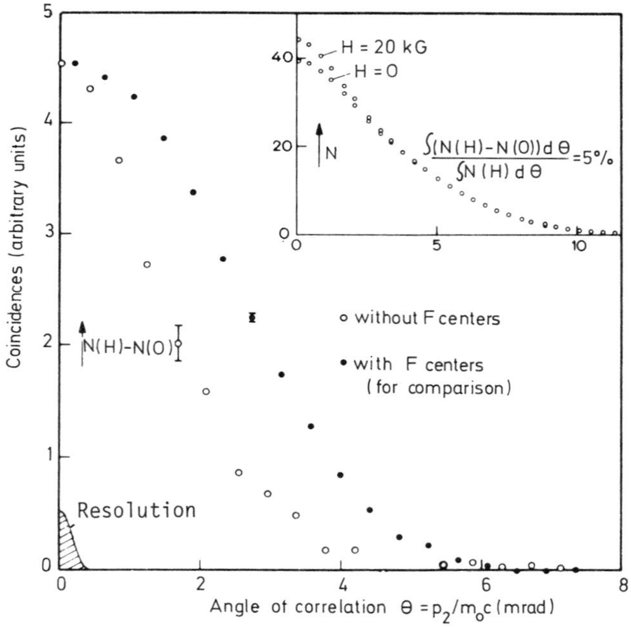

Fig.5.3. Narrow components in the angular distributions for the cases of uncolored KCl (circles) and of additively colored KCl (dots). The insert shows the effect of a static magnetic field on angular distribution for uncolored KCl. [5.26]

Table 5.2. Angular correlation curves for alkali halides containig low defect concentration. Results of decomposition in two Gaussian terms. The narrowest of the two Gaussians has intensity I_N and FWHM Γ_N; the broadest has complementary intensity and FWHM Γ_B

Substance	I_N [%]	Γ_N [mrad]	Γ_B [mrad]
NaCl [a]	5.3 ± 0.3	2.9 ± 0.2	9.3 ± 0.1
KCl [b]	8.3 ± 0.6	3.3 ± 0.1	9.3 ± 0.1

[a] [5.27]
[b] [5.28]

nent. As a systematical analysis is lacking, we content ourselves with listing below the main available information on the correlation curves taken as a whole:

I) As already said, the width of the (spherically averaged) angular distribution depends chiefly on the anion. Exceptions to this rule may be found when the anion is much larger than the cation (e.g., LiCl) [5.44]. This behavior is substantially confirmed by the calculations of NIEMINEN [5.45], based on a positron wave function of the proper symmetry, solution of the one-particle Schrödinger equation for the polarized crystal field.

II) The distributions are considerably narrower than predicted by models based on the assumption that the positrons are all in the same state and annihilate with anion electrons. An agreement may be found only for large angles [5.39,44].

III) The curves display a small but distinct anisotropy, depending on the orientation of the crystal in relation to the axis of the apparatus [5.46-49]. The momentum distribution is broader along [110] direction than along [100], while direction [111] is an intermediate case. Seemingly this effect is not related to anisotropies of the positron wave function. NIEMINEN's calculations [5.45] show that, in the coordinate space, the positron wave function is more compressed along [100] than along [110]; thus, the momentum distribution would be broader along [100] than along [110], in contradiction to the experimental results. Reasonably, HERLACH and HEINRICH [5.47] propose that the origin of the observed effect is due to the anisotropy of the wave function of anion external electrons. This explanation is qualitatively in agreement with the experiment, but the complex structure of the curves prevents quantitative comparisons.

c) Doppler-Broadened Annihilation Line Shape

The method of correlated lifetime-line-shape measurements, used by BRÜMMER et al. [5.14], has given a clear demonstration of the complex structure of the momentum distribution of annihilation photons in uncolored KCl. The above authors obtain significantly different line shapes for annihilation events registered by the time spectrometer either before or after a given instant $\underline{t}$($\underline{t} \simeq 1.35$ ns) from the birth of the positron; the narrowest profile corresponds to events occurring before $\underline{t}$. The low intensity of the source used in this experiment (2 μCi) ensured a low concentration of radiation-induced defects.

Further evidence of the complexity of the annihilation line shape for KCl has been given by DANNEFAER et al. [5.31]. By numerical procedure they decompose the spectrum in a sum of three Gaussians, the narrowest of which has FWHM corresponding to a momentum spread $\Delta p = (4.2 \pm 0.2)mc \times 10^{-3}$. This value is in disagreement with angular correlation results [5.25,27] ($\Delta p \simeq 3 \times 10^{-3}$ mc). The intensity of the narrow component (I_N) is correlated with the sum of the intensities of the longest-living components ($I_3 + I_4 \simeq 3I_N$). For annealed KCl it is found that [$I_N = (7.7 \pm 0.5)\% = (I_3 + I_4)/(2.6 \pm 0.3)$] at room temperature. After subtraction of the narrow component, the obtained momentum distribution is in reasonable agreement with the calculations of HAUTOJÄRVI and NIEMINEN [5.39].

d) Three-Quantum Annihilation

The three quantum annihilation probability has been measured for all chemically stable alkali halides with a triple-coincidence method by BISI et al. [5.10]. It appears that in general this probability exceeds by 15-40% the value 2.7×10^{-3} typical for annihilation in free positron-electron collisions. According to the above authors, it has been impossible to find a correlation between the measured values of the three-quantum yield and the time spectra. We feel that the matter deserves further experimental investigation, with simultaneous or, better, correlated lifetime-triple coincidence measurements. The problem of positronium formation on surfaces, that could give an excess three-quantum yield, has to be carefully considered.

5.2.2 Temperature Effects

The lifetime spectra are known for a few typical alkali halides at temperatures from liquid nitrogen temperature up to and beyond the melting point of the crystals. While dramatic changes of the spectral shape may occur when the crystal melts, in solids the spectra display only moderate variations. Taking, for instance, the case of NaCl, the shortest resolved lifetime τ_1 increases by ~13% when the temperature passes from 295 to 973 K (melting occurs at T_M = 1074 K) while the other spectral parameters remain essentially constant [5.37]. In annealed KCl [5.31] it was observed that there was a ~10% increase in τ_1, and also a slight decrease of the intensity I_3 of the long-living component (essentially compensated by the increase of I_1) when the temperature was lowered from 296 to 90 K. The same specimens of KCl have been studied by the line-shape method; and here the numerical analysis of the energy spectrum gives a narrow component whose intensity is $I_N = 4.7 \pm 0.5\%$ at 90 K and $7.7 \pm 0.5\%$ at 296 K.

The three-quantum annihilation probability P has been measured in a temperature range of from 25 to 450° C for NaCl, NaF and NaBr [5.10]. No change was observed when the temperature was increased; a temporary increase of P, followed by a slow recovery to the original value, may become manifest when the specimen is cooled again.

Interesting temperature effects are revealed by the angular correlation method. Recently, HYODO and TAKAKUSA [5.6] have shown that the shape of the angular correlation curves may drastically change when the temperature is lowered from room temperature to liquid nitrogen temperature. This effect is evident in the data reported in Fig.5.4. It consists of the appearance of a prominent narrow component, whose width (at least in the case of NaF) is essentially determined by the resolution of the angular correlation setup (0.67 mrad); also in the occurrence of small satellite peaks at the projection of the reciprocal-lattice vectors (200) and ($\bar{2}00$) onto the measured momentum direction. The intensity of the narrow component is ~2.8% in the case of NaF shown in the figure. Seemingly, the effect is dependent on the origin and on the actual state of the samples: it has been observed in NaF, NaCl and in a Harshaw specimen of KBr, but not in LiF, KCl, KI or in a Horiba specimen of KBr. Moreover, VOROBEV et al. [5.50] failed to observed any temperature effect with their specimens of KCl, KBr and NaCl. It has also been noted that the intensity of the narrow component is gradually reduced by the positron-source

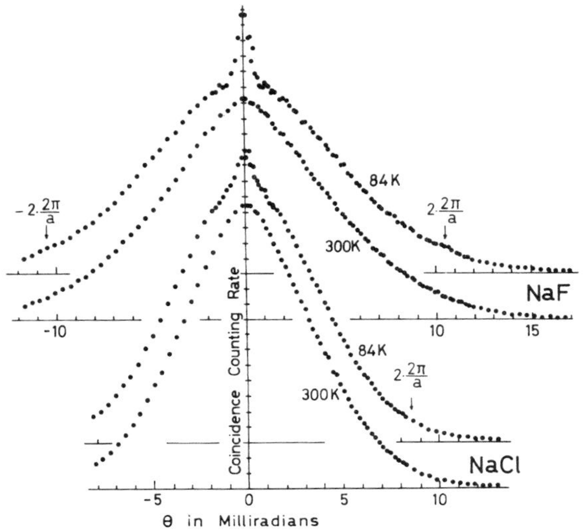

Fig.5.4. Angular correlation of photons from positrons annihilating in NaCl and NaF. The arrows indicate the position of the reciprocal lattice vectors onto the measured momentum direction. [5.6]

irradiation during the measurement. We recall that similar effects have been observed at higher temperatures in quartz [5.51], fluorite [5.51], and ice [5.52].

The discovery of HYODO and TAKAKUSA enables one to draw some immediate conclusions. The narrowness of the central peak indicates that: I) a fraction of positrons annihilate with slow electrons not belonging to the filled valency band or to deeper ion shells; II) the annihilating pair constitutes, or is part of, a delocalized system. The presence of satellite peaks at reciprocal lattice vectors shows that the wave function of the annihilating system has the translational symmetry of the lattice. It follows from the above that this system is typically associated with the ideal crystal structure and is not due to the presence of defects. On the contrary, lattice defects or impurities, originally contained in the crystal or introduced by irradiation, disturb its formation and/or delocalization.

5.2.3 Annihilation in Crystals with High Defect Concentration

The sensitivity of positrons to lattice defects in alkali halides becomes manifest through modifications in lifetime and momentum distributions, which occur when some treatment significantly alters the concentration of defects contained in the crystal. The relative changes of the annihilation parameters, observed in many experiments, are extremely strong in comparison with the molar fraction of defects present in the samples. This fact demonstrates that positrons are attracted and possibly captured by some defect species. In this way, annihilation characteristics do not respond to average properties of the crystal but to the situation locally encountered by the positron in the defect region. After BRANDT [5.53], a positron-defect bound system is named *annihilation center* (A center).

The first experimental evidence of positron trapping at defect locations in alkali halides was obtained with the lifetime method by BRANDT et al. [5.2]; modifications occurring in angular correlation curves were discovered somewhat later by HERLACH and HEINRICH [5.32]. Since then, a large number of papers concerning experimental investigations on positron-defect interaction has been accumulated [5.22,25-28,32,33,43,54-88]. We also have examples of the use of annihilation techniques for the investigation of defect kinetics [5.89-91].

In general, reported effects are quite distinct. In some cases, the analysis of the data has made it possible to correlate the parameters of a well-isolated spectral component with a given defect species. In other experiments, the contemporary formation of different species of defects and the complexity of the annihilation spectrum have prevented the elucidation of existing correlations. Of course, it is impossible to give here even a brief account of all received results and of their proposed explanations. We limit ourselves to summarizing the known phenomenology in a scheme ordered according to the method used for generating or reducing defects into the crystal. In what follows we indicate crystal defects with the notation stated by SONDER and SIBLEY [5.92]. The notation used for A center is but a small modification of the one suggested by BRANDT [5.93]. The rules of this notation are stated here below.

Each center is indicated by the label ${}^{\beta}_{n}A^{\alpha}(x)$, where:

α is an integer, followed by the sign + or -, specifying the total effective charge of the complex; α = 0 is omitted; α = 1+ is indicated +; α = 1- is indicated - .

β is the letter a or the letter c, specifying whether the center is formed around anion or cation vacancies respectively.

n is the number of vacancies participating in a complex; it is omitted if equal to 1.

The chemical symbol of an adjacent impurity, if any, is x.

The number of electrons is not shown on the label, as it is readily calculated given the total charge.

For instance:

${}^{c}A$: cation vacancy + positron;
${}^{a}A$: anion vacancy + positron + two electrons;
${}^{c}A^{-}(Ca^{2+})$: cation vacancy + positron + electron + adjacent Ca^{2+} impurity;
${}^{a}_{2}A^{+}$: two anion vacancies + positron + two electrons.

a) Thermal Defect Generation

It is well known that thermal disorder in pure alkali halides produces Schottky defects (i.e., anion-cation vacancy pairs). In nominally *pure* crystals, which may well contain uncontrolled divalent impurities at concentration of 1 ppm or so, fairly high temperatures are necessary for obtaining thermal generation of cation vacancies in concentrations substantially higher than those already present at room temperature. For a better definition, let us take the example of KCl containing 1 ppm molar fraction of Sr^{2+}: application of the Lidiard-Debye-Hückel theory shows that the cation vacancy concentration remains constant at a level of the order of 1 ppm molar fraction up to $\sim$700 K and then increases with rising temperature; at 1000 K this concentration is $\sim$40 ppm [5.94].

Measurements of lifetime spectra taken at temperatures up to the melting point for the chloride family, already mentioned in Sect.5.2.2, have shown an increase of the short decay constant τ_1, but have failed to demonstrate a correlation of the spectral intensities with the concentration of defects [5.37]. If one assumes that cation vacancies act as effective positron traps, it follows from the experiment that the effect of trapping is already saturated at room temperature or else that it is hidden in the unresolved short-living part of the spectrum.

b) Thermal Quenching

Quenching from high temperatures may freeze into the crystal a nonequilibrium concentration of defects which, in the case of alkali halides, are of the Schottky type. Partial anneal at surfaces and coalescence in aggregates is possible, depending on the cooling rate. Mechanical stress due to rapid cooling may produce dislocations.

DANNEFAER et al. [5.31] have studied annihilation in KCl quenched from temperatures up to 750°C, using lifetime and line-shape methods. They report modifications of the spectra as functions of quenching temperature; the effects are more distinct when the measurements are taken at 90 K. The analysis of the results they present is based on a self-consistent decomposition of lifetime spectra in four exponential components and of line shapes in three Gaussians. The variation of the spectral parameters with quenching temperature is displayed in Fig.5.5 for the four-component fit. The intensity I_N of the narrow component of the energy profile is found to decrease with temperature, while the ratio $(I_3 + I_4)/I_N$ remains nearly constant. A direct comparison with the results of the equilibrium measurements at high temperatures of BERTOLACCINI and DUPASQUIER (see Sect.5.2.4), is not possible, since in this case a two-component fit was adopted for KCl; a qualitative agreement is, however, apparent in the increase of the shortest decay constant τ_1 with rising temperature, observed in both experiments. DANNEFAER and co-workers explain their results assuming that, at concentrations that increase with the quenching temperature, cation vacancies trap positrons in a state whose lifetime turns out to be

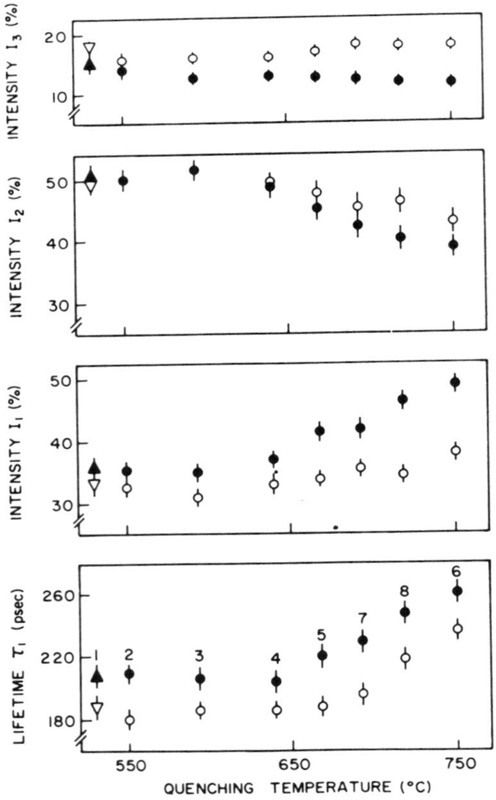

Fig.5.5. Lifetime results for KCl as a function of quenching temperature with τ_2 = 464 ps and τ_3 = 774 ps kept constant. The triangles show results before any quenching. Solid and open symbols are for 90 and 296 K measurements, respectively. τ_4 (not shown) had a mean value of 4.4 ns and I_4 (not shown) the value 0.8% at 90 K and 2% at 296 K. The numbers at the τ_1 values indicate the chronological order of the experiments. [5.31]

~ 340 ps. According to their explanation, the effect of trapping is manifest in the increase of τ_1, which is interpreted as the apparent lifetime of a sum of unresolved components.

c) Additive Coloration

It is well known that substantial F center concentration is introduced in some alkali halides by the addition of alkali metal in stoichiometric excess. Other consequences of additive coloration are: I) effective suppression of cation vacancies; II) formation of colloidal inclusions of metal and F center aggregates in amounts depending on concentration of the excess metal and on temperature. To obtain at room temperature high concentrations of isolated F centers and to reduce the formation of colloids, thermal quenching may be necessary; this may introduce further lattice disorder.

The effects of additive coloration on annihilation characteristics are well known [5.27,28,32,43,54,63]: I) a new long-living component appears in the lifetime spectrum (see Fig.5.6); II) the momentum distribution of annihilating pairs becomes narrower than in uncolored pure crystals. The narrowing of momentum distribution is interpreted as the rise of an independent narrow component that may be isolated by difference with the distribution of an uncolored standard after normalization at large angles. Values of the lifetime τ^* of the mentioned lifetime spectrum component and of the width Γ_N of the narrow component of the angular correlation curve are reported in Table 5.3. The shape of the narrow component in the case of KCl is shown in Fig.5.3. While τ^* and Γ_N are independent on F center concentration, the corresponding intensities I^* and I_N increase with increasing F center concentration. Some

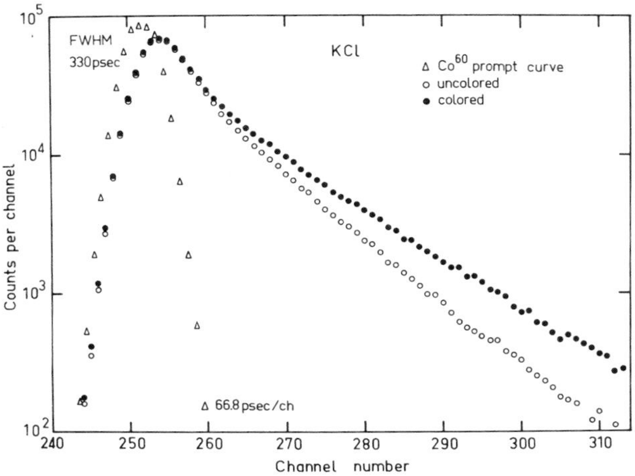

Fig.5.6. Effect of additive coloration on the lifetime spectrum of positrons in KCl. [5.63]

Table 5.3. Annihilation parameters for positrons captured by F centers in additively colored crystals

Substance	τ^*[ns]	Γ_N[mrad]
NaCl	1.09 ± 0.02 [a]; 1.1 ± 0.1 [b]	3.7 ± 0.2 [b]
KCl	0.99 ± 0.02 [a]; 1.04 [c]	5.0 ± 0.2 [d,e]; 5.3 [f]
KBr	1.06 ± 0.02 [a]	5.0 ± 0.2 [e]
KI	1.46 ± 0.04 [a]	4.2 ± 0.2 [e]

[a][5.54,78]; [b][5.27]; [c][5.28]; [d][5.25,96]; [e][5.77]; [f][5.87]

disagreement exists between the data concerning this dependence, as reported in [5.54] and [5.28]; the data of [5.54] and [5.25] are mutually consistent in the hypothesis that I^* and I_N are contributions originating respectively from ortho and para states of a qPs system. The discrepancies are probably due to difficulties in determining the F center concentration for very dark crystals.

The direct association of the above components to F centers and not to other defect species is confirmed by other observations: I) annihilation spectra are not affected by the presence of colloidal X centers [5.54]; II) $F \rightarrow F^-$ conversion and F center aggregation does not alter quenching- and impurity-induced defects, but modifies the spectrum by reduction of the intensity of the long-living component [5.70,72]. The nature of the positron state associated with F centers is indicated by the following further circumstances: I) application of a static magnetic field quenches the long-living component and enhances the central peak of the angular correlation curve [5.23,24,26]; II) the width Γ_N of the narrow component of the angular correlation curve indicates localization of the system in a space region the linear dimensions of which are very close to the lattice constant of the host crystal [5.77,95,96].

All the above elements concur in indicating that positrons are trapped at F centers and form, together with the electron of the F center, a localized qPs system. According to the previously stated notation, this system is indicated as $^aA^+$ center. The experimental identification of $^aA^+$ centers has stimulated several theoretical contributions on which we shall report in Sect.5.3.2.

Lifetime measurements [5.79] for additively colored KCl at temperatures in the range of 7-300 K have shown that: I) the lifetime of the $^aA^+$ center decreases by 5% when the crystal is cooled to 7 K; II) the probability of positron trapping at F centers is temperature independent. The first effect can probably be attributed to the volume decrease of the void where the system is localized. The second result, which is instructive as regards both the behavior of positrons before trapping and the trapping mechanism itself, will be commented on in Sect.5.3.3.

d) $F \rightleftarrows F^-$ Conversion

Irradiation with light in the F band of additively colored KCl crystals cooled to about -100 °C, partially converts F centers into F^- and F^+ centers, according to the reaction $2F \rightarrow F^- + F^+$. Subsequent irradiation with light in the F^- band at a wavelength longer than F band, and at a temperature below 90 K, completely reverts the process.

The most evident consequences of $F \rightarrow F^-$ conversion on annihilation are the following: I) the narrow component of the angular correlation curves narrows even further [5.87]; II) the intensities of the components associated to F centers in lifetime spectra and angular correlation curves are reduced [5.70,72]. This reduction is so strong that it cannot be due only to the decreased number of capturing F centers, but must be caused also by a new competing capture process. Assuming that F^+ centers (i.e., empty vacancies) are not effective as positron traps, it follows that the capture process is on F^- centers. The resulting system is the ${}^{a}A$ center (positron + two electrons + anion vacancy). However, the corresponding component has not been resolved in lifetime spectra. From independent experimental data, the ratio k_{F^-}/k_F of positron capture rates in F^- and in F centers turns out to be 11.5 ± 5.8 according to [5.70], and 8.0 ± 2.0 according to [5.72].

e) Aggregation of F Centers

Illumination at room temperature with F band light of additively colored crystals produces aggregation of F centers. F_2 and F_3 centers are formed. The consequences on annihilation are the following. The narrow component of the angular correlation curves, due to coloration, remains unaltered in intensity, but narrows even further. In an experiment of HERLACH and HEINRICH [5.43] on KCl with 16.10^{17} F centers cm^{-3} and $0.6.10^{17}$ F_2 centers cm^{-3} before illumination, and with $2.8.10^{17}$ F_2 centers cm^{-3} after illumination, the width is reduced by 12%. The result, indicating localization of the annihilating pair in a larger region, is probably attributable to positron capture in F_2 centers to form ${}^{a}_{2}A^+$ centers. The intensity of the long-living lifetime component due to coloration is reduced. On analyzing the results of an experiment in which a substantial amount of F centers is converted into F_3 centers, DANNEFAER et al. [5.80] conclude that F_3 centers compete with other centers in capturing positrons.

The system formed if a positron is captured by an F_3 center is the ${}^{a}_{3}A^+$ center; it has the character of a qPs system and should therefore contribute to lifetime spectra with two separate ortho and para components. The isolation of these components from experimental data is, in practice, an almost impossible task.

f) Doping with Divalent Impurities

Doping of alkali halides with divalent impurities at concentration C increases the concentration x_c of cation vacancies and reduces the concentration x_a of anion vacancies. Omitting the effect of thermal disorder, one obtains $x_c = C$ and $x_a = 0$. A fraction of cation vacancies remains associated to impurities on nearest positive ion sites. The concentration of free cation vacancies is approximately given by

$$x_d \simeq (C/Z)^{1/2} \exp(-E_a/2k_BT)$$

where Z is the cation coordination number (Z = 12 for rocksalt structures) and E_a is the free energy of association between a vacancy and an impurity. Several annihilation experiments have been carried out on divalent-doped crystals, with the aim of demonstrating positron capture in cation vacancies to form ^cA centers. The results can be summarized as follows: I) a new long-living component of lifetime $\tau \simeq 1$ ns appears in lifetime spectra, with an intensity that grows with the concentration of Ca^{2+} in KCl [5.57,62]; the effect is determined by the whole concentration of vacancies x_c and not only by x_d [5.62]; the new component is affected by magnetic quenching [5.62]; II) no effect of Mg^{2+} doping is observed on lifetime spectra for LiF [5.71]; III) no difference is observed in angular correlation curves taken for Ca^{2+} - doped KCl at Ca^{2+} concentrations of 3 ppm and of 340 ppm [5.67].

Apparently, ^cA centers have escaped observation. Even the positive result I), taking into account the magnetic-quenching effect, could be indicative of the formation of $^cA^-$ and/or $^cA^-(Ca^{2+})$ centers, but not of ^cA centers. However, doped crystals become very easily colored by ionizing radiation, and F centers created by the irradiation of the positron source may have played their role.

g) *Defect Creation by Ionizing Radiation*

Defect creation by ionizing radiation is a fairly complicated process. The reader may refer to the review article by SONDER and SIBLEY [5.92] for a coverage of the subject. In order to give a simplified picture of the phenomenon we recall a few facts.

Ionizing radiation may produce disorder in ionic crystals by elastic knock-on collisions or through electronic processes. Only the second mechanism is really important for irradiation with X- and γ-rays and with electrons at energies below, say, 2 MeV. We shall discuss this case first. The energy released by radiation into the crystal in the form of electronic excitation is partially converted by a chain of reactions (*radiolysis*) into kinetic energy capable of producing ion displacement. The mechanism is, however, not effective in displacing positive ions.

The end products of electronic excitation and radiolysis are: I) Frenkel pairs in the halide sublattice, II) trapped electrons, III) trapped holes. Depending on temperature some rearrangement takes place. At temperatures in the liquid helium range the predominating species are vacancy-associated centers (F^-, F, F^+), self-trapped holes (V_K centers), also interstitial halide atoms (H centers). When the temperature is raised, F^- and V_k centers become unstable and disappear, while interstitials cluster together or around impurities. At room temperature the predominating species are F, F_2, F_3 centers, and interstitial halogen clusters of various sizes. A further increase in temperature produces the ionization of F centers and aggregates; at still higher temperatures halogen clusters evaporate; finally, interstitials and vacancies recombine.

Reflecting the multifarious defect structure existing in irradiated crystals, annihilation characteristics display modifications not easily framed in a comprehensive picture. In fact, authors who have investigated this problem often reach conflicting conclusions. However, the various pieces of experimental evidence seem to combine consistently, providing one separates schematically the response of annihilation properties to irradiation in two stages, according to the total dose received by the crystals. For definiteness we shall discuss this point below with special reference to lifetime studies on NaCl and KCl.

At high irradiation doses (*stage II*), the irradiation effects on lifetime spectra are dominated by F centers. As mentioned in Sect.3.3.3, F centers in additively colored crystals capture positrons to form $^aA^+$ centers. The corresponding lifetime component is clearly revealed also in strongly irradiated crystals. In fact, for KCl X-rayed up to 6.10^8 R at room temperature, MALLARD and HSU [5.66] isolate a component with a lifetime of $\tau^* = 1.1$ ns, its intensity I^* being well correlated to F center concentration. It is interesting to note that intensity I^* is definitely smaller in irradiated than in additively colored samples for equal F center concentrations. Ondoubtedly, this is the result of capture processes in other defects (probably F center aggregates) competing with capture at F centers.

The complexity of lifetime spectra, due to the presence of the long-living $^aA^+$ component, is an obstacle in elucidating the modifications occurring in the shorter-living part of the spectrum. Only at low irradiation doses (*stage I*), is the effect of F centers small enough to allow a reliable study of these modifications. BRANDT et al. [5.59] studied NaCl X-rayed up to 2.10^5 R at room temperature. After subtracting a long-living component of intensity $\leqq 2\%$, they were able to observe a distinct decrease of the intensity I_2 of the second component in a three-component fit of the spectrum. Such a decrease is correlated to an increase of the optical absorption in the V band; it is not affected by F center bleaching and disappears only with the complete anneal of the crystal at 450°C. A similar effect, due to V_k centers, has been observed by HSU et al. [5.71] in Ag-doped KCl X-rayed up to 7.10^5 R. In this experiment, irradiation and lifetime spectra measurements were performed at liquid nitrogen temperature to form and maintain stable V_k centers; V_k center production was enhanced by the presence of Ag impurities which act as electron traps and prevent electron-hole recombination. When pure KCl was examined, the effect of irradiation on I_2 was barely visible. The ability of V_k centers to reduce I_2 has been independently proposed by DANNEFAER et al. [5.72] in order to explain modifications occurring in lifetime spectra, with illumination and with changing temperature in the 95-260 K range, in the case of KCl irradiated only by the positron source itself.

The mechanism by which the effects in stage I are determined is at present a matter of conjectures. BRANDT et al. [5.59], as well as DANNEFAER et al. [5.72], offer the following explanation: the component of intensity I_2 is due to positron trapping by negatively charged defects existing before irradiation; free holes created by irradiation, and also V_k centers under illumination, migrate to potential positron traps rendering them inoperative and thus decreasing I_2. On the other hand, the observation that cation vacancies introduced by Ca^{2+}-doping in KCl do not increase I_2, and the result of magnetic quenching experiments indicating that I_2 is the intensity of ortho qPs (see Sect.5.2.4), conflict with the above hypothesis for the origin of I_2. HSU et al. [5.71] propose that V_k centers react with positrons in the process of being thermalized before qPs formation in perfect regions of the crystal. Since the process of qPs formation in perfect crystals is still rather unknown (see Sect.5.3.3), we may well increase this list of conjectures proposing that: I) V_k centers recombine with the electrons of the ionization spur of the positron, and therefore hinder qPs formation; II) V_k, behaving as a paramagnetic free radical, reacts with hot qPs (but not with thermalized qPs, otherwise a change in τ_2 would be observable).

Studies of irradiation effects on annihilation by means of angular correlation and line-shape methods have given less information than lifetime spectra experiments. Generally speaking, defect generation by radiation damage produces a narrowing of momentum distribution or, equivalently, an increase in the number of co-linear photons. To quote an example, BRANDT and PAULIN [5.90] found that in NaCl the number of photon pairs emitted within 1 mrad from co-linearity increases by 10% after X-irradiation at the total dose of 6×10 R (which is well within stage II). Isolation of a narrow component may be obtained by difference with a *zero-irradiation* standard. The width Γ_N of the narrow component so obtained is not too far from the width of the similar component isolated in the case of additively colored crystals: i.e., Γ_N = 5.4 mrad for electron irradiated KCl [5.73], compared with $\Gamma_N = 5.0 \pm 0.2$ mrad for additively colored KCl [5.43,77]. At the same concentration of F centers, the intensity I_N is, however, much larger in irradiated than in additively colored crystals. This effect, which is opposite to that observed for lifetime spectra long-living components, is in agreement with the hypothesis of competitive positron capture in F center aggregates.

To conclude our exposition of experimental results concerning defects produced mainly by radiolytic processes, we wish to underline one point. A critical reading of the literature convinces us that the frequently repeated conjecture that observed irradiation effects on annihilation may be due to positron capture in an increased numer of cationic vacancies is not really founded on experimental evidence. This is in agreement with current views on radiation damage in ionic crystal, according to which cation vacancies are but a small minority in the diversified population of defects produced by radiation (see, for instance, [5.92]).

In principle, the case of irradiation with heavy particles or with very energetic electrons and γ-rays may be different. Here primary and secondary elastic collisions produce Frenkel-type disorder among cations as well as anions. In practice, it is difficult to isolate the effect of cation vacancies from those produced by other defect species. As a matter of fact, WILLIAMS and ACHE [5.33] did not report qualitative differences in effects produced by irradiation either with 30 MeV protons or with ^{60}Co γ-rays at a total dose of 5×10^8 R.

h) Plastic Deformation

In plastically deformed crystals high concentrations of dislocations are to be expected, together with pointlike defects left behind by mobile dislocations.

NIEMINEN et al. [5.41] have measured lifetime spectra for KCl and NaCl at various degrees of plastic deformation under uniaxial compression in the crystallographic direction [100]. Their results show changes in the spectrum at the very first stages of deformation, followed only by modest variations: a new long-living component whose lifetime is ~ 2 ns may then be isolated. The maximum intensity of this component is 2.6% in KCl at 35% deformation, and 3.3% in NaCl at 50% deformation.

i) Mixed Crystals

Solid solutions may be imperfect to a higher degree than their pure components. The KCl-KBr solid solution has been studied by the angular correlation method in the whole 0-100% range of relative concentrations [5.65]. The FWHM of the curve, $\Gamma_{1/2}$, correlates with the estimated density of anion vacancies; at maximum defect concentration $\Gamma_{1/2}$ reaches its minimum.

5.2.4 Magnetic Field Effects

As discussed in Sect.5.1.2, magnetic quenching experiments lead to the identification of lifetime and angular correlation components associated to qPs states of the positron. Furthermore, the quantitative analysis of the effect enables one to calculate the density $|\psi(0)|^2$ of the unpaired-spin electron at the position of the positron, at least to the extent to which the hyperfine energy splitting of ortho and para states remains proportional to $|\psi(0)|^2$. It is convenient to express $|\psi(0)|^2$ in units of density $|\psi_0(0)|^2 = (8\pi a_0)^{-3}$ for the free positronium atom, by giving the ratio[1]

[1]For this ratio, we prefer the symbol κ sometimes encountered in the Russian literature, instead of the more usual symbol α, which is often confusing.

$$\kappa = \frac{|\psi(0)|^2}{|\psi_0(0)|^2} \quad . \tag{5.3}$$

A qPs state characterized by $\kappa < 1$ may be intuitively viewed as a *relaxed* positronium, as it could result from a decrease of the effective reduced mass of from the immersion of the atom in a dielectric continuum. Conversely, $\kappa > 1$ would indicate a *concentrated* atom. Here below we report and comment on the results of magnetic quenching studies of virgin alkali-halide crystals, on colored crystals and on doped ones.

a) Crystals with Low Defect Concentration

In the case of several alkali halides, lifetime spectra contain – at least apparently – two main components and, possibly, a faint long-living tail. With the aim of getting information on the origin of the second main component, such compounds have been systematically studied in a series of experiments by the Milano group by means of the magnetic quenching method. For all cases considered, it was found that the above component is quenched by the magnetic field, as indeed one would expect for a qPs state with the properties of relaxed positronium. In Table 5.4 we report the values of the relaxation parameter κ, (5.3) and of the decay rate λ of the quenchable component, as given in [5.24]. The parameter κ appears empirically correlated to the anion radius (also given in Table 5.4), as shown in Fig.5.7. The formation of qPs states is confirmed by changes occurring in angular distributions on application of the magnetic field. For KCl, HERLACH [5.26] found that, when the field is applied, a number of counts is transferred from the large- to the small-angle region of the angular distribution. The transferred counts form a contribution,

Table 5.4. Experimental values of the relaxation parameter κ for alkali halides with low defect concentration (r_-: anion radius; λ: annihilation rate of the long-living component)

Substance	r_-[a.u.]	$\lambda[10^9\ s^{-1}]$	κ_{LS} [a]	κ_{ACC} [b]
LiF	2.06	3.37	0.09 ± 0.01	
NaF	2.20	3.24	0.10 ± 0.02	
LiCl	3.11	2.30	0.29 ± 0.03	
NaCl	3.11	1.46	0.42 ± 0.07	
KCl	3.11	1.59	0.32 ± 0.07	0.42 ± 0.04
LiBr	3.41	1.80	0.42 ± 0.07	
NaBr	3.41	1.34	0.43 ± 0.06	
KBr	3.41	1.32	0.42 ± 0.08	0.48 ± 0.06
NaI	3.88	1.75	0.64 ± 0.10	

[a] From lifetime measurements [5.24]; [b] From angular correlation measurements [5.86]

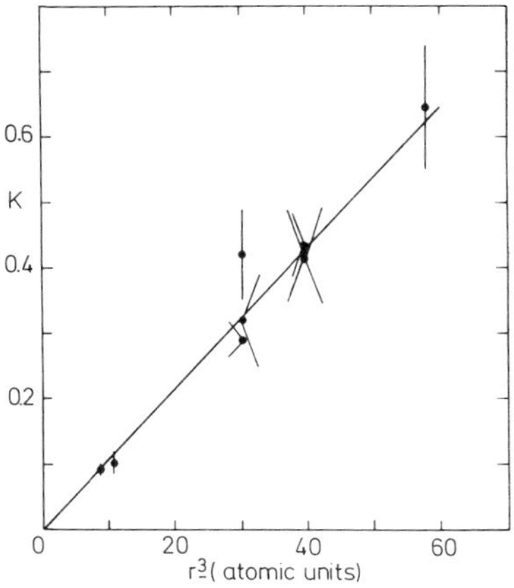

Fig.5.7. Relaxation parameter as a function of the cube of the anion radius. [5.24]

shown in Fig.5.3, whose FWHM is $\Gamma_N \simeq 3.0$ mrad, and whose relative intensity is $I_N \simeq 5\%$ at H = 20 kG. For KCl and KBr also AREFEV et al. [5.86] observed the magnetic enhancement of angular distributions at small angles, and they deduced relaxation parameters in close agreement with those obtained from lifetime measurements. The results of AREFEV et al. are also collated in Table 5.4.

The results of the experiments in [5.23,24,86], indicating $\kappa < 1$, are contested by DANNEFAER et al. [5.3] on the basis of their own data. Let us illustrate this point. Assuming that the intensity I_0 of the quenchable component of the lifetime spectrum is equal to the fraction of positrons bound in ortho qPs state, the intensity I_N of the narrow component in momentum distributions can be predicted as follows:

$$I_N = \frac{1}{3} I_0 \frac{\kappa\lambda_p}{\kappa\lambda_p+\delta} \quad , \tag{5.4}$$

where $\lambda_p = 8.10^9\ s^{-1}$ is the annihilation rate of free parapositronium, and δ is the pick-off annihilation rate, the latter being assumed equal to the measured annihilation rate λ_0 of the ortho state in the crystal. From a four-component fit for KCl, the above authors obtained $I_0 \simeq 3I_N$ when taking for I_N the value resulting from line-shape measurements. This result is consistent only with κ values that are not substantially smaller than 1. However, if one takes I_0, λ_0, and κ as reported in Tables 5.1,4, (5.4) gives $I_N = (9 \pm 1)\%$, within error limits coinciding with $I_N = (7.7 \pm 0.5)\%$ and with $I_N = 8.2 \pm 0.4$, as determined respectively by line-shape and angular correlation methods [5.28,31]. Since the four-component fit of lifetime spectra proposed by DANNEFAER and co-workers is not strictly justified on the basis of a χ^2 test, we do not think that their conclusion is unobjectionable.

b) Additively Colored Crystals

The presence of magnetic field effects on annihilation characteristics for positrons trapped at F centers has been observed with lifetime and angular correlation methods. This gives a strong support to the hypothesis that trapping leads to the formation of $^{a}A^{+}$ centers which have qPs character.

As regards values for relaxation parameter κ, there is disagreement between determinations from lifetime results [5.22,78], which give $\kappa < 1$ as reported in Table 5.5, and angular correlation results indicating $\kappa \geqq 1$ [5.25,28]. The reason for the discrepancy has not been clarified, but some observations may be made: I) angular correlation results are altered by the fact that no attempt has been made to separate the effects of the magnetic quenching of $^{a}A^{+}$ centers from the contributions of qPs not formed at F centers; II) the presence of a narrow component unaffected by the magnetic field (as it would exist if, besides F centers, nonparamagnetic positron traps were present in the crystal) would result, if not properly accounted, in κ-ratios higher than the real ones; III) the result $\kappa \simeq 1$ for colored KCl has been obtained by SMEDSKJAER and DANNEFAER [5.28] from data at fields from 13 to 21 kG, while results of the same authors for fields <13 kG seem to indicate the strong quenching that one expects to find when the hyperfine splitting is reduced, i.e., when $\kappa < 1$.

Table 5.5. Experimental and theoretical values of the relaxation parameter κ for $^{a}A^{+}$ centers. The theory of $^{a}A^{+}$ centers is discussed in Sect.5.3.2

Substance	κ_{exp} [a]	κ_{th}^{H} [b]	κ_{th}^{KS} [b]
NaCl	< 1	0.490	0.242
KCl	0.29 ± 0.04	0.527	0.265
KBr	0.48 ± 0.05	0.544	0.286
KI	0.59 ± 0.07	0.562	0.292

[a] [5.78]; [b] values deduced from calculated lifetimes given in [5.11]; κ_{th}^{H} corresponds to a hydrogenoid potential and κ_{th}^{KS} to a Krumshansl-Schwartz potential

c) Doped Crystals

Magnetic quenching of the longest-living spectrum component has been observed in Ca^{2+}-doped KCl [5.62]. The relaxation parameter deduced from the above measurements turns out to be $\kappa = 0.10 \pm 0.01$; this result helps in differentiating on the origin of the above component in respect to the longest-living component observed in additively doped KCl, which has nearly the same lifetime ($\tau \approx 1$ ns) but a different κ ($\kappa = 0.29 \pm 0.04$) [5.22]. The above supports the attribution of the longest-living component in doped crystals to $^{c}A^{-}$ or $^{c}A^{-}(Ca^{2+})$ centers, or a mixture of both; alternatively, a *correlated* ^{c}A center may be considered (an ^{c}A center with a nearby small exciton [5.20]).

5.3 Positron States in Alkali Halides

The existence of several positron states in alkali halides was realized when the complexity of lifetime spectra was observed for the first time [5.1]. Since then, many models, not all equally successful, have been proposed to describe possible positron states. The wide experimental basis summarized in the previous section has led to the identification of the structure of some states with reasonable certainty. In a few cases details regarding the microscopic properties of the states are also known. For years, the main question of annihilation studies in the field of alkali halides has been: is the complexity observed in annihilation spectra a pureley bulk effect, or is it essentially determined by positron trapping at lattice defects? We think that this question can finally be answered. As we shall see in this section, in ideal crystals at least two positron states are formed; furthermore, a variety of states arises from the association of the positron with defects (A centers). This second kind of state is probably easier to study, as defects can be sometimes introduced into the crystal in controlled amounts; they can also be characterized by optical or ESR measurements.

We present below the information available on both kinds of positron states, linking so far as possible theoretical predictions with observed properties.

5.3.1 Intrinsic States

a) Nearly Free Positrons in a Perfect Lattice

Theoretical and experimental results concur in indicating that positrons in their lowest-energy state behave in the periodic lattice of alkali halides like delocalized particles. We recall from Sect.5.2.1 that the shape of the broad component of angular distributions and annihilation energy lines is essentially determined by the momentum distribution of the external electrons of the halogen ion; the hypothesis of positron localization would result in calculated curves broader than the measured ones.

The reasons why a positron that can be bound to halogen ions in empty space is not bound to anions in the crystal can be intuitively understood if one observes that: I) the size of positronic orbitals around the ion in vacuo is so large that the system cannot fit into the crystal without serious distortion [5.97]; II) the

Madelung potential repels the positron from negative cells. This point of view is confirmed by the calculations of HAUTOJÄRVI and NIEMINEN [5.39], and NIEMINEN [5.45]. By solution of the Schrödinger equation for the positron in a rigid lattice, NIEMINEN obtained Bloch wave functions of the proper translational symmetry, and he concluded that positrons maintain a nearly free-particle behavior with a preference for interstitial regions and for negative ion cells. NIEMINEN also calculated the energy eigenvalues $E(\underline{k} = 0)$ at the bottom of the positron band, the effective mass ratios m_+/m, and the ratios $R = S^+/S^-$ of the positron overlap on cations (S^+) to the overlap on anions (S^-). His results are reported in Table 5.6.

Table 5.6. Values of the positron zero-point energy E(0), effective mass ratios m_+^*/m, and the cation-anion overlap ratio R for some alkali halides [5.45]

Substance	E(0) [eV]	m_+/m	R
LiF	4.65	1.07	0.08
LiCl	3.27	1.19	0.05
NaCl	2.54	1.18	0.10
KCl	1.69	1.25	0.13
KI	1.46	1.38	0.13

In principle, one must consider that positrons, even if not bound to individual anions in a rigid lattice, could become localized by self-trapping in the real polarizable lattice. In fact GOLDANSKII and PROKOPEV [5.98-100] suggested the formation of a *small polaron* state, centered at a negative ion; quite recently, DANNEFAER et al. [5.42] also interpreted their results by assuming the formation of a small polaron centered in an interstitial site. Self-trapping is not considered in Nieminen's theory, but this same author, using the band masses reported in Table 5.6, estimated a value ~ 6 for the coupling constant of the polaron theory. This means that the positron is strongly coupled to longitudinal optical phonons, but nevertheless is still to be considered as an itinerant *large polaron* with effective mass $m_+^\dagger \simeq 30$ m. Polarization effects, of course, will influence the transport properties of the positron. Nieminen's results for m_+^* and $m_+^\dagger$ are consistent with estimates independently made in the case of LiF by FUJIWARA et al. [5.35].

The annihilation rate of nearly free positrons cannot be calculated from Nieminen's wave function, which does not take into account electron-positron correlation. This rate is also difficult to obtain from lifetime measurements for two reasons: I) nearly free positrons contribute to the short-living part of the spectrum which may also contain other unresolved components; II) in case of trapping, the experimental decay rate is not the pure annihilation rate but the sum of the annihilation and trapping rates. However, if one assumes that, after instantaneous thermalization, at the instant t = 0 all positrons populate the fundamental state of the perfect crystal, then the annihilation rate of this state is given by [5.37,101]

$$\lambda_1 = \sum_i I_i/\tau_i \tag{5.5}$$

where I_i and τ_i are observed intensities and decay time constants.

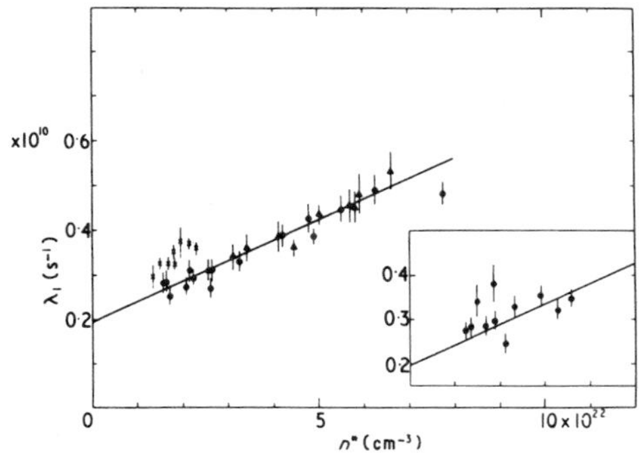

Fig.5.8. Annihilation rate λ_1 as a function of n^* (see text). The insert contains the λ_1 values for salts displaying three components in the lifetime spectrum. (▲) oxides; (×) sulphides, selenides and tellurides; (●) alkali and alkaline-earth halides. [5.101]

The above hypothesis may seem hard to justify, also in view of the formation of other intrinsic positron states; nevertheless, the application of (5.5) to a wide class of binary compounds (metal dioxide, alkaline-earth oxides, sulphides and selenides), besides alkali halides, has led BERTOLACCINI et al. [5.101] to reveal an impressive empirical regularity. As is shown in Fig.5.8, the averaged annihilation rate λ_1 is well correlated to the number n^* of negative ions per unit volume accessible to positrons (cations are considered as impenetrable spheres). The experimental results of KELLY [5.102] for silver halides show that also in the case of these compounds λ_1-points fall on the same straight line. The authors of [5.101] wrote the $\lambda_1 = \lambda_1(n^*)$ linear relation in the form

$$\lambda_1 = \lambda' + \xi \frac{C_s}{4} n^* \quad , \tag{5.6}$$

where C_s is the Dirac singlet annihilation rate per unit electron density, and, by least-squares adjustment, $\lambda' = (0.195 \pm 0.008) \times 10^{10}\ s^{-1}$ and $\xi = 6.09 \pm 0.30$ electrons/ion. They observe that λ' is very close to the annihilation rate of the negative positron ion, and offer the following tentative interpretation of (5.6): delocalized positrons, while sweeping accessible zones of the crystal, annihilate with the eight outer electrons of the anions; however, two out of eight electrons strictly correlate their motion with that of the positron, giving an n^*-independent contribution to the annihilation rate.

b) Quasi-Positronium in Perfect Crystals

The formation of a positronium atom or of a positroniumlike system in alkali halides has for a long time been a debated problem. On the one hand, FERRELL [5.3] showed that a positronium atom could not fit into the interstices of the lattice; moreover, until HERLACH's experiment in 1972 [5.26] the existence of a complex structure in angular distributions for uncolored crystals had not been revealed. On the other hand, as long ago as 1966 GOLDANSKII and PROKOPEV [5.103] were able to show that the early results on magnetic quenching of lifetime spectra and on three-quantum yield could be explained quite simply by the hypothesis of quasi-positronium formation, and they worked out models for this system [5.100,104,105]. At present, qPs formation in ideal crystals is firmly established by mutually consistent independent observations: I) magnetic quenching of the long-living component in lifetime spectra for pure crystals (see Sect.5.2.2); II) isolation of narrow components in momentum distributions for pure crystals (see Sects.5.2.1,2); III) magnetic enhancement of the above narrow components (see Sect.5.2.3); IV) presence of secondary peaks at boundaries of Brillouin zones in angular distributions at low temperatures (see Sect.5.2.2).

The problem as to whether the formation of qPs in alkali halides is energetically possible has been discussed by several authors [5.106,107]. Also to be recalled is the positive answer to the above question obtained by NEAMTAN and VERRALL for LiH [5.108]; and their results can be extended to alkali halides as well. It is easy to realize that Ferrell's argument against positronium formation does not necessarily apply to any qPs system. Let us recall Ferrell's reasoning: to localize an atom of mass 2m in an interstitial space of linear dimensions $d \approx 1$ Å it is necessary to spend the energy E_L, given by

$$E_L \simeq \frac{(\Delta p)^2}{4m} = \frac{\hbar^2}{4md^2} = 16.7 \text{ eV} \quad . \tag{5.7}$$

Since E_L is greater than the binding energy of positronium, $E_B = 6.8$ eV, the atom would become energetically unstable. However, suppose that an electron, raised in the conduction band by some process, is attracted by a nearly free positron; then the two particles will correlated their motion forming a qPs system with a binding energy E_B'. The stability of this system is assured unless the electron recombines with a hole in the valence band (normally the recombination energy E_{gap} is greater than E_B'). And since the center of mass remains free, no localization energy (à la FERRELL) is to be considered.

The above picture of a delocalized qPs, whose center of mass moves in the periodic field of the lattice, is in qualitative agreement with the observation of relaxation parameters $\kappa < 1$, as derived from magnetic quenching measurements, and

$\hbar = h/2\pi$ (normalized Planck's constant)

of side peaks in angular distributions at the projections of reciprocal vectors (200) and ($\bar{2}$00). This can be shown quite simply, as follows.

I) Within the framework of the effective mass approximation, the internal wave function $\psi(\underline{r})$ of the electron-positron pair is the normal hydrogenoid wave function $\psi_0(\underline{r})$ of positronium in vacuo, the only difference being a scaling in distances to account for the modification of the reduced mass and for the dielectric properties of the medium. For the relaxation parameter one readily obtains [5.100]

$$\kappa = \frac{|\psi(0)|^2}{|\psi_0(0)|^2} = 8\left(\frac{\mu}{\varepsilon m}\right)^3 \tag{5.8}$$

where ε is the relative dielectric constant of the medium, and μ is the reduced effective mass $(m_-^{*-1} + m_+^{*-1})^{-1}$, calculated with the band masses of the conduction electron, m_-^*, and of the positron m_+^*. In general m_-^* is substantially smaller than m, and this will give $\kappa < 1$, even if one takes $\varepsilon = 1$. For instance, in the case of KCl ($m_-^* = 0.43$ m; $m_+^* = 1.25$ m), (5.8) gives $\kappa = 0.26$ for $\varepsilon = 1$.

II) An approximate expression of the wave function $\Phi(\underline{R})$ of the center of mass of the qPs system may be obtained treating qPs as a particle characterized by polarizability γ. The corresponding Schrödinger equation is

$$\left[-\frac{\hbar^2}{2(m_e+m_p)}\nabla_{\underline{R}}^2 - \frac{1}{2}\gamma\mathscr{E}^2(\underline{R})\right]\Phi(\underline{R}) = E\Phi(\underline{R}) \quad , \tag{5.9}$$

where $\mathscr{E}(\underline{R})$ is the electrostatic field of the lattice. We do not try to solve (5.9), but observe that $\Phi(\underline{R})$ must have the translational symmetry of the periodical potential-energy term $-\gamma\mathscr{E}^2(\underline{R})/2$. Suppose now that the leading terms in the Fourier expansion of $\mathscr{E}(\underline{R})$ correspond to reciprocal lattice vectors on the first Brillouin zone; in such a case the leading Fourier terms of $\mathscr{E}^2(\underline{R})$, and thus of $\Phi(\underline{R})$, will correspond to reciprocal lattice vectors on the second Brillouin zone.

Combining the ideas derived from the simple model of relaxed positronium with experimental information, a fairly complete characterization of the system becomes possible. We therefore conclude our discussion by listing the characteristic properties of this particular qPs state, and we present below numerical examples for the well-studied case of KCl.

I) Relaxation parameter κ; for KCl the weighted average of experimental data from independent sources gives $\kappa = 0.39$, and the effective mass approximation $\kappa = 0.26$. For other compounds see Table 5.4.

II) Mean radius a'; this can be calculated from the equation

$$a' = a\kappa^{-1/3} \quad , \tag{5.10}$$

where a = 1.06 Å is the positronium mean radius. For KCl a' = 1.45-1.66 Å.

III) Binding energy E'_B; according to the model, this is

$$E'_B = E_B \kappa^{1/3} \quad , \tag{5.11}$$

where E_B = 6.8 is the positronium binding energy. For KCl E'_B = 5.0-4.3 eV.

IV) Total energy E'; taking the zero level of energies in correspondence with the following situation, "all the electrons of the crystal in unexcited valence states plus one positron in its fundamental Bloch state", the total energy level of delocalized qPs is

$$E' = E_{gap} - E'_B + E_{pol} \quad , \tag{5.12}$$

where E_{gap} is the gap energy and E_{pol} is the energy change due to lattice polarization. E_{pol} is in general of the order of 10 meV and can be neglected with respect to $E_{gap} - E'_B$. For KCl, E' = 4.0-4.7 eV.

V) Annihilation rate of the ortho state λ_0; this is essentially determined by the annihilation rate δ with external electrons. One has

$$\lambda_0 = \kappa\lambda_t + \delta \approx \delta \quad , \tag{5.13}$$

neglecting the three-quantum annihilation rate $\kappa\lambda_t$ in comparison with δ. At least for halides for which the component associated to qPs is identified by magnetic quenching measurements, λ_0 is experimentally known. See again Table 5.4. For KCl, $\lambda_0 = 1.6 \times 10^9$ s^{-1}.

VI) Annihilation rate of the para state λ_p; the model gives

$$\lambda_p = \kappa\lambda_s + \delta \tag{5.14}$$

where $\lambda_s = 8 \times 10^9$ s^{-1} is the singlest annihilation rate of positronium. The corresponding component should contribute with intensity $I_p = I_0/3$ to lifetime spectra; however, it has not in fact been resolved. From (5.14) one calculates for KCl $\lambda_p = 4.7 - 3.7 \times 10^9$ s^{-1} and $\tau_p = \lambda_p^{-1}$ = 220-270 ps.

VII) Localization length L; assuming that the center of mass of qPs can move freely in a cubic region, the length L of the edge of the cube is related to the FWHM Γ_N of the angular distribution by the equation

$$L[\text{Å}] \approx 28\, \Gamma_N^{-1} \tag{5.15}$$

where Γ_N is expressed in mrad.

The low temperature measurements of HYODO and TAKAKUSA [5.6] point toward a complete delocalization of qPs ($L \to \infty$). However, this is not true at room temperature, nor is it true at liquid nitrogen temperature for all samples; for KCl at room temperature one has $L \approx 9.3$ Å. The mechanism that leads from complete to partial delocalization is a problem open to further investigation. It is reasonable to think that this ensues from qPs interaction with holes, as suggested in [5.24], or with lattice defects. NIEMINEN has pointed out that a fixed hole would not bind qPs [5.45]; this is also true for any other point-like fixed charge, as can be inferred by the absence of binding in the system $H^+ - e^- - e^+$ [5.109]. It would be interesting to check if Nieminen's conclusion retains its validity even when taking into account the delocalization of the hole on the volume of the anion, or of two anions in the case of holes self-trapped in V_k centers. We suspect that qPs-hole interaction is the clue for understanding some still unclear effects observed in stage I of crystal irradiation (see Sect.5.2.3).

VIII) Three-quantum annihilation probability P *of the ortho state*; here the model gives

$$P = \frac{\kappa\lambda_t}{\kappa\lambda_t+\delta} + \frac{1}{372}\,\frac{\delta}{\kappa\lambda_t+\delta} \quad . \tag{5.16}$$

Experimental P values reported in [5.10] are generally higher than predicted by (5.16) and this indicates contributions from other positronium-like states.

5.3.2 Annihilation Centers

Annihilation centers are bound states of the positron at lattice defects; their existence was predicted by GOLDANSKII and PROKOPEV [5.110]. They are formed upon positron trapping [5.53], a process possible whenever the binding energy of the system is positive vis-à-vis deposition of the positron in its lowest propagating state in a perfect crystal. Several examples are known of positron trapping at negatively charged and neutral defects associated with vacancies. Trapping at interstitial ions is also possible, but this is not expected to give observable effects with ordinary annihilation techniques.

We list here those annihilation centers for which there is reasonable experimental evidence, and briefly discuss their properties. Labelling of the centers is made according to the rules stated in Sect.5.2.3.

a) ^{c}A *Center*

Formed by a positron trapped at an empty cation vacancy, this is the simplest A center; it may be viewed as the antimorph of F centers. It has often been invoked to explain the complexity of annihilation spectra and the effects of irradiation. Having now identified the origin of the long-living intense component in lifetime spectra for uncolored crystals with qPs in a Bloch state intrinsic to the perfect crystal, and also for the reasons given in Sect.5.2.3, no direct evidence for ^{c}A centers seems to remain. However, thermal treatments produce definite changes in the shortest-living spectral component; this is probably a manifestation of the increased trapping probability in thermally generated cation vacancies. According to this hypothesis DANNEFAER et al. deduced from lifetime measurements a mean life of 340 ps for ^{c}A centers [5.31].

To the best of our knowledge, no detailed calculation of the binding energy E_B of the system has been made, although by analogy with F centers one expects this to be of the order of a few eV; FARAZDEL and CADE [5.111] estimated E_B = 4-6 eV.

Due to the very limited relaxation of the external electron shell of anions in nearest-neighbor positrons around an ^{c}A center, no marked narrowing is to be expected in momentum distributions when positrons are trapped at cation vacancies.

^{c}A centers, being closed-shell systems, will not exhibit magnetic quenching effects (see Sect.5.1.2). However, if an exciton is formed close to an ^{c}A center, the positron and the electron of the exciton will correlate their motion; the whole system then assumes qPs characteristics. Such a complex has been named *correlated* ^{c}A *center*, and has been considered in a tentative interpretation of the long-living component that appears in lifetime spectra for Ca^{2+} -doped KCl (see Section 5.2.3) [5.20].

b) $^{a}A^{+}$ *Center*

This center, formerly indicated as $A_+^{\prime}$ center or, in Russian literature, as F_{e+} or $F_+^{\prime}$, is a positron-electron pair trapped by an anion vacancy. It is by far the best known annihilation center, and is easily observed in additively colored and heavily irradiated crystals. Being a qPs system, it can be formed in ortho and in para states, and is affected by static magnetic fields. Its ortho state contributes a long-living component to lifetime spectra, and the para state a narrow component to momentum distributions. Such components are easily resolved, and the accurate determination of their characteristics is possible (see Table 5.3). On the other hand, the short-lived contribution of the para state has never been singled out. The value of the relaxation parameter κ has been determined from magnetic quenching measurements (see Table 5.5 and comments in Sect.5.2.4).

An excellent fit for the narrow component observed in angular correlation experiments on additively colored KCl is provided by a model that represents the center of mass as a particle contained in a rigid cubic cavity. The length of the edge of the cavity was treated as an adjustable parameter, and was evaluated at L = 5.68 Å [5.96].

Calculations for the wave function of the system, and for its binding energy, have been performed by several authors [5.111-114]. Here we report some of their conclusions: I) explicit consideration of the electron-positron correlation is mandatory not only for reasonable lifetime calculations but also for energy calculations; the inclusion of this correlation lowers the calculated energy values by 1-2 eV [5.111]; II) the stability of the system against the dissociation

$^{c}A^{+}$ center → delocalized positron + F center

is in general easily determined; the situation is more critical when considering the dissociation reaction

$^{c}A^{+}$ center → delocalized qPs + F^{+} center;

it becomes then essential to take into account the lowering of the binding energy of the delocalized $e^{+} - e^{-}$ pair, which lowering is due to interaction with the lattice [5.114]; III) the calculated values of the binding energy are markedly influenced by the choice of the model potential and of the trial wave functions used in variational calculations; therefore the regular trends that the theory predicts are more significant than the absolute values; in fact, FARAZDEL and CADE [5.111] showed that the binding energy of the positron to the F center increases with both anion and cation sizes; IV) from the predicted lifetimes for self-annihilation, the calculation of the relaxation parameter κ immediately follows; the values of κ for two choices of the potential of the vacancy deriving from the results of FARAZDEL and CADE are reported together with experimental values in Table 5.5, also here, however, trends are more significant than absolute values.

c) $^{c}A^{-}$ *and* $^{c}A^{-}(Ca^{2+})$ *Centers*

These centers, which are formed by an electron-positron pair trapped by a cation vacancy, and are of a qPs character, are to be considered as a possible origin of the long-living component appearing for Ca^{2+}-doped KCl. However, we must not forget that an alternative interpretation of the same effect is based on the correlated ^{c}A center mentioned in Sect.5.3.2.

d) ^{a}A *Center*

This neutral center comprises two electrons and a positron at an anion vacancy; it is formed by the capture of a delocalized positron in F^{-}center. It has closed-shell character, and is likely to contribute to the short-living part of lifetime spectra and to the small-momentum region in momentum distributions. Although spectral components related to ^{a}A centers have not been experimentally observed, the evidence of the existence of the above centers is quite convincing (see Sect.5.2.3). BEREZIN [5.113] has estimated that the positron affinity to F^{-} center is almost twice as much as the positron affinity to neutral F centers. In point of fact, after the capture of the positron by F^{-} centers, the system will tend to dissociate into an F center plus a delocalized qPs; however, annihilation probably occurs before dissociation.

e) ${}_2^aA^+$ *and* ${}_3^aA^+$ *Centers*

For these complexes, formed respectively by I) two anion vacancies, plus two electrons, plus a positron, and II) three anion vacancies, plus three electrons, plus a positron, there is only indirect evidence (see Sect.5.2.2). Both centers should be characterized by short lifetimes and a low total momentum of the annihilating pair.

5.3.3 Kinetics of State Formation

It is generally assumed (see, for instance, [5.9]) that the kinetics of positron state formation in condensed matter is governed by the rate equations

$$\frac{dn_i}{dt} = -\left(\lambda_i + \sum_{j \neq i} k_{ij}\right) n_i + \sum_{j \neq i} k_{ji} n_j \quad , \tag{5.17}$$

where $n_i = n_i(t)$ is the number of positrons populating state i at time t after the injection into the sample; λ_i is the characteristic annihilation rate of state i; and k_{ij} is the transition rate from state i to state j. On the hypothesis that the slowing down of positrons is instantaneous, rates λ_i and k_{ij} are to be taken as time-independent. All the above assumptions seem to be confirmed by the experimental observation of lifetime spectra which, within instrumental resolution, are composed of a sum of exponentially decaying terms, just as is predicted by the linear equations (5.17) when the coefficients are constant. The rate equations have also to be supplemented by knowledge of initial conditions, or in other words, by a knowledge of the probability of state formation during the slowing down. When interpreting their results on a basis of the above model, authors often introduce a number of hypotheses as to initial conditions and transition rates. The numerical consequences of these hypotheses are somewhat subtle, and the fit of calculated curves to experimental data is not always a discriminating test.

In order to give a better insight into the problem we shall now present a brief outline of the slowing down and state formation processes. As far as the latter is concerned, we shall consider only quantitatively important processes, namely the formation of qPs in perfect crystals and that of A centers by positron capture in defects.

a) Slowing Down

Positrons are injected into the crystal with a distribution of kinetic energy. ranging up to hundreds of keV (for ^{22}Na the maximum energy of the β^+ spectrum is 0.542 MeV). At energies higher than the energy of the forbidden gap (E_{gap}) of the crystal, the positrons are slowed down both by Bhabha scattering on electrons and by elastic collisions with ions. Below E_{gap}, electronic excitation is impossible, and positrons interact mainly with longitudinal optical phonons via the associated polarization wave. In consideration of the above mechanism, BARTENEV et al. [5.115] have carried out a detailed calculation of the duration of the slowing down process. Their estimate for the slowing down time from initial energy down to E_{gap} is $t_1 = 4.15 \times 10^{-12}$ s in the case of NaI. For the slowing down time t_2 in the lower energy interval (from E_{gap} to k_BT), their result is

$$t_2 = \frac{\sqrt{2}\, Ma^2\hbar\omega_{LO}}{12\pi Z^2 e^4\ k_BT\ m_+^{*\frac{1}{2}}}\left[E_{gap}^{3/2} - (k_BT)^{3/2}\right] \quad , \tag{5.18}$$

where M is the reduced mass of an ion pair, a is the lattice constant, Ze is the ion charge, ω_{LO} is the frequency of the longitudinal optical mode and m_+^* is the positron band mass. The above equation holds for $k_BT > \hbar\omega_{LO}$, and, in this limit, t_2 turns out to be $(3\text{-}5) \times 10^{-12}$ s for NaI. These estimates indicate that the total slowing down time is of the order of $\sim 10^{-11}$ s, which is substantially less than the resolving power of life spectrometers, and also shorter than the shortest mean lives observed in ionic crystals.

b) *Quasi-Positronium Formation in Perfect Crystals*

The energy level of qPs in perfect crystals is several eV above the fundamental level of the positron + crystal system. Therefore a thermalized positron in an unexcited crystal cannot form qPs. To account for the formation of this system, two processes have to be considered, namely the *Ore-gap mechanism* and the *spur model*. Let us recall the basic ideas behind these two processes (which are not mutually exclusive) and adapt them to the case in question.

I) *Ore-gap mechanism* [5.116]. To form qPs a positron must have the residual kinetic energy $E_k \geqq E_{gap} - E_B'$ (see Sect.5.3.1); however, if the kinetic energy of the system so formed $(E_k + E_{B'} - E_{gap})$ is greater than E_{gap}, the system will be rapidly dissociated in a free conduction electron plus a free positron. Formation of stable qPs therefore occurs only when, in the process of being slowed down, positrons fall into the energy interval (*Ore-gap*)

$$E_{gap} \geqq E_k \geqq E_{gap} - E_B' \quad . \tag{5.19}$$

An approximate prediction for the fraction f of positrons forming qPs may be obtained by assuming that the positron energy spectrum (in the stochastic sense) is flat in the interval $E_{gap} \geqq E_k \geqq k_BT \sim 0$, and that all the positrons falling into the Ore-gap do in fact form qPs: in this case $f = E_B'/E_{gap}$. With the numerical estimate of E_B' given for KCl in Sect.5.3.1, f = 47-55% is obtained.

II) *Spur model* [5.117]. The positron spur in an ionic solid is the cluster of electron-hole pairs formed when the positron loses the last part of its kinetic energy in the solid. According to the above model, qPs is formed by a positron and one of the conduction electrons in the spur of the positron itself. The formation reaction competes with electron-hole recombination, with electron and positron out-diffusion and with electron trapping by defects or recombination centers. We are unable to give even an approximate estimate of the effectiveness of the spur process, but we recall that the deliberate introduction into KCl of Ag impurities and of V_k centers, which are both effective electron traps, hinders the formation of qPs (see Sect.5.2.3 and [5.71]). This observation is in qualitative accordance with the spur model.

Finally, we note that in real crystal free qPs may also be formed by electron transfer from a shallow donor level to the bound state with a thermalized positron. For instance, the reaction $F^- + e^+ \rightarrow F$ + free qPs is certainly possible from the energy viewpoint.

c) Formation of A *Centers*

Thermalized positrons may fall into states bound to lattice defects, thereby forming A centers, at a rate proportional to the concentration of defects acting as A center precursors. The important parameter of the trapping process is the capture rate per unit-trap-density ν (henceforth simply called *volume rate*). We have some direct experimental information on the volume rate for capture in F centers, i.e.: I) form the measurement of the intensity of the long-living component in the lifetime spectrum for additively colored KCl [5.54,101] it can be deduced that $\nu \sim 10^{-8}\,\mathrm{cm}^3\mathrm{s}^{-1}$; II) again with colored KCl, ν is found temperature-independent in the range of 7-300 K [5.73].

Theoretical expressions for the volume rate in the case of trapping by vacancies in metals have been deduced by several authors (see Chapt.4). The case which we are dealing with is different mainly for two reasons: I) long-range interactions are possible between positron and traps; and II) the binding energy cannot be transferred to the lattice via electron-hole excitation because of the existence of the forbidden energy gap. The theoretical treatment of the problem given by BRANDT [5.93] is such as to permit one to take into consideration the above points. In the following we adhere fairly closely to Brandt's outline.

The volume rate is determined by the rate ν_1 of positron arrival at defects and by the rate ν_2 of transition to the bound state, which together combine to give

$$\nu = (\nu_1^{-1} + \nu_2^{-1})^{-1} \quad . \qquad (5.20)$$

In the classical approach, ν_1 can be obtained by considering the positron as a diffusing particle in thermal equilibrium with the lattice, which particle is driven to the trap by concentration gradients (in a stochastic sense) and by electric forces. In the case of F centers, the integration of the diffusion equation leads to the final result (see [5.73] and literature therein quoted)

$$\nu_1 = 4\pi D_+ \times 1.1 \left(\frac{\gamma e^2}{2\varepsilon^2 k_B T}\right)^{1/4} , \tag{5.21}$$

where D_+ is the positron diffusion coefficient, γ the F center polarizability, and ε the static dielectric constant. With due account for the high polaronic mass of the positron, NIEMINEN's estimate [5.45] for D_+ at room temperature is 0.35 $cm^2 s^{-1}$; at lower temperatures, D_+ increases exponentially. Insertion of the appropriate numbers in (5.21) gives, at room temperature, $\nu_1 \sim 2 \times 10^{-7}$ $cm^3 s^{-1}$. Since this value is substantially greater than the experimental value for ν, the term ν_1^{-1} can be neglected in (5.20) and therefore $\nu \sim \nu_2$; physically this means that the trapping process is not diffusion limited. Assuming that phonon emission is the only important process for energy transfer, the physical factors determining ν_2 are: I) the wavelengths $\lambda\!\!\!^{-}$ and $\lambda\!\!\!^{-}_B$ of the positron before and after trapping; II) the binding energy E_B of the trapped state; III) the coupling to the phonon field, as represented by the damping width Γ related to longitudinal optical phonon frequency $(2\pi\Gamma \approx \hbar\omega_{LO})$. BRANDT has derived an approximate expression for ν_2, which can be put in the form

$$\nu_2 = 4\pi r_B \cdot \frac{\hbar}{\pi m_+^*} \cdot \rho , \tag{5.22}$$

where $r_B = \pi\lambda\!\!\!^{-}_B$, m_+^* is the positron band mass, and ρ is the factor representing the probability of the energy transfer to the crystal. The expression for ρ is

$$\rho = \left(1 + \frac{2}{\pi} \cdot \frac{\lambda\!\!\!^{-}_B}{\lambda} \cdot \frac{E_B}{\Gamma}\right)^{-1} . \tag{5.23}$$

Considering the distortion of the positron wave function due to the long-range potential of the trap even prior to capture, it seems inappropriate to insert the thermal wavelength for $\lambda\!\!\!^{-}$ in (5.23), but rather a temperature-independent value of $\lambda\!\!\!^{-} \sim 2\text{-}5\lambda\!\!\!^{-}_B$. By setting $E_B/\Gamma \sim 10^2$ we obtain $\rho = 3\text{-}8 \times 10^{-2}$, and, for ν_2, the temperature-independent value $0.4\text{-}1 \times 10^{-8}$ cm^2 s^{-1}. Considering the approximations on which (5.22,23) are based, and the wide margin of uncertainty for the numerical values introduced, the nearly perfect agreement of this result with the experimental value is probably accidental. Nevertheless, the mechanism of trapping and the role of the controlling factors seem to be correctly accounted for.

5.4 Annihilation in Other Ionic Compounds

In this section we report on present knowledge in the field of positron interaction with ionic solids not belonging to the alkali halide family. A certain degree of ionic bond is to be found in any nonelemental substance, but practical considerations demand that we limit our discussion to the case of compounds that are not entirely different in their general properties from alkali halides, namely, to insulating binary compounds of predominant ionic character. With this reservation, we hope that the basic concepts referring to those positron states which we discussed in the previous section may retain their validity. A synthetic but accurate account of work performed before 1971 in the field of our present concern, also an extended reference list, are given in the review article by BARTENEV et al. [5.7]. Therefore we shall not mention specific contributions already quoted in the above paper, unless strictly relevant to the context of this presentation.

5.4.1 Hydrides of Alkali and Alkaline-Earth Metals

All compounds of the above families which have been studied (LiH, NaH, KH, MgH_2, CaH_2, SrH_2, BaH_2) display complex lifetime spectra with an intense long-living component [5.118]. Instead, no structures have been revealed in angular correlation in the few cases where measurements were performed (LiH, NaH, CaH_2) [5.119,120].

Turning to an interpretation of complex spectra, we note the following. Analogous with alkali halides, and also considering the large dimensions of system H^-e^+ in vacuo, it seems reasonable that thermalized positrons behave in hydride crystals as nearly free particles, and without forming bound states around the negative ion. At least in the case of LiH, the above contention is substantiated by the calculations of BRANDT et al. [5.121]. In addition, these authors were able to demonstrate that experimental momentum distributions are correctly reproduced by calculations when proper account is taken of the phase relation of the wave functions of the annihilating pair in two adjacent Li and H cells. The theory of NEAMTAN and VERRALL [5.108] demonstrates that delocalized qPs can be formed in LiH, and indeed this may be the origin of the intense long-living component observed in lifetime spectra. Yet no experimental evidence of the existence of qPs in hydrides has so far been produced, nor have any investigations on the effect of lattice imperfections been performed.

5.4.2 Copper, Silver, Gold, Thallium Halides

The existence of various positron states in the above compounds is demonstrated by the complexity of the lifetime spectra [5.122]. In AgCl the intense long-living component is strongly quenched by Cd^{2+}-doping [5.102] and since the excess charge introduced by Ca^{2+}-doping is compensated by Ag^+ vacancies, this quenching effect is likely to be the result of positron trapping at cation vacancies in competition with the formation of a longer-living state. This interpretation implies that ^{c}A centers are short-living; such, in fact, is the prediction that KELLY [5.102] obtained from a theoretical evaluation of probability

$$p = \int_{|\underline{r}|>r_v} |\psi_+(\underline{r})|^2 \, d\underline{r}$$

that a positron bound in the ^{c}A center is to be found outside the radius r_v of the vacancy. Kelly's result ($p = 0.98$) indicates that the annihilation characteristics of ^{c}A centers are essentially the same as for bulk material. The effectiveness of trapping processes has been revealed also in AgBr, for which the average lifetime $\bar{\tau} = \sum_i I_i\tau_i$ appears to be correlated to the occurrence of molecular ions $AgBr^+$ and Br_2^+ [5.123] and to the mechanical deformation of the sample [5.124].

On the subject of intrinsic positron states no direct experimental information is available. We note, however, that trapping at defects indicates that positrons in their fundamental state for perfect regions of the crystal are mobile enough to reach traps.

5.4.3 Alkaline-Earth Halides

For these materials complex lifetime spectra have been observed [5.125]. Moreover, the formation of delocalized qPs is clearly proven in the case of CaF_2 single crystals by the occurrence of narrow peaks in the angular distribution [5.51]. On the contrary, the angular distributions for $MgCl_2$ and $BaCl_2$ do not display any evident structure [5.9]. RAMASAMY et al. [5.126] have applied the so-called *optical positron model* to deduce from angular correlation data the effective ionic charge ηe for the two latter compounds and for certain other halides of the RX_2 type. Here we shall comment briefly on the optical positron model which was originally developed by GOLDANSKII and PROKOPEV [5.127] for alkali halides, and has often been applied to other ionic compounds. This model assumes that the positron is bound to negative ions by the potential energy $-\eta e^2/r$, and that its wave function has the following form

$$\psi_+ \propto r \exp(-3\eta r/2a_0) \quad . \tag{5.24}$$

On taking a Slater wave function to describe electrons in the external anion shell, the calculated FWHM of the momentum distribution of the annihilating pair is given by

$$\Gamma_{1/2} = k(\tfrac{3}{2}\eta + \beta) \quad , \tag{5.25}$$

where k is a numerical constant and β the Slater exponent. Equation (5.25) is then used to calculate η from the experimentally determined $\Gamma_{1/2}$. In our opinion (but see also [5.39,121]) this model, which neither considers correct boundary conditions nor Madelung repulsion acting on the positron in the negative crystal cell, cannot give a physically sound representation of positron behavior in an ionic crystal. Nevertheless, the form of (5.24) may represent a fair fit for the true positron wave function in the region where positron-electron overlap is intense. In spite of this, doubt remains as to the true significance of the η values calculated from (5.25)

5.4.4 Alkaline-Earth Oxides

An abundant formation of positronium occurs on the surface of powder grains of alkaline-earth oxides. Discussion of this phenomenon is beyond the limits of the present report, which concerns only positron bulk states. In single crystals and well-sintered samples the long-living component due to the surface formation of positronium disappears, or is reduced to below 1%. The remainder of the spectrum has been reduced to two components by BERTOLACCINI et al. [5.101]. The longest-living component has a lifetime of $5\text{-}7 \times 10^2$ ps, and a moderately feeble intensity (4% in MgO, 6.7% in CaO). The momentum distributions for single crystals do not manifest any narrow component [5.128]. BARTENEV et al. [5.106] have made a theoretical analysis of the stability of qPs in several alkaline-earth oxides. Their conclusion is that interaction with the lattice strongly reduces the binding of the $e^+ - e^-$ pair, so that the Ore-gap (see Sect.5.3.2) is small or even absent in BaO and MgO, while for CaO qPs formation is possible. On the other hand, the formation of qPs is revealed in Ag^{3+}-doped MgO by the appearance in the angular distribution of a narrow component (Γ_N = 4.5 mrad) and by magnetic field effects; the reasonable interpretation of these results points twoard qPs formation at cation vacancies [5.129].

Acknowledgements. The author takes this occasion to acknowledge the valuable guidance and encouragement given to him by Profs. A. Bisi and L. Zappa during many years of work in the field of positron annihilation.

References

5.1 A. Bisi, A. Fiorentini, L. Zappa: Phys. Rev. *131*, 1023 (1963)
5.2 W. Brandt, H.F. Waung, P.W. Levy: Proc. Int. Symposium on Color Centers in Alkali Halides (Roma 1968) p.48
5.3 R.A. Ferrell: Rev. Mod. Phys. *28*, 308 (1956)

5.4 P.R. Wallace: *Solid State Physics*, Vol.10, ed. by F. Seitz, D. Turnbull (Academic Press, New York, London 1960) P.1
5.5 J. Green, J.Lee: *Positronium Chemistry* (Academic Press, New York, London 1964)
5.6 T. Hyodo, Y. Takakusa: J. Phys. Soc. Jpn. *42*, 1065 (1977)
5.7 G.M. Bartenev, A.D. Tsyganov, E.P. Prokopev, L.A. Varisov: Usp. Fiz. Nauk *103*, 339 (1971)
5.8 S.A. Vorobev: *Prokhoždenie Beta-Častiz Čeres Kristally* (Atomisdat, Moskva 1975)
5.9 R.N. West: *Positron Studies of Condensed Matter* (Taylor and Francis, London 1974)
5.10 A. Bisi, C. Bussolati, S. Cova, L. Zappa: Phys. Rev. *141*, 348 (1966)
5.11 A. Gainotti, E. Germagnoli, G. Schianchi, L. Zecchina: Phys. Lett. *13*, 9 (1964); Nuovo Cimento *32*, 1 (1964)
5.12 C. Bussolati, L. Zappa: Phys. Rev. *136*, A657 (1964)
5.13 I.K. MacKenzie, B.T.A.McKee: Bull. Am. Phys. Soc. *12*, 687 (1967)
5.14 O. Brümmer, G. Brauer, V. Andreitscheff, L. Kaubler: Phys. Status Solidi (b) *71*, 59 (1975)
5.15 I.K. MacKenzie, B.T.A. McKee: Appl. Phys. *10*, 245 (1976)
5.16 I.K. MacKenzie, P. Sen: Phys. Rev. Lett. *37*, 1296 (1976)
5.17 P. Sen, I.K. MacKenzie: Nucl. Instrum. Methods *141*, 293 (1977)
5.18 T.A. Pond, R.H. Dicke: Phys. Rev. *85*, 489 (1952)
5.19 O. Halpern: Phys. Rev. *94*, 904 (1954)
5.20 W. Brandt, A. Dupasquier: Bull. Am. Soc. *17*, 284 (1972)
5.21 H.A. Bethe, E.E. Salpeter: Quantum Mechanics of One- and Two-Electron Atoms (Springer, Berlin, Göttingen, Heidelberg 1957) p.88
5.22 A. Bisi, A. Dupasquier, L. Zappa: J. Phys. C *4*, L33 (1971)
5.23 A. Bisi, A. Dupasquier, L. Zappa: J. Phys. C *4*, L311 (1971)
5.24 A. Bisi, A. Dupasquier, L. Zappa: J. Phys. C *6*, 1125 (1973)
5.25 D. Herlach, F. Heinrich: Helv. Phys. Acta *45*, 10 (1972)
5.26 D. Herlach: Helv. Phys. Acta *45*, 894 (1972)
5.27 S. Dannefaer, L.S. Smedskjaer: J. Phys. C *6*, 3536 (1973)
5.28 L. Smedskjaer, S. Dannefaer: J. Phys. C *7*, 2603 (1974)
5.29 L.A. Page, M. Heinberg: Phys. Rev. *106*, 1220 (1957)
5.30 P. Kirkegaard, M. Eldrup: Computer Phys. Commun. *3*, 240 (1972); *7*, 401 (1974)
5.31 S. Dannefaer, G.W. Dean, B.G. Hogg: Phys. Rev. B *13*, 3715 (1976)
5.32 D. Herlach, F. Heinrich: Phys. Lett. A *31*, 47 (1970)
5.33 T.L. Williams, H.J. Ache: J. Chem. Phys. *51*, 3536 (1969)
5.34 H.W. Etzel: Unpublished results quoted by I.H. Schulman, W.D. Compton: *Color Centers in Solids* (Pergamon Press, Oxford 1962) Chap.VII
5.35 K. Fujiwara, T. Hyodo, Y. Takakusa, O. Sueoka: J. Phys. Soc. Jpn. *39*, 403 (1975)
5.36 C. Bussolati, A. Dupasquier, L. Zappa: Nuovo Cimento B *52*, 529 (1967)
5.37 M. Bertolaccini, A. Dupasquier: Phys. Rev. B *1*, 2896 (1970)
5.38 C.S. Tumosa, J.B. Nicholas, H.J. Ache: J. Phys. Chem. *75*, 2030 (1971)
5.39 P. Hautojärvi, R. Nieminen: Phys. Status Solidi (b) *56*, 421 (1973)
5.40 F.H. Hsu, W.C. Mallard, J.K. Fu: Appl. Phys. *4*, 75 (1974)
5.41 R. Nieminen, P. Hautojärvi, P. Jauho: Appl. Phys. *5*, 41 (1974)
5.42 S. Dannefaer, G.W. Dean, D.P. Kerr, B.G. Hogg: To be published
5.43 D. Herlach, F. Heinrich: Helv. Phys. Acta *43*, 489 (1970)
5.44 A.T. Stewart, N.K. Pope: Phys. Rev. *120*, 2033 (1960)
5.45 R. Nieminen: J. Phys. C *8*, 2077 (1975)
5.46 B. Rozenfeld, W. Swiatkowski, J. Wesolowski: Acta Phys. Pol. *29*, 429 (1966)
5.47 D. Herlach, F. Heinrich: Helv. Phys. Acta *42*, 601 (1969)
5.48 K.P. Arefev, S.A. Vorobev: Krist. Tech. *7*, 841 (1972)
5.49 P. Ramasamy, T. Nagarajan: Physica B *81*, 305 (1976)
5.50 A.A. Vorobev, Yu.M. Annenkov, K.P. Arefev, S.A. Vorobev: Phys. Status Solidi (b) *57*, K27 (1973)
5.51 W. Brandt, G. Coussot, R. Paulin: Phys. Rev. Lett. *23*, 522 (1969)
5.52 O.E. Mogensen, G. Kvajič, M. Eldrup, M. Milosevič-Kvajič: Phys. Rev. B *4*, 71 (1971)

5.53 W. Brandt: In *Positron Annihilation*, ed. by A.T. Stewart, L.O. Roellig (Academic Press, New York, London 1967) p.179
5.54 A. Dupasquier: Lett. Nuovo Cimento *4*, 13 (1970)
5.55 K.P. Singh, R.M. Singru, M.S. Tomar, C.N.R. Rao: Phys. Lett. A *32*, 10 (1970)
5.56 W. Brandt, H.F. Waung, P.W. Levy: Phys. Rev. Lett. *26*, 496 (1971)
5.57 W. Brandt, H.F. Waung: Phys. Rev. B *3*, 3432 (1971)
5.58 C.S. Tumosa, J.B. Nicholas, H.J. Ache: J. Phys. Chem. *75*, 2030 (1971)
5.59 W. Brandt, G. Coussot, R. Paulin: Phys. Lett. A *35*, 175 (1971)
5.60 G.M. Bartenev, A.D. Tsyganov, A.Z. Varisov: Izv. Vuzov, Ser. Fiz. *5*, 60 (1971)
5.61 G.M. Bartenev, A.Z. Varisov, V.I. Goldanskii, E.P. Prokopev, A.D. Tsyganov: Unpublished results quoted in Ref. [1.7]
5.62 W. Brandt, A. Dupasquier, G. Dürr: Phys. Rev. B *6*, 3156 (1972)
5.63 P. Hautojärvi, P. Jauho: *Physics of Impurity Centers in Crystals* (Estonian Acad. Sci., Tallin 1972) p.645
5.64 K.P. Arefev, S.A. Vorobev: Phys. Letters A *39*, 381 (1972)
5.65 A.A. Vorobev, K.P. Arefev, S.A. Vorobev: Phys. Status Solidi (b) *53*, K133 (1972)
5.66 W.C. Mallard, F.H. Hsu: Lett. A *38*, 164 (1972)
5.67 O. Brümmer, W. Schülke, G. Brauer, G. Dlubek: Jahresbericht ZfK *243*, 104 (1972)
5.68 A. Balogh, U. Dézsi, D. Horváth, Z. Kajcsos: Phys. Lett. A *45*, 299 (1973)
5.69 T. Nagarajan, P. Ramasamy: Nuclear Physics and Solid State Symposium, Bangalore, Vol.16C, 102 (1973)
5.70 L. Bosi, A. Dupasquier, L. Zappa: J. Phys. (Paris) C *9*, 295 (1973)
5.71 F.H. Hsu, W.C. Mallard, J.H. Hadley Jr.: Appl. Phys. *4*, 83 (1974)
5.72 S. Dannefaer, G. Trumpy, R.M.J. Cotterill: J. Phys. C *7*, 1261 (1974)
5.73 A.A. Vorobev, K.P. Arefev, S.A. Vorobev: Appl. Phys. *3*, 241 (1974)
5.74 P.K. Tseng, P.K.L. Chang, G.J. Jan: Phys. Lett. A *48*, 211 (1974)
5.75 A. Balogh, I. Dészi, D. Horvath, Z. Kajcsos: Appl. Phys. *6*, 21 (1975)
5.76 P. Ramasamy, T. Nagarajan, V. Devanathan: J. Phys. Chem. Sol. *36*, 859 (1975)
5.77 O. Brümmer, G. Brauer, V. Andrejtscheff: Phys. Status Solidi (b) *70*, 683 (1975)
5.78 A. Bisi, L. Bosi, A. Dupasquier, L. Zappa: Phys. Status Solidi (b) *69*, 515 (1975)
5.79 L. Bosi, A. Dupasquier, L. Zappa: Phys. Rev. B *11*, 2485 (1975)
5.80 S. Dannefaer, D.P. Kerr, B.G. Hogg: J. Phys. C *8*, 2667 (1975)
5.81 A.A. Vorobev, K.P. Arefev, S.A. Vorobev: Zh. Eks. Teor. Fiz. *68*, 1486 (1975) [Sov. Phys. JETP *41*, 743 (1975)]
5.82 A.A. Vorobev, K.P. Arefev, S.A. Vorobev: Appl. Phys. *10*, 227 (1976)
5.83 K.P. Arefev, S.A. Vorobev, B.G. Starodubov: Kristallografiya *21*, 422 (1976)
5.84 K.P. Arefev, S.A. Vorobev, B.G. Starodubov: *Khimiceskaya svyaz'v kristallakh i ikh fiziceskie svoistva 2* (Nauka i Tekhnika, Minsk 1976) p.29
5.85 A.A. Vorobev, K.P. Arefev, S.A. Vorobev, G.I. Etin: *Khimiceskaya svyaz'v kristallakh i ikh fiziceskie svoistva 2* (Nauka i Tekhnika, Minsk 1976) p.44
5.86 K.P. Arefev, S.A. Vorobev, V.P. Kileev: Fis. Tverd. Tela *18*, 669 (1976) [Sov. Phys. Solid State *18*, 386 (1976)]
5.87 T. Nagarajan, P. Ramasamy, S. Ramasamy: Phys. Status Solidi *82* , 75 (1976)
5.88 T. Nagarajan, P. Ramasamy: J. Phys. Chem. Sol. *37*, 549 (1976)
5.89 J.B. Nicholas, C.S. Tumosa, H.J. Ache: J. Chem. Phys. *58*, 2902 (1973)
5.90 W. Brandt, R. Paulin: Phys. Rev. B *8*, 4125 (1973)
5.91 Yu. M. Annenkov, K.P. Arefev, T.S. Frangulyan, V.G. Starodubov, S.A. Vorobev: Appl. Phys. *7*, 83 (1975)
5.92 E. Sonder, W.A. Sibley: *Point Defects in Solids*, Vol.1, ed. by J.H. Crawford Jr., L.M. Slifkin (Plenum Press, New York, London 1972) Chap.4
5.93 W. Brandt: Appl. Phys. *5*, 1 (1974)
5.94 R.G. Fuller: *Point Defects in Ionic Solids*, Vol.1, ed. by J.H. Crawford Jr., L.M. Slifkin (Plenum Press, New York, London 1972) Chap.2
5.95 D. Herlach, F. Heinrich: Helv. Phys. Acta *44*, 561 (1971)
5.96 D. Herlach, F. Heinrich: Helv. Phys. Acta *43*, 491 (1970)
5.97 P.E. Cade, A. Farazdel: J. Chem. Phys. *66*, 2598, 2612 (1977)

5.98 V.I. Goldanskii, E.P. Prokopev: Fiz. Tverd. Tela *6*, 3301 (1964) [Sov. Phys.-Solid State *6*, 2641 (1965)]
5.99 E.P. Prokopev: Fiz. Tverd. Tela *8*, 464 (1966) [Sov. Phys.-Solid State *8*, 368 (1966)]
5.100 V.I. Goldanskii, E.P. Prokopev: Fiz. Tverd. Tela *13*, 2955 (1971) [Sov. Phys.-Solid State *13*, 2481 (1972)]
5.101 M. Bertolaccini, A. Bisi, G. Gambarini, L. Zappa: J. Phys. C *4*, 734 (1971)
5.102 T.M. Kelly: Photogr. Sci. Eng. *17*, 201 (1973)
5.103 V.I. Goldanskii, E.P. Prokopev: Zh. Eksp. Teor. Fiz. Pis. Red. *4*, 422 (1966) [Sov. Phys.-JETP Lett. *4*, 284 (1966)]
5.104 E.P. Prokopev: Fiz. Tverd. Tela *9*, 1266 (1967) [Sov. Phys.-Solid State *9*, 993 (1967)]
5.105 E.P. Prokopev: Fiz. Tverd. Tela *19*, 472 (1977) [Sov. Phys.-Solid State *19*, 271 (1977)]
5.106 G.M. Bartenev, M.N. Pletnev, E.P. Prokopev, A.D. Tsyganov: Fiz. Tverd. Tela *12*, 2733 (1970) [Sov. Phys.-Solid State *12*, 2201 (1971)]
5.107 G.M. Bartenev, A.Z. Varisov, A.V. Ivanova, M.N. Pletnev, E.P. Prokopev, A.D. Tsyganov: Fiz. Tverd. Tela *14*, 715 (1972) [Sov. Phys.-Solid State *14*, 608 (1972)]
5.108 S.M. Neamtan, R.I. Verrall: Phys. Rev. A *134*, 1254 (1964)
5.109 I. Aronson, C.J. Kleinmann, L. Spruch: Phys. Rev. A *4*, 841 (1971)
5.110 V.I. Goldanskii, E.P. Prokopev: Fiz. Tverd. Tela *6*, 3301 (1964)
5.111 A. Farazdel, P.E. Cade: Phys. Rev. B *9*, 2036 (1974)
5.112 A.A. Berezin, R.A. Evarestov: Phys. Status Solidi (b) *48*, 133 (1971)
5.113 A.A. Berezin: Phys. Status Solidi (b) *50*, 71 (1972)
5.114 P. Hautojärvi, R. Nieminen, P. Jauho: Phys. Status Solidi (b) *57*, 115 (1973)
5.115 G.M. Bartenev, A.D. Tsyganov, E.P. Prokopev, A.Z. Varisov: Izv. Vuzov (Fizika) *4*, 68 (1971)
5.116 Aa. Ore: Univers. Bergen Årbook, Naturvidenskap. Rekke *9* (1949)
5.117 O.E. Mogensen: J. Chem. Phys. *60*, 998 (1974)
5.118 A. Gainotti, C. Ghezzi, M. Manfredi, L. Zecchina: Nuovo Cimento B *56*, 47 (1968)
5.119 A.T. Stewart, R.H. March: Phys. Rev. *122*, 75 (1961)
5.120 L.D. Burton, W.F. Huang, W.G. Bos: J. Chem. Phys. *59*, 5205 (1973)
5.121 W. Brandt, L. Eder, S. Lundqvist: Phys. Rev. *142*, 165 (1966)
5.122 S. Cova, A. Dupasquier, M. Manfredi: Nuovo Cimento *47*, 263 (1967)
5.123 J.T. Muheim, H. Surbeck: Helv. Phys. Acta *48*, 446 (1975)
5.124 H. Surbeck: Helv. Phys. Acta, to be published
5.125 G. Gambarini, L. Zappa: Phys. Lett. A *27*, 498 (1968)
5.126 P. Ramasamy, T. Nagarajan, S. Ramasamy: Phys. Lett. A *59*, 397 (1976)
5.127 V.I. Goldanskii, E.P. Prokopev: Fiz. Tverd. Tela *8*, 515 (1966) [Sov. Phys.-Solid State *8*, 409 (1966)]
5.128 A.D. Mokrushin, A.D. Tsyganov: Fiz. Tverd. Tela *11*, 3177 (1969) [Sov. Phys.-Solid State *11*, 2575 (1970)]
5.129 A.A. Vorobev, K.P. Arefev, V.P. Arefev: Fiz. Tverd. Tela *18*, 896 (1975) [Sov. Phys.-Solid State *18*, 518 (1976)]

Additional References with Titles

Chapter 1

C.F. Coleman: Positron annihilation — a potential new ndt technique. NDT International *10*, 227 (1977)

R.M. Singru, K.B. Lal, S.J. Tao, R.M. Lambrecht: Positron-annihilation data tables, table III. Doppler Broadening of annihilation-radiation lineshape. At. Data Nucl. Data Tables *20*, 475 (1977)

Chapter 2

A.J. Arko, Z. Fisk, F.M. Mueller: de Haas-van Alphen effect and Fermi surface of Nb_3Sb. Phys. Rev. B*16*, 1387 (1977)

A.J. Arko, D.H. Lowndes, F.A. Muller, L.W. Roeland, J. Wolfrat, A.T. van Kessel, H.W. Myron, F.M. Mueller, G.W. Webb: de Haas-van Alphen effect in the high-T_c A15 superconductors Nb_3Sn and V_3Si. Phys. Rev. Lett. *40*, 1590 (1978)

S. Berko, M. Haghgooie, J.J. Mader: Electronic momentum densities by two-dimensional angular correlation of annihilation radiation, in *Transition Metals 1977*. Inst. Phys. Conf. Ser. *39*, 94 (1978)

J. Callaway, C.S. Wang: Energy bands in ferromagnetic iron. Phys. Rev. B*16*, 2095 (1977)

N.C. Debnath, S. Chatterjee: ℓ-dependent pseudopotential in the theory of band structure of disordered solids — an application to binary substitutional alloys MgLi, MgCd and MgIn. J. Phys. C*11*, 3403 (1978)

H.P. Fischer, K.H. Steinmetz, E. Nembach: A new high-resolution, high-intensity apparatus for the study of Fermi surfaces of alloys by positron annihilation. Phys. Status Solidi (a) *44*, 669 (1977)

C.D. Gelatt, Jr., H. Ehrenreich, J.A. Weiss: Transition-metal hydrides: Electronic structure and the heats of formation. Phys. Rev. B*17*, 1940 (1978)

V.I. Goldanskii, K. Petersen, V.P. Shantarovich, A.V. Shishkin: Another method of deconvoluting positron annihilation spectra obtained by the solid-state detector. Appl. Phys. *16*, 413 (1978)

B.L. Gyorffy, G.M. Stocks: Electronic states in random substitutional alloys: The CPA and beyond, in *Electrons in Finite and Infinite Structures*, ed. by P. Phariseau, L. Scheire (Plenum Press, New York 1977) pp.144-235

T. Hyodo, O. Sueoka: Study of Umklapp annihilation of positrons in copper by means of rotating-specimen method. J. Phys. Soc. Jpn. *43*, 1137 (1977)

A.P. Jeavons: The high-density proportional chamber and its applications. Nucl. Instrum. Methods *156*, 67 (1978)

D.G. Kanhere, R.M. Singru: Systematics of the electron momentum distributions in some 3d transition metals. J. Phys. F *7*, 2603 (1977)

B.M. Klein, L.L. Boyer, D.A. Papaconstantopoulos: Self-consistent APW bandstructure of V_3Ga. J. Phys. F *8*, 617 (1978)

A.A. Manuel, Ø. Fischer, M. Peter, A.P. Jeavons: An application of proportional chambers to the measurement of the electronic properties of solids by positron annihilation. Nucl. Instrum. Methods *156*, 67 (1978)
A.A. Manuel, S. Samoilov, Ø. Fischer, M. Peter, A.P. Jeavons: Fermi surface topology by the two dimensional angular correlation of annihilation radiation: Copper, a test case. J. Phys. (Paris) *39*, C6-1084 (1978)
P.G. Mattocks, R.C. Young: de Haas-van Alphen effect and Fermi surface of yttrium. J. Phys. F *8*, 1417 (1978)
P.E. Mijnarends, R.M. Singru: Point-geometry angular correlation curves for Cu: A study of enhancement in positron annihilation (To be published)
D.A. Papaconstatopoulos, B.M. Klein, J.S. Faulkner, L.L. Boyer: Coherent-potential-approximation calculations for PdH_x. Phys. Rev. B*18*, 2784 (1978)
P. Pattison: Derivation of the radial distribution function from experimental Compton profiles. Solid State Commun. *24*, 721 (1977)
S. Samoilov, M. Weger: Experimental proof for a quasi-one-dimensional band structure of V_3Si. Solid State Commun. *24*, 821 (1977)
D.J. Sellmyer: Electronic structure of metallic compounds and alloys: Experimental aspects, in *Solid State Physics*, Vol.33, ed. by H. Ehrenreich, F. Seitz, D. Turnbull (Academic Press, New York 1978) pp.83-248
G.M. Stocks, W.M. Temmerman, B.L. Gyorffy: Complete solution of the Korringa-Kohn-Rostoker coherent-potential-approximation equations: Cu-Ni alloys. Phys. Rev. Lett. *41*, 339 (1978)
M.J. Stott, R.N. West: The positron density distribution in metals: Temperature effects. J. Phys. F *8*, 635 (1978)
Y. Tsuchiya, S. Tamaki: Positron annihilation and charge transfer in simple alloys. J. Phys. F *8*, L29 (1978)
A.T. van Kessel, H.W. Myron, F.M. Mueller: Electronic structure of Nb_3Sn. Phys. Rev. Lett. *41*, 181 (1978)
R.C. Young: Fermi surface studies of pure crystalline materials. Rep. Prog. Phys. *40*, 1123 (1977)

Chapter 3

K.P. Aref'ev, S.A. Vorob'ev, E.P. Prokop'ev, A.A. Tsoi: Positron annihilation in electron irradiated silicon. Sov. Phys-Solid State *19*, 8 (1977)
L.J. Cheng, M.L. Swanson: Positron lifetimes in quenched aluminium and 0.09 at % Mn crystals. Appl. Phys. *6*, 273 (1975)
W. Fahs, U. Holzhauer, S. Mantl, F.W. Richter, R. Sturn: Annihilation of positrons in electron irradiated silicon crystals. Phys. Status Solidi (b) *89*, 69 (1978)
H. Fukushima, M. Doyama: Positrons as the probe of phase transformation in Cu-Mn alloy. J. Phys. F *8*, 205 (1978)
J.A. Jackman, C.W. Schulte, J.L. Campbell: Comparison of positron annihilation behaviour with lattice expansion in Ni, Fe and V at low temperatures. J. Phys. F *8*, L13 (1978)
J.A. Jackman, C.W. Schulte, J.L. Campbell: Temperature-dependent positron trapping in nickel after 5 MeV electron irradiation. J. Phys. F *8*, 1845 (1978)
B. Lengeler, S. Mantl, W. Trifthäuser: Interaction of hydrogen and vacancies in copper investigated by positron annihilation. J. Phys. F *8*, 1691 (1978)
B.T.A. McKee, T. McMullen: Some effects of temperature on positron annihilation characteristics in metals. J. Phys. F *8*, 1175 (1978)
P. Rice-Evans, I. Chaglar, K. El Khangi: Positron annihilation in Indium, zinc, cadmium and gold in the temperature range down to 4 K. Phys. Rev. Lett. *40*, 716 (1978)
P. Rice-Evans, I. Chaglar, K. El Khangi: On the trapping rate of positrons in deformed lead over the temperature range 4-100 K. Phys. Lett. *64* A (1978)
Y. Tsuchiya, S. Tamaki: Positron annihilation and charge transfer in simple alloys. J. Phys. F *8*, L29 (1978)
Y. Tsuchiya, S. Tamaki, F. Iton: Positron annihilation in liquid mercury-indium and mercury-thallium. J. Phys. Soc. Jpn. *44*, 866 (1978)

Chapter 4

A.K. Gupta, P. Jena, K.S. Singwi: Nonlinear electron-density distribution around point defects in simple metals. Phys. Rev. B *18*, 2712 (1978)

G.M. Hood, B.T.A. McKee: Some systematics of positron-vacancy interactions in metals. J. Phys. F *8*, 1457 (1978)

C. Koenig: On the positron localization in ordered and disordered metallic alloys. Phys. Status Solidi (b) *88*, 569 (1978)

E. Kuramoto, E. Kitajima, M. Hasegawa: Positron annihilation in niobium containing microvoids. Radiat. Eff. *37*, 241 (1978)

C.S. Lam, Y.P. Varshni: Binding energy of positrons to F and F' color centers in alkali halides. Phys. Status Solidi (b) *89*, 103 (1978)

M. Manninen, R.M. Nieminen: Electronic structure of vacancies and vacancy clusters in simple metals. J. Phys. F *8*, 2243 (1978)

A.P. Mills, Jr., P.M. Platzman, B.L. Brown: Slow-positron emission from metal surfaces. Phys. Rev. Lett. *41*, 1076 (1978)

R.M. Nieminen, J. Laakkonen, P. Hautojärvi, A. Vehanen: Temperature dependence of positron trapping at voids in metals. Phys. Rev. B *19* (February 1979)

S.W. Tam, R.W. Siegel: On the effect of vacancy migration upon the annihilation of a trapped positron in metals. J. Phys. F *7*, 877 (1977)

Chapter 5

K.P. Aref'ev, V.P. Aref'ev, S.A. Vorob'ev: Irradiation damage of alkali halides crystals during positron bombardment. Radiat. Eff. *36*, 141 (1978)

K.P. Aref'ev, V.M. Lisitsyn, V.G. Starodubov, M.I. Kalinin: Positron spectroscopy studies of charge transformations of rare-earth ions in CaF_2. Fiz. Tverd. Tela *19*, 3593 (1977)

K.P. Aref'ev, E.P. Prokop'ev: Kinetics of annihilation decay of positron states in ionic crystals. Izvest. Vyssh. Uch. Zav., Fiz. *9*, 50 (1977)

S.Y. Chuang, P.K. Tseng, G.J. Jan, C.C. Dai: Temperature dependence of positron annihilation in KBr single crystals. Phys. Lett. A*65*, 438 (1978)

T. Hyodo, Y. Takakusa: Direct observation of delocalized positronium in KCl. J. Phys. Soc. Jpn. *45*, 795 (1978)

D.P. Kerr, S. Dannefaer, G.W. Dean, G.B. Hogg: Positron annihilation in the alkali halides. Can. J. Phys. *56*, 1453 (1978)

H.A. Kurtz, K.D. Jordan: Ab initio study of the positron affinity of LiH. J. Phys. B *11*, L479 (1978)

C.S. Lam, Y.P. Varshni: Binding energies of positron to F and F' colour centres in alkali halides. Phys. Status Solidi *89*, 103 (1978)

T. Nagarajan, S. Ramasamy, Y.V.G.S. Murti, N. Sucheta: Study of C centres in NaCl by positron angular correlation. Phys. Lett. A*64*, 141 (1977)

E.P. Prokop'ev: Positronium and its properties in semiconductors and alkali halides crystals. Khim. Vys. Energ. *12* (1978) [English transl.: High Energy Chem. , 145 (1978)]

Y. Takakusa, T. Hyodo: On the existence of delocalized positronium in NaF at room temperature. J. Phys. Soc. Jpn. *45*, 353 (1978)

Y. Yoshizawa, K. Shizuma, T. Fujita, M. Nishi: Doppler broadening measurement of positron annihilation in alkali halides. J. Phys. Soc. Jpn. *44*, 204 (1978)

Subject Index

Springer Series in

Solid-State Sciences

Volume 1
C. P. Slichter

Principles of Magnetic Resonance

2nd, revised and expanded edition 1978
115 figures, X, 397 pages
ISBN 3-540-08476-2

Contents: Elements of Resonance. – Basic Theory. – Magnetic Dipolar Broadening of Rigid Lattices. – Magnetic Interactions of Nuclei with Electrons. – Spin-Lattice Relaxation and Motional Narrowing of Resonance Lines. – Spin Temperature in Magnetism and in Magnetic Resonance. – Double Resonance. – Advanced Concepts in Pulsed Magnetic Resonance. – Electric Quadrupole Effects. – Electron Spin Resonance. – Summary. – Problems. – Appendices.

Volume 2
O. Madelung

Introduction to Solid-State Theory

Translated from the German by B. C. Taylor
1978. 144 figures. XI, 486 pages
ISBN 3-540-08516-5

Contents: Fundamentals. – The One-Electron Approximation. – Elementary Excitations. – Electron-Phonon Interaction: Transport Phenomena. – Electron-Electron Interaction by Exchange of Virtual Phonons: Superconductivity. – Interaction with Photons: Optics. – Phonon-Phonon Interaction: Thermal Properties. – Local Description of Solid-State Properties. – Localized States. – Disorder. – Appendix: The Occupation Number Representation.

Volume 3
Z. G. Pinsker

Dynamical Scattering of X-Rays in Crystals

1978. 124 figures, 12 tables. XII, 511 pages
ISBN 3-540-08564-5

Contents: Wave Equation and Its Solution for Transparent Infinite Crystal. – Transmission of X-Rays Through a Transparent Crystal Plate. Laue Reflection. – X-Ray Scattering in Absorbing Crystal. Laue Reflection. – Poynting's Vectors and the Propagation of X-Ray Wave Energy. – Dynamical Theory in Incident-Spherical-Wave Approximation. – Bragg Reflection of X-Ray I. Basic Definitions. Coefficients of Absorption: Diffraction in Finite Crystal. – Bragg Reflection of X-Rays. II. Reflection and Transmission Coefficients and Their Integrated Values. – X-Ray Spectrometers Used in Dynamical Scattering Investigation. Some Results of Experimental Verification of the Theory. – X-Ray Interferometry. Moire Patterns in X-Ray Diffraction. – Generalized Dynamical Theory of X-Ray Scattering in Perfect and Deformed Crystals. – Dynamical Scattering in the Case of Three Strong Waves and More. – Appendices.

Volume 4

Inelastic Electron Tunneling Spectroscopy

Proceedings of the International Conference and Symposium on Electron Tunneling, University of Missouri-Columbia, USA, May 25–27, 1977
Editor: T. Wolfram
1978. 126 figures, 7 tables. VIII, 242 pages
ISBN 3-540-08691-9

Contents: Review of Inelastic Electron Tunneling. – Applications of Inelastic Electron Tunneling. – Theoretical Aspects of Electron Tunneling. – Discussions and Comments. – Molecular Adsorption on Non-Metallic Surfaces. – New Applications of IETS. – Elastic Tunneling.

Volume 5
F. Rosenberger

Fundamentals of Crystal Growth

Macroscopic Equilibrium and Transport Concepts
ISBN 3-540-09023-1

Contents: Thermodynamics. – Phase Transitions and Phase Diagrams. – Crystal Growth and Phase Diagrams. – Mass Transport and Heat Transfer. – Segregation.

Volume 6
R. P. Huebener

Magnetic Flux Structures in Superconductors

1979. 99 figures, 5 tables. Approx. 270 pages
ISBN 3-540-09213-7

Volume 7
E. N. Economou

Green's Functions in Quantum Physics

1979. 49 figures, 2 tables. IX, 251 pages
ISBN 3-540-09154-8

Contents: Green's Function in Mathematical Physics. Green's Functions in One-Body Quantum Problems. Green's Functions in Many-Body Systems. – Appendices.

Volume 8

Solitons and Condensed Matter Physics

Proceedings of the Symposium on Nonlinear (Soliton) Structure and Dynamics in Condensed Matter. Oxford, England, June 27–29, 1978
Editors: A. R. Bishop, T. Schneider
1978. 120 figures, 4 tables. XI, 341 pages
ISBN 3-540-09138-6

Contents: Introduction. – Mathematical Aspects. Statistical Mechanics and Solid-State Physics. – Summary.

Springer-Verlag
Berlin Heidelberg New York